MW01552559

“ᑐᒃᑐᑦ ᑎᐱᖕᒋᑦ ᐊᔾᔨᒋᔭᓐᓃᖅᑕᖕᒋᑦ ᒫᓐᓇ”

“THE CARIBOU TASTE DIFFERENT NOW”

NUNAVUT
ARCTIC COLLEGE
MEDIA

SILATTUQSARVINGMI
ᓯᓚᑦᑐᖅᓴᕐᕕᖕᒥ
TUSAQATTAUTILIRIJIIT
ᑐᓴᖃᑦᑕᐅᑎᓕᕆᔩᑦ

“ᑐᒃᑐᑦ ᑎᐱᖕᒋᑦ ᐊᔾᔨᒋᔪᓐᓃᖅᑕᖕᒋᑦ ᒫᓐᓇ”

ᐃᓄᐃᑦ ᐃᓐᓇᐃᑦ ᖃᐅᔨᓯᒪᔭᖕᒋᑦ ᓯᓚᐅᑉ ᐊᓯᔾᔨᖅᓂᖓᓂ

“THE CARIBOU TASTE DIFFERENT NOW”

Inuit Elders Observe Climate Change

ᐋᖅᑭᒋᐊᕆᔨ ᔫᓴᐃ ᒐᕆᓐ-ᓚᔪᐊ, ᐊᓚᐃᓐ ᑭᐅᕆᐊᐃ, ᐊᒻᒪ ᓗᐊᕋ ᓰᒍᐊᖅᑦ ᑰᓚᐅᖅ

Edited by José Gérin-Lajoie, Alain Cuerrier, and Laura Siegwart Collier

ᐊᐱᖅᓱᖅᑏᑦ

Interviewers

ᔫᓴᐃ ᒐᕆᓐ-ᓚᔪᐊ:
ᑲᖏᖅᓱᐊᓗᔾᔪᐊᖅ, ᑲᖏᖅᓱᔪᐊᖅ,
ᐅᒥᐅᔭᖅ, ᐸᖕᓂᖅᑑᖅ,
ᖃᒪᓂ’ᑐᐊᖅ, ᖁᖅᓗᖅᑐᖅ

José Gérin-Lajoie:
Kangiqsualujjuaq, Kangiqsujuaq,
Umiujaq, Pangnirtung,
Pond Inlet, Baker Lake, Kugluktuk

ᑳᒪᓐ ᓯᐱᑦ: ᖃᒪᓂ’ᑐᐊᖅ

Carmen Spiech: Baker Lake

ᐊᓚᐃᓐ ᑭᐅᕆᐊᐃ:
ᑲᖏᖅᓱᐊᓗᔾᔪᐊᖅ,
ᑲᖏᖅᓱᔪᐊᖅ, ᓇᐃᓂ

Alain Cuerrier:
Kangiqsualujjuaq,
Kangiqsujuaq, Nain

ᓗᐊᕋ ᓰᒍᐊᖅᑦ ᑰᓚᐅᖅ: ᓇᐃᓂ

Laura Siegwart Collier: Nain

ᐊᓂᑕ ᕚᓪᔅ: ᓇᐃᓂ

Anita Fells: Nain

ᖃᐅᔨᓴᖅᑐᑦ

Researchers

ᐊᓚᐃᓐ ᑭᐅᕆᐊᐃ, ᓴᐃᕋ ᑎᕈᓯ,
ᐋᔅᓕ ᑕᐅᓂᖕ, ᔫᓴᐃ ᒐᕆᓐ-ᓚᔪᐊ,
ᒍᕋᐃᒃ ᕼᐊᓐᕆ, ᓗᐃᔅ ᕼᐅᒪᓄᑦ,
ᐃᔅᑕ ᓚᕚᔅᒃ, ᓗᐊᕋ ᓰᒍᐊᖅᑦ ᑰᓚᐅᖅ

Alain Cuerrier, Sarah Desrosiers,
Ashleigh Downing, José Gérin-Lajoie,
Greg Henry, Luise Hermanutz, Esther
Lévesque, Laura Siegwart Collier

Published by Nunavut Arctic College Media
www.nacmedia.ca
Box 600, Iqaluit, NU, X0A 0H0

Text copyright © 2016 by José Gérin-Lajoie, Alain Cuerrier, and Laura Siegwart Collier
Design and layout by Inhabit Media © 2016 Nunavut Arctic College Media

Cover photograph © Freezingpictures/Dreamstime.com
Photographs copyright as indicated

All rights reserved. The use of any part of this publication reproduced, transmitted in any form or by any means, electronic, mechanical, photocopying, recording, or otherwise, or stored in a retrievable system, without written consent of the publisher, is an infringement of copyright law.

ISBN: 978-1-897568-39-2

Printed in Canada.

Every reasonable effort has been made to contact copyright holders. Nunavut Arctic College Media would be pleased to have any errors or omissions brought to its attention.

Funded by International Polar Year 2007–2008; ArcticNet; Université du Québec à Trois-Rivières Fondation de l'Université du Québec à Trois-Rivières; University of British Columbia; Memorial University of Newfoundland; Jardin botanique de Montréal; Institut de recherche en biologie végétale; Nunatsiavut Government; World Wildlife Federation of Canada

Library and Archives Canada Cataloguing in Publication

"Tuktut tipingit ajjigijunniiqtangit manna" : Inuit innait qaujisimajangit silaup asijjirningani / Aaqkigiaqtaujuq Juusi gaarinlajuai, Alain kuariu, amma Laura siguat kuuliu ; Apiqsuqtiit,Juusi giurinlajui: kangiqsualujjuaq, Kangiqsujuaq, umiujaq, Pangniqtuuq, Qamani'tuaq, Qurluqtuq, Kaaman sipit: Qamani'tuaq, Alain kuriu: Kangiqsualujjuaq, Kangiqsujuaq naini, Laura siguat kuuliu: Naini, Anita vaals: Naini ; Qaujisaqtut, Alain kuriu, Saira tiruusi, Asli tauning, Juusi giurinlajui, Guraik Hanri, Luis Humanut, Ista lavask, Laura Siguat Kuuliu = "The caribou taste different now" : Inuit elders observe climate change / edited by José GérinLajoie, Alain Cuerrier, and Laura Siegwart Collier ; interviewers, José Gérin-Lajoie: Kangiqsualujjuaq, Kangiqsujuaq, Umiujaq, Pangnirtung, Pond Inlet, Baker Lake, Kugluktuk, Carmen Spiech: Baker Lake, Alain Cuerrier: Kangiqsualujjuaq, Kangiqsujuaq, Nain, Laura Siegwart Collier: Nain, Anita Fells: Nain ; researchers, Alain Cuerrier, Sarah Desrosiers [and six others].

Text in Inuktitut and English.
ISBN 9781897568392 (paperback)

1. Climatic changes--Canada, Northern. 2. Climatic changes--Social aspects--Canada, Northern. 3. Climatic changes—Effect of human beings on--Canada, Northern. 4. Inuit--Canada, Northern. 5. Inuit elders--Canada. 6. Traditional ecological knowledge--Canada, Northern. 7. Canada, Northern--Climate. 8. Canada, Northern--Environmental conditions. I. GérinLajoie, José, editor II. Cuerrier, Alain, 1962-, editor III. Siegwart Collier, Laura, editor IV. Title: Caribou taste different now. V. Title: Tuktut tipingit ajjigijunniiqtangit manna. VI. Title: Tuktut tipingit ajjigijunniiqtangit manna. English.

GF512.N6T84 2016 304.2'509719 C20169016625

ᐃᓗᓕᖅᑎᑦ

Contents

ᓯᕗᓕᐊᕈᑦ

ᖁᕕᐊᓱᒻᒪᕆᒃᐳᖓ ᐃᓱᒪᕆᔭᒃᑲ ᑕᕝᕙᓂ ᓯᕗᓕᐊᕈᒻᒥ ᐃᓚᓕᐅᔾᔭᐅᔪᓐᓇᕋᒪ.

ᖃᐅᔨᒪᔪᐃᓐᓇᐅᔪᒍᑦ ᓯᓚᐅᑉ ᐊᓯᔾᔨᖅᐸᓪᓕᐊᓂᖓᓄᑦ ᐅᑭᐅᖅᑕᖅᑐᖅ ᓄᓇᕐᔪᐊᕐᓗ ᐊᑦᑕᕐᓇᖅᑐᒦᓕᕐᓂᖏᓐᓂᒃ. ᐅᑭᐅᓄᑦ ᐊᒥᓱᓄᑦ, ᐃᓄᐃᑦ ᐅᖃᖃᑦᑕᖅᓯᒪᖓᒪᑕ ᐃᓄᐃᑦ ᖃᐅᔨᒪᔭᑐᕐᑲᖏᑦ ᑕᐃᒪᙵᓂᐊᓗᒃᓗ ᐊᑐᖅᑕᐅᖃᑦᑕᖅᓯᒪᔪᑦ ᐃᓱᒪᕆᔭᐅᔭᓇᐊᖅᑲᖅᒍᑎᒃᓗ ᓯᓚᐅᑉ ᐊᓯᔾᔨᖅᐸᓪᓕᐊᓂᖓ ᐅᖅᑲᒪᔭᐅᑕᐅᑎᓪᓗᒍ ᓱᖏᐅᑎᔭᐅᕙᓪᓕᐊᑎᓪᓗᒍᓗ. ᑕᒪᓐᓇ ᖃᐅᔨᓴᕐᓂᖅ ᑕᒪᑐᒥᖓ ᐊᑑᑎᖅᑲᓕᕈᑕᐅᓂᐊᖅᑐᖅ. ᐃᓕᖅᑯᓯᑦᑎᓐᓂᒃ ᐃᓚᒦᓄᓪᓗ ᐊᑐᖅᑕᐅᕙᒃᑐᓂᒃ ᓴᖅᑭᑎᑦᑎᔾᔭᑕᐅᓂᐊᖅᑐᖅ, ᐊᒻᒪ ᓇᓗᓇᐃᖅᓯᔾᔭᑕᐅᓗᓂ ᖃᓄᖅ ᓯᓚᐅᑉ ᐊᓯᔾᔨᖅᐸᓪᓕᐊᓂᖓ ᐊᒃᑐᐃᓯᒪᓂᖅᑲᖅᒪᖔᑦ ᓂᕿᕙᒃᑕᑦᑎᓐᓂᒃ.

ᐸᐅᕐᖓᖅᑕᕈᒻᒪᒻᒪᕆᐅᔪᖓ, ᓄᓇᓕᓐᓂ ᑰᔾᔪᐊᕐᒥ ᓄᓇᕕᖕᒥ ᐅᑭᐊᒃᓵᑕᒫᖅ ᐸᐅᕐᖓᖅᑕᖅᐸᒃᑐᖓ. ᐸᐅᕐᖓᖅᑕᕐᓂᖅ, ᓂᕆᔭᒃᓴᓂᒃ ᐊᕝᕗᕐᓂᖅ ᓯᕗᓕᖓᓂᑦ ᐱᓯᒪᔭᕋ ᖁᕕᐊᕆᓂᖅᐹᓇᓪᓗᒍ.

ᐅᑭᐅᒐᓴᓄᑦ ᐸᐅᕐᖓᖅᑕᓕᕌᕐᒐᒪ ᐃᓱᒪᖅᑲᑦᑕᖅᓯᒪᔪᕐᒐ ᖃᓄᖅ ᐃᓅᓯᕗᑦ ᐃᓕᖅᑯᓯᕗᓪᓗ ᐊᒃᑐᖅᑕᐅᓂᖅᑲᕐᓂᐊᕐᒪᖔᑦ ᓯᓚᐅᑉ ᐊᓯᔾᔨᖅᐸᓪᓕᐊᕐᑎᓐᓇᕐᓂᕐᒐᓄᑦ. ᓯᕐᑭᓐᓂᑦᑎᐊᖅᑲᑦᑕᕐᓂᐊᕋᓗᐊᖅᐹ— ᒪᕐᑯᓗᐊᖅᑲᑦᑕᕐᓂᐊᖅᐸᓘᓐᓃᑦ? ᐃᓚᐃᑎᒍᑦ “ᐸᐅᕐᖓᖅᑕᕈᒻᒪᒻᒪᕆᐅᔪᒍᑦ”

Foreword

I am very pleased to be adding a few thoughts to this publication through this foreword.

We all know that climate change continues to be a growing threat in the Arctic and around the world. For many years now, Inuit have stated that our traditional knowledge and ancient practices are extremely relevant and need to be considered in the discussion on climate change mitigation and adaptation. This research puts that idea into practice. It also reflects our key cultural and family practices, and how climate change is affecting what we eat.

I am a very passionate berry picker, scouring the hills around my hometown of Kuujjuaq, Nunavik, every autumn. This gathering of food has been passed down from my ancestors and is my most cherished and satisfying pastime.

For years, at just about every patch I have witnessed and wondered about how this essential part of our life and culture will be affected as our climate continues to change. Will we get enough sun—or too much rain? Certainly all of us who pride ourselves as “serious

ᐅᔾᔨᕈᓱᕐᑲᑦᑕᖅᓯᒪᔪᑦ ᐊᓯᔾᔨᖅᐸᓪᓕᐊᓂᐅᔪᓂᒃ.

ᑖᓐᓇ ᐅᕐᑲᓕᒫᒐᖅ ᑎᑎᕋᐅᓯᕐᑲᖅᓯᒪᔪᖅ ᐅᑭᐅᖅᑕᖅᑐᕐᒥᐅᑦ ᐃᓚᖏᓐᓂ ᓄᓇᓕᖕᓂ ᐅᔾᔨᕆᔭᐅᓯᒪᓕᖅᑐᓂᒃ. ᐊᑐᐊᒐᑦᑎᐊᕙᐅᓂᐊᖅᑐᖅ—ᐊᒻᒪ ᐃᓄᐃᑦ ᐃᑲᔫᑎᕐᑲᕈᓐᓇᕐᓂᖏᑦ ᓯᓚᐅᑉ ᐊᓯᔾᔨᖅᐸᓪᓕᐊᓂᖓᓄᑦ ᐃᓕᓴᕆᔭᐅᓕᕐᓗᑎᒃ.

ᒥᐊᓕ ᓴᐃᒪᓐ
ᐊᖏᔪᖅᑳᖑᓚᐅᖅᓯᒪᔪᖅ ᐃᓄᐃᑦ
ᐅᑭᐅᖅᑕᖅᑐᒥ ᑲᑎᒪᔨᖏᓐᓄᑦ ᐊᒻᒪ
ᐃᓄᐃᑦ ᑕᐱᕇᑦ ᑲᓇᑕᒥ
ᑲᓇᑕᓕᒫᒥ ᐱᓇᓱᐊᖅᓯᒪᓂᖓᓄᑦ
ᑯᐃᓐᒧᑦ ᐃᓕᓴᕆᔭᐅᔾᔪᑎᓕᒃ

pickers" have seen our own seasons change over time.

This book documents what others in selected communities across the North have observed. It is a very useful guide—and I hope it is the beginning of recognizing the contribution Inuit have made and continue to make to the science of climate change.

Mary Simon
Former President of Inuit Circumpolar Council and Inuit Tapiriit Kanatami
Officer of the Order of Canada

ᖃᐅᔨᔭᐅᖁᔭᒍᑦ ᐅᖃᐅᓯᖅᓂᒃ ᒧᒥᑦᑎᓂᐅᑉ ᒥᒃᓵᓄᑦ

ᐅᓇ ᐅᖃᓕᒫᒐᖅ ᓴᖅᑭᑕᐅᖕᒪᑦ ᐃᓄᖕᓂᒃ ᐊᐱᖅᓱᖅᑕᐅᓂᑯᓂᒃ ᑲᑎᖅᓱᒐᐅᓯᒪᔪᓂᒃ ᓇᑯᐃᓐᓈᖅᓯᒪᔪᓂᒃ ᑲᓇᑕᐅᑉ ᐅᑭᐅᖅᑕᖅᑐᖕᒐᓂᑦ. ᐊᐱᖅᓱᒐᐅᓯᒪᔪᑦ ᓇᑯᐃᓐᓈᖅᓯᒪᓂᖕᒋᑦ ᕿᑭᖅᑖᓗᖕᒥ, ᑭᕙᓪᓕᕐᒥ, ᕿᑎᕐᒥᐅᓂ, ᓄᓇᕕᖕᒥ ᐊᒻᒪᓗ ᓄᓇᑦᓯᐊᕗᒻᒥ ᐅᖃᐅᓯᖃᑎᒌᖕᒋᑦᑐᑦ. ᑖᓐᓇ ᐱᓕᕆᐊᖑᓚᐅᖅᑐᖅ ᐊᖕᒋᔫᓚᐅᕐᒪᑦ ᓴᖅᑮᓂᐊᖅᑎᓪᓗᑕ ᐃᓱᒪᓕᐅᕆᐊᖃᖅᓯᒪᓚᐅᕋᑦᑕ.

ᐅᖃᓕᒫᒐᓕᐊᕆᖃᑦᑕᕐᓂᐊᖅᑕᕗᑦ ᐅᖃᐅᓯᖃᑎᒌᖕᒋᑦᑐᓂᒃ ᓴᖅᑭᔮᖅᑎᑦᑎᖃᑦᑕᕐᓂᐊᖅᑑᒐᓗᐊᑦ, ᑖᓐᓇ ᓴᖅᑭᑎᑕᕗᑦ ᑐᑭᖕᒐᒻᒪᑦ ᕿᑭᖅᑖᓗᖕᒦᑦᑐᓄᑦ ᐃᓕᓐᓂᐊᖅᑎᓄᑦ. ᑕᐃᒪᐃᖕᒪᑦ, ᐃᓄᒃᑏᑑᖅᑐᖕᒋᑦ ᕿᑭᖅᑖᓗᖕᒥᐅᔭᖅᑐᑦ.

ᐊᐱᖅᓱᖅᑐᖃᖅᑎᓪᓗᒍ, ᕿᑭᖅᑖᓗᖕᒥ ᐊᐱᖅᓱᖅᑕᐅᔪᑦ ᐃᓄᒃᑏᑑᖅᑐᓂᒃ ᑎᑎᕋᖅᑕᐅᓯᒪᔪᑦ. ᐊᐱᖅᓱᖅᑕᐅᓂᑯᑦ ᕿᑭᖅᑖᓘᑦ ᓯᓚᑖᓂ ᐅᖃᓪᓚᒃᑎᓪᓗᒋᑦ ᖃᓪᓗᓈᑎᑑᓕᖅᑎᑕᐅᕙᓪᓕᐊᓐᓂᕐᒪᑕ ᑐᓵᔨᒧᑦ. ᑕᐃᒪ ᑕᐃᒃᑯᐊ ᖃᓪᓗᓈᑎᑑᖅᑐᕕᓃᑦ ᕿᑭᖅᑖᓗᖕᒥᐅᔭᓕᖅᑎᑕᐅᓯᒪᓪᓗᑎᒃ. ᑕᐃᒪᐃᑦᑕᐃᓕᒪᓘᖕᒍᔪᒍᑦ, ᑭᓱᐊᓂ ᐅᖃᐅᓯᖅᑎᒌᖖᒋᓗᐊᖅᑎᓪᓗᒋᑦ ᓂᐱᓕᐅᖅᑕᐅᓯᒪᔪᑦ ᐃᓖᓐᓂᒃᑯᑦ ᑕᐃᒪᐃᒋᐊᖃᓘᖕᒍᒻᒪᑦ. ᑕᐃᒪᐃᓚᐅᓘᖕᒍᔪᒍᑦ ᐅᖃᓕᒫᒐᓕᐅᖅᑎᓪᓗᑕ ᐅᑭᐅᖅᑕᖅᑐᒥ.

Notes on Translation

This book is published from a compilation of Elder interviews conducted across Canada's Arctic. The interviewees come from the Baffin, Kivalliq, Qitirmiut, Nunavik, and Nunatsiavut Regions, representing different dialects. This broad fieldwork makes for unique and specific perspectives, and difficult publishing decisions.

Although we plan future editions to reflect the various dialects represented, the translation of the present volume is intended to serve students in the Baffin Region. Therefore, the Inuktitut in this text is in the Baffin dialect.

In the field, the interviews from the Baffin Region were transcribed in Inuktitut. The interviews from the other regions were interpreted in real time through a local translator from the respective dialects into English. This was then translated into the Baffin dialect. This is a practice that we work to avoid; however, translating multi-dialectic oral material into printed form makes this a necessity at times. This is a reality and challenge of both fieldwork and publishing in the Arctic.

ᐱᓕᕆᓯᒪᔪᒍᑦ ᑖᒃᑯᐊ ᑐᑭᓪᓚᑦᑖᖅᑎᑦ ᑐᙵᕕᖏᓐᓇᕆᐊᖅᑐᑎᑦ ᐊᕙᑎᒥ ᓄᓇᒥᓪᓗ ᖃᐅᔨᒪᔭᐅᔪᑦ ᓯᓚᐅᑉ ᐊᓯᔾᔨᕐᓂᖓᓂᒃ ᓇᐅᑦᑎᖅᓱᒐᐅᓯᒪᔪᓂᒃ ᐅᑭᐅᖅᑕᖅᑐᒥ. ᐊᑏᑐᖅ ᐅᕐᑲᓕᒫᒐᖅ ᑖᓐᓇ ᐃᓕᖕᓄᑦ ᐊᑑᑎᖅᑲᑦᑎᐊᕐᓂᐊᖅᐳᖅ ᐅᕐᑲᓕᒫᖅᑎᐅᔪᑎᑦ ᐅᖅᑲᓪᓚᒃᑎᐅᔪᑎᓪᓗ.

ᖁᔭᓐᓇᒦᕐᔪᐊᖅ ᐅᕐᑲᓕᒫᕋᕕᑦ.

ᓗᐃᓯ ᕙᓚᕼᐅᕐᑎ
ᑐᑭᒧᐊᒃᑎᑦᑎᔨ ᐃᓄᐃᑦ ᐅᖃᐅᓯᖓᓂᒃ
ᐃᓕᖅᑯᓯᖓᓂᓗ ᐃᓕᓐᓂᐊᖅᑐᓄᑦ
ᓄᓇᕗᑦ ᓯᓚᑦᑐᖅᓴᕐᕕᒃ

We have worked to ensure that little is lost in translation through our efforts to represent various geographical observations of climate change across Canada's Arctic. We hope this book holds meaning for all readers and speakers.

Thank you for reading.

Louise Flaherty
Director of Language and Culture
Nunavut Arctic College

ᐃᓕᓴᕆᔭᐅᔪᑦ

ᖁᔭᓕᔭᕆᐊᖃᒻᒪᕆᒃᐸᕗᑦ 145-ᖑᔪᑦ 8-ᖑᔪᓂᑦ ᓄᓇᓕᖕᓂᑦ ᐊᐱᖅᓱᖅᑕᐅᓚᐅᖅᑐᑦ ᑲᓇᑕᐅᑉ ᐅᑭᐅᖅᑕᖅᑐᖕᒐᓂ: ᐅᒥᐅᔭᖅ, ᑲᖏᖅᓱᔪᐊᖅ ᐊᒻᒪ ᑲᖏᖅᓱᐊᓗᒃᔪᐊᖅ ᓄᓇᕕᖕᒥ (ᑯᐸᐃᒃ ᑕᕐᕿᖕᒐᓂ); ᐸᖕᓂᖅᑑᖅ, ᒥᑦᑎᒪᑕᓕᒃ, ᖃᒪᓂ'ᑐᐊᖅ, ᐊᒻᒪ ᖁᕐᓗᖅᑐᖅ ᓄᓇᕗᒻᒥ; ᐊᒻᒪᑦᑕᐅᖅ ᓇᐃᓂ ᓄᓇᑦᓯᐊᕗᒻᒥ (ᓛᐸᑐᐊᓕᒥ). ᐅᐱᒋᒻᒪᕆᒃᓱᑎᒍᑦ ᖁᔭᓕᕙᕗᑦ ᖃᐅᔨᒪᔭᕐᒥᓂᒃ ᐊᑐᖅᑲᑦᑕᖅᓯᒪᔭᕐᒥᓂᓪᓗ ᐅᓂᒃᑳᕈᓐᓇᓚᐅᕐᒪᑕ. ᑐᙵᓇᕐᓂᖕᒋᓐᓄᑦ ᑕᒪᓐᓇ ᐱᓕᕆᐊᕆᔪᓐᓇᓚᐅᖅᐸᕗᑦ. ᖁᔭᒋᔪᒪᒋᕙᕗᑦ ᑐᓵᔩᑦ ᐊᐱᖅᓱᖅᑎᑦᓗᑕ ᐃᑲᔪᕐᓂᖅᒪᕆᓚᐅᖅᑐᑦ: ᓘᑲᓯ ᑐᒃᑐ ᐅᒥᐅᔭᕐᒥ; ᐹᓴ ᐊᖓᒃ ᐊᒻᒪ ᒥᐊᓕ ᐋᑕᒥ ᑲᖏᖅᓱᔪᐊᕐᒥ; ᒥᐊᓕ ᐊᓈᓇᒃ, ᓅᑕ ᐊᓈᓇᒃ, ᐋᓂ ᐹᕋᓐ, ᖃᕆᔦ ᐃᑦᑐᖅ, ᐊᒻᒪ ᓵᓚᑦ ᒧᐊᒐᓐ ᑲᖏᖅᓱᐊᓗᒃᔪᐊᕐᒥ; ᔫᓇ ᕿᓚᖅᐳᖅ, ᐃᐅᓇᒃ ᔫᒥ, ᐊᒻᒪ ᓅᑕ ᓄᑭᑯᐊᖅ ᐸᕐᓂᖅᑑᕐᒥ; ᑑᒥᓇ ᑯᓈᒃ, ᔪᐊᓇ ᐃᓄᐊᓗᒃ ᑯᓄᒃ, ᐃᓕᓴᐱ ᖃᔮᖅ, ᐊᒻᒪ ᓘᓯ ᖁᐊᓴ ᒥᑦᑎᒪᑕᓕᖕᒥ; ᕕᐅᓚ ᐊᕚᓚ, Hᐊᑎ ᒪᓂᖅ, ᐊᒻᒪ ᑑᒪᓯ ᒪᓂᖅ ᖃᒪᓂ'ᑐᐊᕐᒥ; ᑑᕆᔅ ᐃᓚᑎᐊᖅ ᖁᕐᓗᖅᑐᕐᒥ; ᐊᒻᒪ ᐅᐃᓪᓴ ᔭᖃᕈᔅ ᓇᐃᓂᒥ. ᖁᔭᒋᕙᕗᑦ ᐊᒡᓂᑯ-ᐄᒍ ᐃᑲᔪᕐᓂᖅᑲᓚᐅᕐᓂᖏᓐᓄᑦ ᖃᒪᓂ'ᑐᐊᕐᒥ ᐱᓕᕆᐊᕐᑲᕆᐊᕐᑲᖅᑎᑦᓗᑕ. ᑭᖑᓪᓕᖅᐹᕐᒥ, ᖁᔭᓕᕗᒍᑦ ᐃᓄᓕᒫᓄᑦ ᑐᙵᓇᖅᐸᓚᐅᖅᑐᓄᑦ ᑐᔪᕐᒥᔭᕆᐊᕐᑲᖄᖕᒐᑦᑕ ᓄᓇᓕᖕᓂ. ᑖᓐᓇ ᐃᑲᔫᑎᕐᑲᕐᓂᐊᕐᑯᓇᖁᓗᐊᖅᐳᖅ ᐅᕐᑲᓖᖅᑐᓄᑦ, ᐱᓗᐊᖅᑐᒥᒃ

Acknowledgements

Very special thanks are due to all 145 participants with whom we conducted interviews in eight communities across the Canadian Arctic: Umiujaq, Kangiqsujuaq, and Kangiqsualujjuaq in Nunavik (northern Quebec); Pangnirtung, Pond Inlet, Baker Lake, and Kugluktuk in Nunavut; and finally Nain in Nunatsiavut (northern Labrador). It is with great respect that we thank them for sharing with us their knowledge of the land and their experiences. Their generosity and welcoming manner made this work possible. We also want to express our gratitude to the interpreters for their essential role in assisting us with the interviews: Lucassie Tooktoo in Umiujaq; Pasha Arngak and Mary Adams in Kangiqsujuaq; Mary Annanack, Rita Annanack, Annie Baron, Harriet Etok, and Charlotte Morgan in Kangiqsualujjuaq; Jonah Kilabuk, Eric Joamie, and Rita Nookiguak in Pangnirtung; Domina Koonark, Joanna Innualuk Kunnuk, Elisapie Kyak, and Lucy Quassa in Pond Inlet; Vera Avaala, Hattie Mannik, and Thomas Mannik in Baker Lake; Doris Elatiak in

ᐅᑭᐅᖅᑕᖅᑐᕐᒥᐅᓄᑦ, ᐊᕙᑎᐅᑉ ᐊᓯᔾᔨᖅᐸᓪᓕᐊᓂᖓᓄᑦ ᓄᓇᓕᖕᓂ, ᐅᑭᐅᖅᑕᖅᑐᕐᒧ.

Kugluktuk; and Wilson Jararuse in Nain. We thank Agnico Eagle for their in-kind support when we had validation work to do in Baker Lake. Finally, we are grateful to all the people who warmly hosted us in their homes during our stays in the communities. We hope this tool will increase awareness among readers, especially Northerners, about the environmental changes that are taking place around the communities, but also throughout the land.

ᐱᒋᐊᕐᓂᖓ

2007-ᖑᑎᓪᓗᒍ ᖃᐅᔨᓴᖅᑎᑦ ᐃᓕᓐᓂᐊᖅᑎᓪᓗ ᐳᓇᑎᔅ ᑲᓚᒻᐱᐊ ᓯᓚᑦᑐᖅᓴᕐᕕᒡᔪᐊᖓᓂᑦ, ᑐᐊ ᓇᕕᐊ ᓯᓚᑦᑐᖅᓴᕐᕕᒡᔪᐊᖓᓂᑦ ᑯᐸᐃᒃᒥ, ᒪᓐᑐᓇᐊᒥ ᐱᕈᖅᑐᓕᓇᖕᕕᖓ, ᐊᒻᒪ ᒪᒨᓇᐅ ᓯᓚᑦᑐᖅᓴᕐᕕᒡᔪᐊᖓ ᓂᐅᕗᐅᓐᓚᓐᒥᑦ ᖃᐅᔨᓴᖅᑎᑦ ᑲᑎᓚᐅᖅᑐᑦ ᑐᑭᓯᓇᓱᐊᖅᓱᑎᒃ ᖃᓄᖅ ᐊᕙᑎᑦᑎᓐᓂ ᐊᓯᔾᔨᖅᐸᓪᓕᐊᔪᑦ ᑲᓇᑕᐅᑉ ᐅᑭᐅᖅᑕᖅᑐᖓᓐᓂ ᐊᒃᑐᐃᓂᖃᕐᒪᖔᑕ ᓄᓇᒥ ᐱᕈᖅᐸᒃᑐᓂᒃ. ᑐᑭᓯᔪᒪᓗᐊᙳᐊᖅᐸᕗᑦ ᐊᓯᔾᔨᖅᐸᓪᓕᐊᓂᐅᑉ ᐊᒃᑐᐃᓂᖃᖅᓯᒪᓂᖓ ᓄᓇᒥ ᐱᕈᖅᐸᒃᑐᓂᒃ, ᓲᕐᓗ ᑭᒍᑕᖏᕐᓇᐃᑦ ᐸᐅᕐᖓᐃᑦᓗ, ᑭᖕᒥᖕᓇᐃᑦ, ᐊᖅᐹᑦ ᑕᒪᒃᑯᐊ ᐱᕈᖅᑐᑦ ᐱᒻᒪᕆᐅᓂᖏᓐᓄᑦ ᐆᒪᔪᓄᑦ, ᐃᓄᖕᓄᑦ, ᐅᑭᐅᖅᑕᖅᑐᕐᒥᐅᓪᓗ ᐃᓕᖅᑯᓯᖏᓐᓄᑦ. ᓄᓇᒥ ᖃᐅᔨᓴᕐᓂᒃᑯᑦ ᓄᓇᓕᖕᓂᓪᓗ ᕿᒥᕐᕈᓂᒃᑯᑦ, ᐱᓕᓇᐊᖃᖃᑎᖃᓚᐅᖅᐳᒍᑦ 145-ᓂᒃ ᐃᓄᑐᖃᕐᓂᒃ ᖃᐅᔨᒪᔨᑕᕐᓂᒡᓗ 8-ᖑᔪᓂᑦ ᓄᓇᓕᖕᓂᑦ ᐱᖓᓱᓂᑦ ᐃᓄᐃᑦ ᓄᓇᖏᓐᓂ ᑲᓇᑕᒥ (ᓄᓇᕗᑦ, ᓄᓇᕕᒃ, ᐊᒻᒪ ᓄᓇᑦᓯᐊᕗᑦ) ᑐᑭᓯᓇᓱᐊᖅᓱᑕ ᐊᓯᔾᔨᖅᓯᒪᔪᓂᒃ ᐊᓯᔾᔨᖅᐸᓪᓕᐊᖏᓐᓇᕐᓂᐊᖅᑐᓂᓪᓗ ᓄᓇᒥ ᐊᕙᑎᐅᔪᒥᓪᓗ. ᐊᐱᖅᓱᕐᓂᒃᑯᑦ ᕿᒥᕐᕈᓂᒃᑯᓪᓗ, ᐃᓄᑐᖃᐃᑦ ᖃᐅᔨᒪᔭᑕᓪᓗ ᐅᓂᒃᑳᖃᑦᑕᓚᐅᖅᑐᑦ ᐊᓯᔾᔨᖅᐸᓪᓕᐊᓯᒪᔪᓂᒃ ᐅᔾᔨᕆᓯᒪᔭᒥᖕᓂᒃ, ᖃᓄᕐᓗ ᐊᓯᔾᔨᖅᐸᓪᓕᐊᓂᖅ ᐊᒃᑐᐃᓂᖃᕐᒪᖔᑦ ᐊᕙᑎᒥᖕᓄᑦ, ᓄᓇᓕᖏᓐᓄᑦ, ᐃᓅᓯᖏᓐᓄᑦ.

Introduction

In 2007, a group of researchers and students from the University of British Columbia, Université du Québec à Trois-Rivières, Jardin botanique de Montréal, and Memorial University of Newfoundland formed a research team with the objective of understanding how environmental changes across the Canadian Arctic are affecting tundra vegetation. We were particularly interested in understanding the impacts of change on the ecology of tundra berry plants, such as blueberry/bilberry (*Vaccinium uliginosum*), redberry/cranberry (*Vaccinium vitis-idaea*), blackberry/crowberry (*Empetrum nigrum*), and cloudberry/bakeapple (*Rubus chamaemorus*) because of the importance of berries to wildlife, human health, and traditional ways of life in the North. In combination with field research and community-based monitoring, we partnered with 145 Elders and knowledge holders from eight Northern communities across three of the four Inuit regions of Canada (Nunavut, Nunavik, and Nunatsiavut) to understand historic and future changes in tundra vegetation

ᑕᕝᕙᓂ ᐅᖃᓕᒫᒐᖅᒥ, ᐱᓕᕆᐊᖃᖃᑎᒌᒍᑕᐅᓯᒪᔪᑦ ᐊᒥᓲᓂ ᓄᓇᓕᖕᓂ ᐊᓯᔾᔨᖅᐸᓪᓕᐊᓯᒪᔪᑦ ᐅᔾᔨᕆᔭᐅᓯᒪᓂᖏᑦ ᐅᓂᒃᑳᕆᔭᐅᓯᒪᔪᑦ. ᑕᒪᑐᒪᓂ, ᐊᔾᔨᒌᖕᓂᒃ ᐊᓯᔾᔨᖅᓂᐅᑉ ᒥᒃᓲᓄᑦ ᐅᔾᔨᕆᔭᐅᓯᒪᔪᓂᒃ ᓄᓇᓕᖕᓂᑦ ᐊᐱᖅᓱᖅᑕᐅᔪᑦ ᓇᑉᐸᖏᑕ ᐅᖓᑖᓃᑦᑐᓂᒃ ᓇᓗᓇᐃᖅᓯᓯᒪᕗᒍᑦ (ᑕᑯᒋᐊᒃᑭᑦ ᐃᓚᒋᐊᕈᑕᐃᑦ). ᑕᒪᐃᓐᓂᓗ ᓄᓇᓕᖕᓂ ᐊᓯᔾᔨᖅᐸᓪᓕᐊᓂᐅᑉ ᒥᒃᓵᓄᑦ ᐅᓂᒃᑲᐅᓯᐅᓯᒪᔪᓂᒃ ᓇᐃᓈᖅᓯᒪᔪᓂᒃ ᑎᑎᕋᐅᓯᖃᖅᓯᒪᓪᓗᑕ ᓇᓗᓇᐃᖅᓯᓪᓗᑕ ᐊᕙᑎᒧᑦ ᐊᓯᔾᔨᖅᐸᓪᓕᐊᓯᒪᔪᓂᒃ. ᐊᐱᖅᓱᖅᑕᐅᔪᑦ ᑭᒡᒍᑎᖏᑦ ᑕᒪᐃᓐᓂ ᓄᓇᓂ ᐊᕕᒃᑐᖅᓯᒪᔪᓂ ᐃᓄᑐᖃᐃᑦ ᐅᔾᔨᕆᓯᒪᔭᖏᑦ ᓄᓇᓕᖕᒥᓂ ᑎᑎᕋᐅᓯᐅᓯᒪᔪᑦ.

ᖃᐅᔩᓴᓂᒃᑲᑦᑕ, ᐃᓄᑐᖃᐃᑦ ᓄᓇᓕᖕᒥᐅᓪᓗ ᖁᔭᒋᓯᒪᕙᕗᑦ ᖃᐅᔨᒪᔭᖅᒥᓄᑦ ᐃᑲᔫᑎᖃᖅᓂᖏᓐᓂᒃ ᑕᒪᐃᓐᓄᑦ ᓄᓇᓕᖕᓄᑦ ᐊᑭᓐᓇᖅᒥᐅᑕᓂᒃ ᓴᖅᑭᑦᑎᓪᓗᑕ. ᑖᒃᑯᐊ ᐊᑭᓐᓇᖅᒥᐅᑕᑦ ᑐᓂᔭᐅᓚᐅᖅᑐᑦ ᐃᓄᑐᖃᖅᓄᑦ ᖃᐅᔨᒪᔨᑕᖅᓄᓪᓗ (ᐅᕝᕙᓘᓐᓃᑦ ᐃᓚᖏᓐᓄᑦ, ᐊᐱᖅᓱᖅᑕᐅᓚᐅᖅᑎᓪᓗᒋᑦ ᐃᓅᔪᓐᓃᖅᓯᒪᓕᖅᑐᓄᑦ) ᓄᓇᓕᖕᓂᓪᓗ ᐊᑭᓐᓇᖅᒥᐅᑕᐃᑦ ᐃᓕᐅᖃᖅᑐᒋᑦ ᖃᐅᔨᒪᓂᑎᒃ ᓄᓇᓕᖕᒥᐅᓄᑦ ᐊᓯᖏᓐᓄᓪᓗ ᐃᓄᖕᓄᑦ ᑲᓇᑕᒥᐅᑕᐅᖃᑎᖏᓐᓄᑦ ᑐᓴᐅᒪᔭᐅᔪᓐᓇᖅᓂᐊᖅᒪᑕ.

ᑕᕝᕙᓂ ᐅᖃᓕᒫᒐᖅᒥ ᑎᑎᕋᐅᓯᓇᓯᒪᔭᕗᑦ ᓄᓇᓕᖕᓄᑦ ᐊᑐᓂ ᑐᖄᖓᔪᑦ, ᐃᓚᖏᓪᓗ ᓄᓇᓂ ᐊᕕᒃᑐᖅᓯᒪᔪᓂ ᐊᔾᔨᒌᖕᓂᖃᖅᑐᑎᒃ. ᓄᓇᓕᖕᒥᐅᑦ ᐅᖃᐅᓯᓇᓯᒪᔭᖏᑦ ᑕᕝᕙᓂ ᐅᖃᓕᒫᒐᖅᒥ ᐃᑲᔫᑎᖃᖁᓐᓇᖅᓗᐊᖅᐳᑦ

and the environment. Through interviews and mapping consultations, Elders and local knowledge holders shared their observations of change, and the impacts of these changes on their environment, community, and well-being.

In this book, we share the results of this collaborative work by presenting a cross-community summary of the changes observed. Here, we focus on commonly observed changes identified by more than half of the participants in each community (see Appendix). We then summarize the changes observed within each individual community by describing major perceptions of environmental change. Direct quotations from the interviews conducted in each region provide a record of each Elder's observations of the changes in her or his community.

After completing our research, we paid tribute to the participating Elders and community members by representing the knowledge they had shared through thematic posters created for each community. Copies of these posters were given to participating Elders and knowledge holders (or their families, for those who had

ᑐᕋᐅᒪᔭᐅᓕᖅᑎᓐᓇᓱᐊᖅᑎᑦᑐᑎᒍᑦ
ᓯᑕᐅᑉ ᐊᓯᔾᔨᖅᐸᓪᓕᐊᓂᖓᑕ
ᐊᒃᑐᐃᓂᕆᓯᒪᔪᖕᒋᓐᓂᒃ ᐅᑭᐅᖅᑕᖅᑐᕐᒥ,
ᐃᑲᔫᑎᕐᑲᕐᑯᒪᓗᐊᕆᕙᑦ
ᓄᓇᓖᑦ ᓄᓇᐃᑦᑐ ᐊᕕᒃᑐᖅᓯᒪᔪᑦ
ᓱᖏᐅᑎᕙᑦᑕᐊᓇᓱᐊᖅᑎᑦᑐᒋᑦ
ᒫᓐᓇᒧᑦ ᐊᓯᔾᔨᖅᓯᒪᓕᖅᑐᓂᒃ
ᐊᒃᓱᕉᑕᐅᔪᓂᓪᓗ
ᐊᓯᔾᔨᖅᐸᓪᓕᐊᓂᐊᖅᑐᓂᓪᓗ
ᓯᕗᓂᒃᓴᑎᓐᓂ.

passed away since) and distributed throughout each community as a legacy of their knowledge, and a means of communicating local and regional environmental changes experienced by Inuit across Canada.

The perspectives we present in this book are specific to each community, apart from some regional trends. We hope that the local knowledge shared within this book will increase the reader's awareness of climate change impacts across the North, and that it will contribute to local and regional adaptation strategies to address current and future environmental challenges.

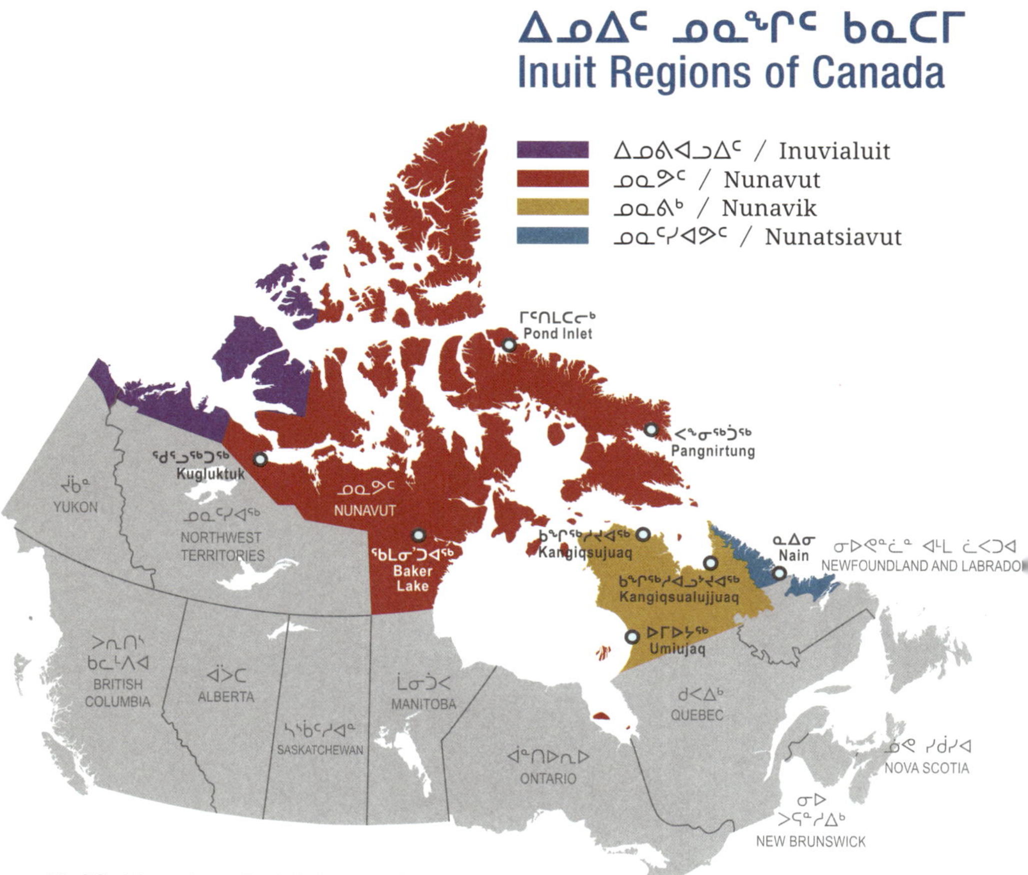

Modified from: Inuit Tapiriit Kanatami (2008). Maps of Inuit Nunangat (Inuit Regions of Canada). Retrieved from www.itk.ca

ᐊᔾᔨᙳᐊᖅ 1. ᓄᓇᓖᑦ ᐃᓄᐃᑦ ᓄᓇᖏᓐᓂ ᑲᓇᑕᒥ ᐱᑕᖃᐅᖅᑲᖅᑕᔪᑦ. ᐊᒻᒪᓗᖅᑐᑦ ᓇᓗᓇᐃᖅᓯᔪᑦ ᐃᓄᐃᑦ ᓄᓇᓕᖏᓐᓂᒃ ᓄᓇᐃᑦ ᐊᕕᑦᑐᖅᓯᒪᓂᖏᓐᓃᑦᑐᓂᒃ.

Figure 1. Locations of participating communities across the Inuit regions of Canada. The dots reflect all Inuit communities within each region.

ᐊᑐᖅᑕᐅᔪᑦ

2007-ᒧ 2010-ᒧ ᐊᑯᓐᖓᓂ, ᐊᐱᖅᓱᓚᐅᖅᐳᒍᑦ 145-ᓂᒃ ᐃᓄᑐᖃᕐᓂᒃ ᖃᐅᔨᒪᔨᑕᕐᓂᓪᓗ 8-ᖑᔪᓂᑦ ᓄᓇᓕᖕᓂ ᑎᓴᒪᐅᔪᓂᑦ ᐃᓄᐃᑦ ᓄᓇᖕᒋᓐᓂ ᑲᓇᑕᒥ; ᖁᕐᓗᖅᑐᖅ (18 ᐃᓄᑐᖃᐃᑦ ᐊᐱᖅᓱᖅᑕᐅᔪᑦ), ᖃᒪᓂ'ᑐᐊᖅ (24), ᒥᑦᑎᒪᑕᓕᒃ (15), ᐊᒻᒪ ᐸᖕᓂᖅᑑᖅ (19) ᓄᓇᕗᒻᒥ; ᐅᒥᐅᔭᖅ (19), ᑲᖏᖅᓱᐊᓗᔾᔪᐊᖅ (9), ᐊᒻᒪ ᑲᖏᖅᓱᔪᐊᖅ (17) ᓄᓇᕕᖕᒥ; ᐊᒻᒪ ᓇᐃᓂ (23) ᓄᓇᑦᓯᐊᕗᒻᒥ (ᑕᑯᓗᒍ ᐊᔾᔨᖖᒍᐊᖅ 1). ᑕᒪᕐᒥᒃ 8 ᓄᓇᑖᑦ ᑲᓇᑕᒥ ᐅᑭᐅᖅᑕᖅᑐᕐᒥ ᓯᒡᔭᕐᒦᑦᑐᑦ, ᑭᑭᔅᓗᒍ ᖃᒪᓂ'ᑐᐊᖅ, ᓄᓇᐃᓐᓇᕐᒦᑦᑐᖅ ᑲᓇᑕᐅᑉ ᕿᑎᑦᓚᓇᑲᓴᖕᖓᓂ. ᐃᓚᒋᔭᐅᓚᐅᖅᑐᑦ 145 ᐃᓄᐃᑦ, ᐊᕝᕙᕐᒥ ᐊᒥᓲᓂᖅᓴᐃᑦ ᐊᕐᓇᐃᑦ ᐊᕝᕙᕐᒥ ᖃᔅᔫᓐᓇᐅᓂᖅᓴᐃᑦ ᐊᖑᑏᑦ, ᐅᑭᐅᖃᖅᑐᑦ 44-ᒥᑦ 92-ᒧᑦ. ᑐᓵᔨᖃᖅᓗᑕ, ᐊᐱᖅᓱᖅᑕᐅᔪᑦ ᐅᖃᐅᓯᖃᖃᑦᑕᓚᐅᖅᑐᑦ ᖃᐅᔨᒪᔭᕐᒥᓂᒃ ᐊᓯᔾᔨᖅᐸᑦᑕᐊᓂᐅᑉ ᒥᒃᓵᓄᑦ ᐅᑯᓇᓂ 6-ᓄᑦ ᐊᕕᒃᑕᐅᓯᒪᔪᓂᒃ: ᐸᐅᖕᖓᐃᑦ ᐊᓯᖏᑦᓗ ᐱᕈᖅᐸᒃᑐᑦ, ᓂᕐᔪᑎᑦ, ᐊᕐᕌᒍᒥ ᓯᓚᓂᒃ, ᓯᓚᐅᑉ ᖃᓄᐃᓐᓂᖕᒋᓐᓂᒃ, ᐊᒻᒪᓗ ᐃᓕᖅᑯᓯᕆᔭᐅᔪᓄᑦ ᐊᒃᑐᐃᓂᐅᔪᓂᒃ. ᐊᐱᖅᑯᑎᑦ ᐊᔾᔨᒌᖑᓚᐅᖅᑐᑦ ᑕᒪᐃᓐᓄᑦ ᐊᐱᖅᓱᖅᑕᐅᔪᓄᑦ, ᐃᓚᓕᐅᔾᔨᓯᒪᓪᓗᑎᒃ ᐊᔾᔨᖖᒍᐊᓂᒃ ᓄᓇᖖᒍᐊᓂᓪᓗ ᑕᒪᐃᓐᓂ ᓄᓇᓂ ᐊᕕᒃᑐᖅᓯᒪᔪᓂ. ᐱᖓᓱᑦ ᑎᓴᒪᓪᓘᓐᓃᑦ ᐊᑭᓐᓇᕐᒥᐅᑕᑦ ᐋᖅᑭᒃᓱᖅᑕᐅᓚᐅᖅᑐᑦ ᑕᒪᐃᓐᓄᑦ ᓄᓇᓕᖕᓄᑦ ᓂᕈᐊᖅᓯᓪᓗᑎᒃ ᐊᐱᖅᓱᖅᑕᐅᔪᑦ ᑭᒡᒍᑎᖏᓐᓂᑦ, ᑭᓯᑉ

Methods

Between 2007 and 2010, we interviewed 145 Elders and local knowledge holders from eight communities in three of the four Inuit regions of Canada: Kugluktuk (18 Elders interviewed), Baker Lake (24), Pond Inlet (15), and Pangnirtung (19) in Nunavut; Umiujaq (19), Kangiqsualujjuaq (9), and Kangiqsujuaq (17) in Nunavik; and Nain (23) in Nunatsiavut (see Figure 1). All eight Inuit communities are located in coastal sub-Arctic or Arctic Canada, with the exception of Baker Lake, which is inland, near the geographic centre of Canada. The participants included 145 individuals, roughly two-thirds women and one-third men, who ranged in age from 44 to 92 years. With the help of interpreters, interview participants discussed their knowledge of change across six broad categories: berries, other plants, animals, seasons, climate/weather, and impacts on traditional ways of life. We used the same questionnaire for each interview, accompanied by related photographs and maps of each region. Three or four posters were designed for

ᒥᒃᓵᓅᖓᔪᑦ ᑲᑎᖅᓱᖅᑕᐅᓪᓗᑎᒃ, ᐊᔾᔨᙳᐊᓂᒡᓗ ᐃᓚᓕᐅᔾᔨᓯᒪᓪᓗᑎᒃ. ᐊᑭᓐᓇᕐᒥᐅᑕᒃᓴᑦ ᐊᐱᖅᓱᖅᑕᐅᔪᓄᑦ ᐱᓕᕆᐊᖃᖅᑕᐅᔪᓄᑦ ᐊᖏᖃᑕᐅᓯᒪᓚᐅᖅᑐᑦ.

each community by selecting interviewees' quotes, grouped by theme, and illustrated with photographs. The poster content was validated with each participant.

ᐃᓚᖓ I
ᓄᓇᓕᖕᓂ ᐊᖏᑎᐅᑉ ᐊᓯᔾᔨᖅᐸᓪᓕᐊᓂᖕᒐᓄᑦ ᐅᖅᑲᐅᑕᐅᓯᒪᔪᓂᑦ ᓇᐃᓈᖅᓯᒪᔪᖅ

Part I
Cross-Community Summary of Environmental Change

ᐊᐱᖅᓱᕐᓂᑦᑎᒍᑦ ᐃᓄᑦᖃᐃᑦ ᑲᓇᑕᒥᓪᓗ ᐅᑭᐅᖅᑕᖅᑐᕐᒥ ᓄᓇᓖᑦ ᖃᐅᔨᓯᒪᓂᖅᐳᒍᑦ ᐅᔾᔨᕆᔭᐅᔪᓂᒃ ᐊᓯᔾᔨᖅᐸᓪᓕᐊᓂᐅᔪᓂᒃ ᐊᕙᑎᑦᑎᓂ. ᐊᓯᔾᔨᕐᓂᐅᔪᑦ ᐊᔾᔨᒌᒃᑐᐃᓐᓇᐅᙱᑦᑐᑦ ᓄᓇᓕᓕᒫᓂ; ᑭᓱᐊᓂ, ᐊᒥᓲᓂᒃ ᐱᒪᕆᐅᔪᓂᒃ ᐊᔾᔨᒌᙱᓂᖃᕆᓪᓗᑎᒃ. 6-ᙳᔪᓂᒃ ᐊᕙᒃᑐᖅᑕᐅᓯᒪᔪᓂ ᐊᐱᖅᑯᑕᐅᔪᓂ, ᑭᒡᒍᑕᐅᔪᑦ ᐊᓯᔾᔨᕐᓂᐅᑉ ᒥᒃᓵᓄᑦ ᓱᓕᒻ ᐱᔾᔪᑎᖃᓚᐅᖅᐳᑦ, ᐊᒻᒪ ᓄᓇᒥ ᐱᕈᖅᐸᒃᑐᓂᒃ ᐊᓯᔾᔨᕐᓂᐅᔪᓂᒃ. ᐃᓚᖏᑦ ᓄᓇᓖᑦ ᖃᐅᔨᓯᒪᓂᖅᑐᑦ ᐱᒻᒪᕆᐅᔪᓂᒃ ᐊᓯᔾᔨᕐᓂᐅᔪᓂᒃ ᐊᕙᑎᓕᒫᒥᓅᑦᑐᓂᒃ. ᐱᕐᑲᓯᐅᑦᑑᒃ ᒫᕐᕈᒃ ᐅᑭᐅᖅᑕᖅᑐᒥ ᓄᓇᓖᒃ ᓇᐃᓂᒥ (ᓄᓇᑦᓯᐊᕗᒻᒥ) ᐊᒻᒪ ᑲᖏᖅᓱᐊᓗᔾᔪᐊᕐᒥ (ᐅᖓᕙᒥ, ᓄᓇᕕᖕᒥ), ᐊᐱᖅᓲᖅᑕᐅᔪᑦ ᓇᑉᐸᖏᑦ ᐅᖓᑎᖏᓐᓂᓪᓚᓐᓃᑦ ᓇᓗᓇᐃᖅᓯᓪᓗᑎᒃ ᐊᓯᔾᔨᕐᓂᐅᔪᓂᒃ ᑕᒪᐃᓐᓂᓕᒫᖅ ᐊᕙᒃᑐᖅᑕᑦᑎᓐᓂ ᐊᐱᖅᑯᑎᓕᐊᕆᓯᒪᔭᑦᑎᓐᓂᒃ. ᐊᒻᒪ, ᖁᑦᑎᙵᓂᐸᐅᔪᓂ ᓄᓇᓕᙵᓂ ᖃᐅᔨᓴᖅᕕᐅᖃᑕᐅᓯᒪᔪᓂ, ᐸᖕᓂᖅᑑᖅᒥ ᒥᑦᑎᒪᑕᓕᖕᒥᓗ (ᖄᑭᖅᑖᓗᖕᒥ ᓄᓇᕗᒻᒥ), ᖃᑦᑐᓈᓄᑦᑐ ᖃᓂᓐᓂᖅᐸᐅᔪᒥ ᓄᓇᓕᖕᒥ, ᐅᒥᐅᔭᕐᒥ (ᑕᓯᐅᔭᕐᔪᐊᕐᒥ, ᓄᓇᕕᖕᒥ), ᐊᒥᓱᙵᓐᓂᖅᐹᓂᒃ ᐊᓯᔾᔨᖅᓯᒪᔪᓂᒃ ᐊᕙᑎᒥᙵ ᐅᔾᔨᕈᓱᒃᓯᒪᓪᓗᑎᒃ.

ᐊᐱᕆᔭᐅᒐᒥᒃ ᐱᕈᖅᑐᓂᒃ ᐊᓯᔾᔨᖅᓯᒪᓂᐅᔪᓂᒃ, ᑕᒪᐃᓐᓂᑲᓴᒃ ᓄᓇᓕᙵ ᓇᓗᓇᐃᖅᓯᓚᐅᖅᐳᑦ ᐱᕈᖅᑐᑦ ᓲᕐᓗ ᐅᖅᐲᑦ, ᐅᖅᐱᐊᓛᑦ ᐊᒻᒪ ᐊᕚᓚᕿᐊᑦ ᐱᕈᖅᓂᖅᓴᐅᕙᓕᕐᓂᖏᓐᓂᒃ. ᑲᖏᖅᓱᐊᓗᔾᔪᐊᕐᒥᐅᑦ ᐊᒻᒪ ᓇᐃᓂᒥᐅᑦ ᐅᔾᔨᕈᓱᒃᓯᒪᓕᖅᑐᑦ ᓇᐹᖅᑐᙱᑦ ᐅᓄᖅᓯᕙᓪᓕᐊᓂᖏᓐᓂᒃ.

Our interviews found that Elders and residents from Canadian Arctic communities are observing significant changes in their local environment. The changes are not uniform across all communities; however, there are many important similarities. Of the six environmental categories addressed in the questionnaire, the most frequent observations of change were related to changes in climate and weather, followed by changes in vegetation. Generally speaking, some communities are experiencing important changes across all aspects of their natural environment. This includes the two most easterly and sub-Arctic communities of Nain (Nunatsiavut) and Kangiqsualujjuaq (Ungava Bay, Nunavik), where half or more interviewees identified changes in every category of our questionnaire. Alternatively, the most northerly communities, Pangnirtung and Pond Inlet (Baffin Island, Nunavut), and the most southerly community, Umiujaq (Hudson Bay, Nunavik), demonstrated the fewest observations of environmental change.

When asked about changes in vegetation, the majority of participants in most

ᑲᖏᖅᓱᐊᓗᔾᔪᐊᕐᒥ, ᐃᓄᑐᖃᐃᑦ ᓇᓗᓇᐃᖅᓯᒋᓪᓗᑎᒃ ᐅᖅᐲᑦ, ᑭᖕᒥᖕᓇᐃᑦ ᐊᖅᐲᑦ ᐸᐅᕐᖓᐃᑦᑐ ᐱᕈᖅᓵᓕᓂᖅᓴᐅᕙᓕᕐᓂᖏᓐᓂ ᐊᕐᖀᒍᓂ ᐊᓂᒍᓵᖅᑐᓂ. ᑕᒪᐃᓐᓂᑦ 8-ᖑᔪᓂᑦ ᓄᓇᓕᖕᓂᑦ, ᐃᓄᑐᖃᐃᑦ ᓇᐃᓂᒥ ᐅᔾᔨᕈᓱᒃᓯᒪᕗᑦ ᐸᐅᕐᖓᐃᑦ ᐅᓄᕈᓐᓃᖅᐸᓪᓕᐊᓯᒪᓂᖏᓐᓂᒃ. ᐅᔾᔨᕈᓱᒃᓯᒪᓪᓕᑎᒃᑐ ᐸᐅᕐᖓᐃᑦ ᒥᑭᓐᓂᖅᓴᐅᕙᓕᕐᓂᖏᓐᓂᒃ ᑎᐱᖕᒋᑦᑐ ᐊᓯᔾᔨᖅᓯᒪᓕᕐᓂᖏᓐᓂᒃ. ᑭᒡᑐᒧᖕᒫᖅ, ᐃᓄᑐᖃᐃᑦ ᑲᖏᖅᓱᐊᓗᔾᔪᐊᕐᒥ ᐅᔾᔨᕈᓱᒃᓯᒪᓪᓗᑎᒃ ᐊᖅᐲᑦ ᐅᓄᕐᓂᖅᓴᐅᓕᕐᓂᖏᓐᓂᒃ ᕿᓚᒻᒥᐅᓂᖅᓴᕐᓗ ᐊᕝᕗᒐᒃᓴᐅᕙᓕᕐᓂᖏᓐᓂᒃ. ᒥᑦᑎᒪᑕᓕᖕᒥ, ᐃᓄᑐᖃᐃᑦ ᐅᔾᔨᕈᓱᒃᓯᒪᔪᑦ ᑭᒍᑕᖏᕐᓇᐃᑦ ᐱᕈᕐᓂᖅᓴᐅᕙᓕᕐᓂᖏᓐᓂᒃ.

ᓄᓇᓖᑦ ᐅᔾᔨᕈᓱᒃᓯᒪᓕᕐᒥᔪᑦ ᓂᕐᔪᑏᑦ ᐅᓄᕐᓂᖏᓐᓂᒃ. ᑲᖏᖅᓱᔪᐊᕐᒥ ᑲᖏᖅᓱᐊᓗᔾᔪᐊᕐᒥᓗ, ᐃᓄᑐᖃᐃᑦ ᐅᔾᔨᕈᓱᒃᓯᒪᔪᑦ ᓂᕐᔪᑏᑦ ᐅᓄᕐᓂᖅᓴᐅᓕᕐᓂᖏᓐᓂᒃ ᓴᖅᑭᒐᔪᖕᓂᖅᓴᐅᓕᕐᓂᖏᓐᓂᓪᓗ, ᓲᕐᓗ ᐊᒃᓚᐃᑦ ᐃᑦᓛᕈᓰᑦᑐ. ᒥᑦᑎᒪᑕᓕᖕᒥ ᓇᐃᓂᒥᓪᓗ ᐅᔾᔨᕈᓱᒃᓯᒪᔪᑦ ᐅᓄᖅᓯᕚᓪᓕᖅᓯᒪᔪᓂᒃ ᖃᔅᓲᓐᓇᐅᓂᖅᓴᐅᓕᖅᓯᒪᔪᓂᓪᓗ ᐆᒪᔪᓂᒃ. ᐆᒃᑑᑎᒋᓗᒍ, ᐃᓄᑐᖃᐃᑦ ᒥᑦᑎᒪᑕᓕᖕᒥ ᐅᖃᖅᓯᒪᔪᑦ ᓇᑦᑏᑦ ᐅᓄᕈᓐᓃᖅᐹᓪᓕᕐᓂᖏᓐᓂᒃ, ᐊᒻᒪ ᓇᐃᓂᒥ, ᐊᒃᓚᐃᑦ ᑐᒃᑐᕙᐃᑦᑐ ᓄᓇᓕᖕᒋᑕ ᖃᓂᒋᔭᖕᒋᓐᓂ ᐅᓄᖅᓯᕚᓪᓕᖅᓯᒪᓂᖏᓐᓂᒃ, ᑐᒃᑐᐃᑦᑐ ᐱᑕᖃᙱᓐᓂᖅᓴᐅᓕᕐᓂᖏᓐᓂᒃ. ᓇᐃᓂᒥᐅᑦ ᐅᖃᖅᓯᒪᖕᒥᔪᑦ ᑎᖕᒥᔾᔭᑦ ᐅᓄᕐᓂᖅᓴᐅᓕᕐᓂᖏᓐᓂᒃ, ᐊᒥᓱᑦᑐ ᓄᓇᓖᑦ (50 ᐳᓴᓐᑎ ᑐᖕᒑᓂᐅᒐᓗᐊᖅ)

communities identified that the abundance of shrubs, such as willow, dwarf birch, and alder, is increasing. Residents from Kangiqsualujjuaq and Nain also observed that trees such as larch and spruce are increasing in abundance. In Kangiqsualujjuaq, Elders further identified that plants such as willow, cloudberry/bakeapple, and blackberry/crowberry appear to be blooming earlier than in past years. Among all the eight communities, Elders in Nain are observing the most significant declines in berry abundance, mainly in cloudberries/ bakeapples and blueberries. They have further recognized a decrease in berry size and changes to their taste. In contrast, Elders in Kangiqsualujjuaq are observing increases in abundance and earlier ripening times for cloudberry/bakeapple. In Pond Inlet, Elders have observed increases in blueberry abundance.

Communities are also experiencing significant changes in mammal abundance. In Kangiqsujuaq and Kangiqsualujjuaq, Elders have observed increases in animal abundance and occurrence, such as in black bear and porcupine. In Pond Inlet and Nain, Elders have seen both

ᐅᕐᑲᖅᓯᒪᓪᓗᑎᒃ ᓄᑖᓂᒃ ᑎᖕᒥᐊᓂᒃ ᓄᓇᓕᖕᒋᑦ ᑎᑭᑕᐅᕙᓕᕐᓂᖕᒋᓐᓂᒃ (ᓲᕐᓗ ᑎᖕᒥᐊᑦ ᕿᐱᓐ ᑲᖕᒋᖅᓱᐊᕐᒥ ᐅᒥᐅᔭᕐᒥᓗ). ᑕᒪᐃᓐᓂ ᓄᓇᓕᖕᓂ, ᐊᔾᔨᒌᖕᓂᕐᑲᓗᐊᓚᐅᖖᒋᑦᑐᖅ ᕐᑯᐱᕐᕈᓂᒃ; ᑭᓯᐊᓂ ᕐᑯᕐᓗᖅᑐᕐᒥ, ᐃᓄᑐᕐᑲᐃᑦ ᐊᐱᖅᓱᖅᑕᐅᔪᑦ ᓇᑉᐸᖕᒋᑕ ᐅᖕᒐᑎᐊᓃᑦᑐᑦ ᐅᕐᑲᖅᓯᒪᓪᓗᑎᒃ ᐊᓇᖖᒌᑦ ᐅᓄᕈᓐᓃᖅᐹᓪᓕᖅᓯᒪᓂᖕᒋᓐᓂᒃ. ᐃᓚᖕᒋᑦ ᓄᓇᓖᑦ ᑕᑯᕐᑲᑦᑕᖅᓯᒪᓕᕐᓂᖁᖅᑐᑦ ᓄᑖᓂᒃ ᑕᑯᓚᐅᖅᓯᒪᖖᒋᑕᒥᖕᓂᒃ ᕐᑯᐱᕐᕈᓂᒃ ᓲᕐᓗ ᑕᕐᕿᓕᑮᓴᕐᓂᒃ, ᐃᒍᑦᑕᕐᓂᒃ, ᑭᑦᑐᓯᕐᓂᓪᓗ ᓄᓇᕕᖕᒥ.

ᐊᔾᔨᐸᓗᖕᒋᓐᓂᒃ ᐊᓯᔾᔨᕐᓂᐅᓯᒪᓕᖅᑐᑦ ᐱᕈᖅᑐᓄᑦ, ᐸᐅᕐᖓᕐᓄᑦ, ᓂᕐᔪᑎᓄᓪᓗ, ᓄᓇᓖᑦ ᐅᔾᔨᕈᓱᒃᓯᒪᓕᕐᒥᔪᑦ ᓯᓚ ᐅᖅᑰᓯᓴᓕᓂᖅᓴᐅᕙᓕᕐᓂᖕᒐᓂᒃ ᐊᐅᔭᑯᑖᖑᓂᖅᓴᐅᕙᓕᕐᓂᖕᒐᓄᓪᓗ. ᐃᓄᑐᕐᑲᐃᑦ ᐊᐱᖅᓱᖅᑕᐅᔪᑦ ᓇᑉᐸᖕᒋᓐᓂᒃ ᐅᓄᕐᓂᖅᓴᐃᑦ ᕐᑯᕐᓗᖅᑐᕐᒥ ᓇᐃᓂᒥᓪᓗ ᐅᔾᔨᕈᓱᒃᓯᒪᓕᖅᑐᑦ ᓯᓚᐅᑉ ᐊᓯᔾᔨᖅᓯᒪᓕᕐᓂᖕᒋᓐᓂᒃ ᐅᑭᐅᓕᒫᒥ, ᐃᓄᑐᕐᑲᐃᓪᓗ ᑲᖕᒋᖅᓱᐊᕐᒥ ᓇᓗᓇᐃᖅᓯᓯᒪᓪᓗᑎᒃ ᐊᐅᔭᒃᑯᑦ ᐊᓯᔾᔨᖅᓯᒪᓂᐅᓕᖅᑐᓂᒃ. ᑕᒪᐃᓐᓂᑲᓴᑦ, ᓄᓇᓕᓕᒫᑦ ᓇᓗᓇᐃᖅᓯᓯᒪᔪᑦ ᓯᓚᐅᑉ ᐊᓯᔾᔨᖅᑕᕐᓂᖅᓴᐅᓕᖅᓯᒪᓂᖕᒋᓐᓂᒃ ᐊᒻᒪ ᓇᓚᐅᑦᑖᒐᕐᓂᖕᒋᓐᓂᖅᓴᐅᓕᕐᓂᖕᒐᓂᒃ ᓯᓚ ᕐᑲᓄᐃᑦᑑᓂᐊᕐᒪᖖᒌᑦ. ᑭᓯᐊᓂ ᐊᖕᒋᕐᑲᑎᒌᒃᑐᐃᓐᓇᑦ (ᐊᐱᖅᓱᖅᑕᐅᔪᑦ ᓇᑉᐸᖕᒋᑕ ᐅᖕᒑᓃᑦᑐᑦ) ᐱᖕᒐᓲᓂᑦ ᓄᓇᓕᖕᓂᑦ (ᕐᑲᒪᓂ'ᑐᐊᖅ, ᑲᖕᒋᖅᓱᐊᖅ, ᐊᒻᒪ ᓇᐃᓂ). ᐃᓄᐃᑦ ᐅᕐᑲᖅᓯᒪᔪᑦ ᓯᓚᐅᑉ ᐊᓯᔾᔨᖅᓯᒪᓕᕐᓂᖕᒐᓄᑦ ᓄᒃᑕᕐᓂᖅ ᐊᐅᓪᓛᕐᓂᖅ ᐊᒃᑐᖅᑕᐅᓯᒪᓕᕐᓂᖕᒐᓄᑦ,

increases and decreases in different species. For example, Elders in Pond Inlet described decreases in ringed seal abundance, and in Nain, black bears and moose are becoming increasingly abundant near the community, whereas caribou have become scarcer. Residents in Nain have also noted that birds are more abundant now, and many communities (albeit less than 50%) have noted new bird species around their communities (e.g., robins in Kangiqsujuaq and Umiujaq). For all communities, there was little consensus on changes in biting insects; however, in Kugluktuk, more than half of the Elders interviewed noticed a decrease in black fly abundance. Some communities are also observing new insect species, such as new species of butterflies, bumblebees, and dragonflies in Nunavik.

Concurrent with changes in vegetation, berries, and animals, communities are experiencing changes in the timing and length of seasons. More than half of the Elders interviewed in Kugluktuk and Nain observed changes in all four seasons, whereas Elders in Kangiqsujuaq identified changes in summer. In general, all communities identified greater variability

ᐃᓕᖅᑯᓯᖕᖏᓐᓄᑦ ᐊᖑᓇᓱᖕᓂᖕᖏᓐᓄᓪᓗ
ᐊᒃᑐᐃᓯᒪᓂᖕᒐᓂᒃ.

ᐊᐱᕆᔭᐅᒐᒥᒃ ᓯᓚᐅᑉ ᒥᒃᓵᓄᑦ,
ᐃᓄᑐᖃᐃᑦ ᑲᖏᖅᓱᐊᓗᔾᔪᐊᕐᒥ
ᐅᖃᓚᐅᖅᑐᑦ ᐅᑭᐅᓕᒫᖅ ᓯᓚ
ᐅᖅᑰᓂᖅᓴᐅᕙᓕᕐᓂᖕᒐᓂᒃ,
ᐃᓄᑐᖃᐃᓪᓕ ᓇᐃᓂᒥ ᐅᖃᖅᖢᑎᒃ
ᐅᐱᕐᖓᒃᑯᑦ ᐊᐅᔭᒃᑯᓪᓗ
ᓂᓪᓕᓱᖕᓂᖅᓴᐅᕙᓕᕐᓂᖕᒐᓂᒃ. 6-ᖑᔪᑦ
8-ᓂᑦ ᓄᓇᓕᖕᓂᑦ ᐅᓂᒃᑳᖃᑦᑕᓚᐅᖅᑐᑦ
ᐊᓄᓈ ᐊᓯᔾᔨᖅᓯᒪᓕᕐᓂᖕᒐᓂᒃ, ᓲᕐᓗ
ᐊᓄᓈᐅᑉ ᓴᕿᕙᖕᓂᖕᖓ, ᐊᒻᒪᓗ
ᐊᓄᕌᕐᔪᐊᕐᓂᖅᓴᐅᕙᓕᕐᓂᖕᒐᓂᒃ
ᐊᓄᕌᖓᔪᖕᓂᖅᓴᐅᓕᕐᓂᖕᒐᓂᓪᓗ.
ᒪᕐᑯᒃᑕᕐᓃᑦ ᐊᔾᔨᒌᙱᑦᑐᑦ
ᓄᓇᓂ ᐊᒥᒃᑐᖅᓯᒪᔪᓂ.
ᒪᕐᕉᒃ ᐅᐊᓕᓂᖅᐸᓯᓖᖑᔫᒃ
ᖁᕐᓗᖅᑐᖅ ᐊᒻᒪ ᖃᒪᓂ'ᑐᐊᖅ,
ᒪᕐᑯᒐᔪᙱᓐᓂᖅᓴᐅᓂᕋᖅᖢᑎᒃ,
ᒪᕐᕉᒃ ᑲᓇᖕᓂᖅᐸᓯᓛᒃ
ᓇᐃᓂ ᐊᒻᒪ ᑲᖏᖅᓱᔪᐊᖅ,
ᒪᕐᑯᒐᔪᖕᓂᖅᓴᐅᓕᕐᓂᕋᖅᖢᑎᒃ.
ᐊᒥᓲᓂᖅᓴᐃᑦ ᓄᓇᓖᑦ
ᐅᔾᔨᕈᓱᖕᓂᕋᓚᐅᖅᑐᑦ
ᐊᐳᑎᖅᑲᙱᓐᓂᖅᓴᐅᕙᓕᕐᓂᖕᒐᓂᒃ.
ᖁᑦᑎᒃᑐᕐᒥᐅᑕᐅᓂᖅᐹᒃ
ᑕᐃᒪᐃᓐᓂᕋᙱᑦᑐᑑᑦᓗᑎᒃ
(ᒥᑦᑎᒪᑕᓕᒃ ᐊᒻᒪ ᐸᖕᓂᖅᑑᖅ),
ᑕᒪᐃᓐᓄᑦ ᐊᔾᔨᒌᙱᔾᔪᑕᐅᓪᓗᓂ, ᐊᒻᒪ
ᖃᓪᓗᓈᑦ ᓄᓇᖓᓐᓄᑦ ᖃᓂᓐᓂᖅᐹᖅ
(ᐅᒥᐅᔭᖅ), ᐊᐱᖅᓱᖅᑕᐅᔪᑦ
ᐊᒥᓲᓂᖅᓴᑦ ᐊᐳᑎᖅᑲᖅᐸᖕᓂᖓ
ᐊᓯᔾᔨᖅᓯᒪᙱᓐᓂᕋᖅᑕᐅᓪᓗᓂ.
ᑲᖏᖅᓱᔪᐊᕐᒥᐅᑦ
ᑲᖏᖅᓱᐊᓗᔾᔪᐊᕐᒥᐅᓪᓗ ᐅᖃᓚᐅᖅᖢᑎᒃ
ᐊᐳᑦ ᐊᐅᓵᓕᓂᖅᓴᐅᕙᓕᕐᓂᖕᒐᓂᒃ.
ᐊᐱᖅᓱᖅᑕᐅᔪᑦ ᐅᖃᖃᑦᑕᓚᐅᖅᑐᑦ
ᐊᓄᓈᐅᑉ ᐊᓯᔾᔨᖅᓯᒪᓕᕐᓂᖕᒐᓄᑦ
ᐊᐳᑎᐅᑉ ᐃᔾᔪᓂᖕᖏᑦ

in their weather, making it more difficult to predict. However, consensus (over half of interviewees) was obtained in only three communities (Baker Lake, Kangiqsujuaq, and Nain). People expressed that this increased variability has affected travel on the land, thus having important effects on traditional activities.

When asked about changes in specific aspects of the weather, Elders in Kangiqsualujjuaq described experiencing warmer weather throughout the year, whereas Elders in Nain are experiencing cooler weather in spring and summer. Six out of eight communities identified changes in the wind, such as shifts in direction, as well as greater intensity and frequency. Rainfall appears to vary regionally. The two most westerly communities, Kugluktuk and Baker Lake, noted less rainfall, whereas the two most easterly communities, Nain and Kangiqsujuaq, noted more rainfall. The majority of communities observed a decrease in snow abundance. Exceptions include the most northerly communities (Pond Inlet and Pangnirtung), where consensus was not achieved, and the most southerly community

ᐊᔾᐳᒌᒃᐸᙱᓕᕐᓂᖏᓐᓄᑦ ᐊᐅᓪᓛᖅᑲᑦᑕᓇᐊᒃᓴᖅ ᐊᔭᕐᓇᕐᓂᖅᓴᐅᕙᓕᕐᓂᖓᓂᒃ ᓄᒃᑕᕐᕕᐅᕐᑲᑦᑕᖅᑐᑐᕐᑲᐃᑦᓗ ᐅᐸᒐᕐᓂᙱᓐᓂᖅᓴᐅᓕᕐᓂᖏᓐᓂᒃ. ᓄᓇᓕᓕᒫᑲᓴᐃᑦ ᐅᔾᔨᕆᓱᒃᓯᒪᓚᐅᖅᑐᖅ ᓯᓚᐅᑉ ᐊᓯᔾᔨᕐᓂᖓᓂ; ᑕᐃᒪᐃᓚᐅᖏᑦᑑᒃ ᓄᓇᓖᒃ ᒪᕐᕉᒃ, ᕿᑭᖅᑖᓗᖕᒦᑦᑑᒃ, ᐸᖕᓂᖅᑑᒥ ᒥᑦᑎᒪᑕᓕᖕᒥᓪᓗ. ᑰᑦ, ᑰᒐᓛ, ᑕᓲᓪᓗ ᐃᒪᑭᓐᓂᖅᓴᐅᓕᕐᓂᖏᓐᓂᒃ, ᓄᓇᓗ ᐸᓂᖅᓯᒪᓂᖅᓴᐅᓕᖅᓱᓂ. ᐆᒃᑑᑎᒋᓗᒍ, 2008-ᖑᑎᓪᓗᒍ ᑲᖏᖅᓱᐊᓗᔾᔪᐊᕐᒥ ᓄᓇᖓ ᐃᑭᓕᓚᐅᖅᓯᒪᖕᒪᑦ, ᐊᐱᖅᓱᖅᑕᐅᔪᓪᓗ ᐅᕐᑲᓚᐅᖅᓱᑎᒃ ᓄᓇᒥᒃ ᐃᑭᑦᑐᒥᒃ ᑕᑯᓚᐅᖅᓯᒪᙱᑦᑎᐊᕐᓂᕋᖅᓱᑎᒃ ᑕᒪᑐᒪ ᓯᕗᕐᖓᒍᑦ.

ᓯᑰᑦᑕᐅᖅ ᖃᓄᐃᓐᓂᖓ ᐊᓯᔾᔨᖅᐸᓪᓕᐊᓯᒪᓕᕐᒥᔪᖅ ᑕᒪᐃᓐᓂ ᓄᓇᓕᖕᓂ, ᐊᒥᓲᓂᖅᓴᖅ ᐃᓄᑐᖅᑲᑦ ᐊᐱᖅᓱᖅᑕᐅᔪᑦ ᖁᕐᓗᖅᑐᕐᒥ ᓇᐃᓂᒥᓪᓗ ᐅᕐᑲᓚᐅᖅᑐᑦ ᓯᑯᓇᓰᕐᓂᖅᓴᐅᕙᓕᕐᓂᖓᓂᒃ. ᑐᓴᕈᒥᓇᓗᐊᖅᑐᖅ ᑲᖏᖅᓱᔪᐊᕐᒥ, ᐊᕐᕌᒍᑕᒫᑦ ᔮᓄᐊᕆ 1-ᒥ ᖃᒧᑕᐅᔭᒃᑯᑦ ᕿᒧᒃᓯᒃᑯᓪᓘᓐᓃᑦ ᓱᒃᑲᓴᐅᔾᔭᐊᖅᐸᒃᑐᑦ 2011-ᖑᑎᓪᓗᒍ ᐅᒥᐊᒃᑯᑦ ᓱᒃᑲᓴᐅᔾᔭᐊᓇᐊᕐᑲᓚᐅᖅᓯᒪᔪᑦ ᓯᑯᕐᑲᙱᓗᐊᕐᓂᖓᓄᑦ. ᐱᖓᓐᓂᖅᐸᓯᒃᑐᑦ ᓄᓇᓖᑦ ᐅᕐᑲᓚᐅᖅᑐᑦ ᓯᑯᐃᖅᓰᓕᓂᖅᓴᐅᕙᓕᕐᓂᖓᓂᒃ. ᑲᖏᖅᓱᔪᐊᕐᒥ, ᑲᖏᖅᓱᐊᓗᔾᔪᐊᕐᒥ ᓇᐃᓂᒥᓗ ᐃᓄᑐᖅᑲᐃᑦ ᐅᔾᔨᕆᓚᐅᕐᒥᔪᑦ ᓯᑯ ᓵᓐᓂᖅᓴᐅᕙᓕᕐᓂᖓᓂᒃ. ᐃᓄᑐᖅᑲᐃᑦ ᐅᔾᔨᕆᓗᐊᓚᐅᙱᑦᑐᑦ ᑕᓲᑦ ᑰᓪᓗ ᓯᑯᕙᖕᓂᖏᑦ ᐊᓯᔾᔨᖅᓯᒪᒋᐊᖏᓐᓂᒃ,

(Umiujaq), where a majority of interviewees observed no change in snow abundance. Residents of Kangiqsujuaq and Kangiqsualujjuaq further noted they are experiencing earlier snowmelt. Participants noted that observed wind changes have affected snow deposition patterns, thus influencing the accessibility of traditional travelling routes. In most communities, a majority of interviewees observed hydrological changes; this was not the case in only two communities, the ones on Baffin Island, Pangnirtung and Pond Inlet. These changes were mostly related to lower water levels in rivers, small streams, lakes, and ponds, causing the drying of the tundra. As an example, a tundra fire was observed in 2008 in Kangiqsualujjuaq, something the interviewees had never observed before. Conditions in the sea ice are also changing across communities, with the majority of participants from Kugluktuk and Nain observing later sea ice freeze-up. An interesting example of this is in Kangiqsujuaq, where the traditional January 1 snowmobile/dogsled race had to be converted to the first ever boat race in 2011

ᑭᓯᐊᓂ ᑲᖏᖅᓱᔪᐊᕐᒥᐅᑦ ᐅᔾᔨᕈᓱᒃᓯᒪᓂᕋᖅᖢᑎᒃ, ᕿᓚᒻᒥᐅᓂᖅᓴᖅ ᑕᓲᑦ ᓯᑯᐃᖅᐸᓕᕐᓂᖏᓐᓄᑦ. ᑕᑦᑕᒪᐅᔪᑦ 8-ᖑᔪᓂᑦ ᓄᓇᓕᖕᓂᑦ ᓇᓗᓇᐃᖅᓯᓚᐅᖅᑐᑦ ᓄᓇᒥ ᖁᐊᖑᐃᓐᓇᖅᐸᒃᑑᒐᓗᐊᑦ ᐊᐅᒃᐸᓕᕐᓂᖏᓐᓂᒃ, ᐱᖓᓲᑦᑐ ᑖᒃᑯᓇᖖᒐᑦ ᑕᑦᑕᒪᓂᑦ ᓄᓇᓕᖕᓂᑦ ᐅᖃᖅᑐᑎᒃ ᓄᓇᖏᑦ ᑎᓯᐊᖃᑦᑕᓕᕐᓂᖏᓐᓂᒃ ᐊᐅᒃᐸᑦᓕᐊᓂᕐᒍᑦ.

ᑭᒡᒍᑕᐅᔪᑦ ᕿᒥᕐᕈᓪᓗᒋᑦ ᐃᓚᖏᑦ ᐊᓯᔾᔨᖅᓯᒪᓂᐅᔪᑦ, ᓲᕐᓗ ᐱᕈᖅᑐᖅᑲᕐᓂᖅᓴᐅᓕᕐᓂᖅ, ᐊᐳᑎᑭᓐᓂᖅᓴᐅᓕᕐᓂᖅ, ᐊᓄᕌᕐᓂᖅᓴᐅᕙᓕᕐᓂᖓᓂᒃ, ᖁᐊᖑᐃᓐᓇᖅᐸᒃᑐᑦ ᐊᐅᒃᐸᓕᕐᓂᖏᓐᓂᒃ, ᐃᒪᑭᓐᓂᖅᓴᐅᓕᕐᓂᖓᓂᒃ, ᓯᑯᐃᖅᓵᓕᕙᓕᕐᓂᖓᓂᒃ, ᓄᑖᓂᒡᓗ ᐱᕈᖅᑐᓂᒃ ᐆᒪᔪᓂᒡᓗ (ᑎᖕᒥᐊᑦ ᖁᐱᕐᕈᓪᓗᐊᑕᑦ), ᐊᔾᔨᒌᐸᓗᖕᓂᖏᓐᓂᒃ ᐅᑭᐅᖅᑕᖅᑐᕐᒥ ᓄᓇᓕᖕᓂ, ᐅᔾᔨᕆᔭᐅᓚᐅᖅᑐᑦ ᓄᓇᓖᑦ ᐊᐱᖅᓱᖅᑕᐅᔪᑦ ᓇᑉᐸᖏᑕ ᐅᖓᑎᖏᓐᓃᑦᑐᓄᑦ. ᐊᓯᖏᑦ ᐊᓯᔾᔨᕐᓂᐅᔪᑦ ᓄᓇᓕᖕᓂᑐᐊᖅ ᐅᔾᔨᕆᔭᐅᔪᑦ ᓲᕐᓗ ᐸᐅᕐᖓᑭᓐᓂᖅᓴᐅᓕᕐᓂᖓ ᓇᐃᓂᒥ ᐊᒻᒪ ᐊᓇᖖᒋᖅᑲᖏᓐᓂᖅᓴᐅᓕᕐᓂᖓ ᖁᕐᓗᖅᑐᕐᒥ. ᑕᒪᓐᓇ ᓇᓗᓇᐃᖅᓯᕗᖅ ᐱᒻᒪᕆᐅᓂᖓᓂᒃ ᐊᒥᓱᓂᒃ ᓄᓇᓕᖕᓂᒃ ᑕᒪᐃᓐᓂ ᓄᓇᓂ ᐊᕕᒃᑐᖅᓯᒪᔪᓂ ᖃᐅᔨᓴᕆᐊᖃᕐᓂᖅ, ᐊᕙᑎᑦᑕ ᐊᓯᔾᔨᖅᐸᑦᓕᐊᓂᖓ ᐊᔾᔨᒌᒃᐸᖖᒋᓐᓂᖏᓐᓄᑦ ᓄᓇᓕᖕᓂ ᓄᓇᓂᓗ ᐊᕕᒃᑐᖅᓯᒪᔪᓂ ᐊᔾᔨᒌᖖᒋᑦᑐᓂ.

because of a lack of sea ice. All easterly communities noted earlier sea ice breakup than in previous years. In Kangiqsujuaq, Kangiqsualujjuaq, and Nain, Elders are also noticing decreases in sea ice thickness. Elders observed very few changes in lake and river ice conditions, with the exception of Elders in Kangiqsujuaq, who identified earlier lake ice breakup. Five out of eight communities identified increased permafrost thawing, and three of those five further noted increased erosion in their territory.

Overall, the results of this comparison show that some changes, such as more shrubs, less snow, more wind, thawing permafrost, lower water levels, earlier sea ice breakup, and new plant and animal species (mostly birds and insects), are common across Northern communities, as seen by more than half of the communities surveyed. Other changes seem to be more locally driven, such as declines in berry abundance and quality in Nain and fewer black flies in Kugluktuk. This emphasizes the importance of consulting with numerous communities within each region, as environmental changes can be driven by both local and regional factors.

ᐃᓚᖓ 1: ᓄᓇᓕᖕᓂ ᐊᕙᑎᐅᑉ ᐊᓯᔾᔨᖅᐸᑦᓕᐊᓂᖓᓄᑦ ᐅᖃᐅᑕᐅᓯᒪᔪᓂᑦ ᓇᐃᓈᖅᓯᒪᔪᖅ

ᐃᓚᖓ II
ᐊᑐᓂ ᓄᓇᓖᑦ ᐅᓂᒃᑳᓚᐊᓇᓯᒪᔭᖏᑦ ᐊᕙᑎᒥ ᐊᓯᔾᔨᖅᐸᓪᓕᐊᓂᖑᑦ

Part II
Individual Community Summaries of Environmental Change

ᓄᓇᕗᑦ Nunavut

ᖃᒡᓗᒃᑐᒃ
Kugluktuk

ᖁᕐᓗᖅᑐᖅ

Kugluktuk

ᖁᕐᓗᖅᑐᖅ ᓄᓇᓕ ᑰᖓᑕ ᐹᖓᓂ, ᑕᐃ'ᔪᓘᑦᑐᕈᒥ, ᐃᓚᒋᓪᓗᓂᐅᒃ ᑕᓪᓗᓇᑎᐅᑉ ᐃᒪᖓᑕ (N 67.83° W 115.08°; ᖁᑦᑎᖓᓂᖓ 23 m). ᐅᑭᐅᖅᑕᖅᑑᓂᕐᒦᑦᑐᖅ ᐊᒃᑎᑦᑐᒥ ᓇᐹᖅᑐᖃᖅᑐᒍᑦ ᖃᓂᑦᑐᒥ, ᓄᓇᖓ ᐅᖅᐱᖃᐅᖅᑐᖅ ᐃᑭᐊᖓᓗ ᓄᓇᐅᑉ ᖁᐊᖑᓪᓗᓂ. ᓄᓇᓕᖓ ᐃᓄᖃᖅᑐᒋᔭᖅ 1450 (ᑲᓇᑕᒥ ᓈᓴᐃᓅᑦ 2011 ᓈᖅᓴᖅᑕᐅᓂᖏᑦ).

ᖁᕐᓗᖅᑐᕐᒥᐅᑦ ᐊᖏᖃᑎᒌᒍᓐᓇᑦᑎᐊᓚᐅᖏᑦᑐᑦ ᐱᕈᖅᑐᑦ ᐊᓯᔾᔨᕐᓂᖏᑕ ᒥᒃᓵᓄᑦ ᐊᕝᕙᖏᓐᓂᒃ (<50%), ᑭᓯᐊᓂ ᐊᖏᖃᑎᒌᖃᑦᑕᓚᐅᖅᑐᑦ ᐅᔾᔨᕈᓱᖕᓂᖏᓐᓂᒃ ᐊᕙᑎᐅᑉ ᐊᓯᔾᔨᕐᓂᖏᓐᓂᒃ ᐊᔾᔩᖏᑦᑐᓂ ᓄᓇᓕᖓᓐᓂ. ᐊᐱᖅᓱᖅᑕᐅᔪᑦ ᐊᒥᓲᓂᖅᓴᐃᑦ ᐅᔾᔨᕈᓱᒃᓯᒪᔪᑦ ᐅᖅᐱᓕᓪᓚᓇᐊᓘᓕᓇᐊᖓ. ᐊᓯᔾᔨᕐᓂᕋᖅᑕᐅᓚᐅᕐᒥᔪᑦ ᐱᕈᖅᑐᑦ ᐱᕈᖅᓴᕋᐃᑦᑐᓐᓃᓇᐊᖏᑦ, ᐃᕕᒃᓱᒃᑲᑦ ᐊᒥᓲᓕᓇᐊᖏᑦ, ᑲᖕᒎᔭᐃᓪᓗ ᖃᔅᓯᓘᓇᕈᕐᓂᖅᓴᐅᕆᐊᖏᑦ.

ᐊᐱᖅᓱᖅᑕᐅᓚᐅᖅᑐᑦ ᐅᖃᓚᐅᕐᒥᔪᑦ ᒥᑐᒋᐊᑦ ᑭᑦᑐᓇᐊᑦ ᐃᒍᑦᑕᐃᓪᓗ ᖃᔅᓯᓘᓇᕈᓇᐊᖓ ᖁᕐᓗᖅᑐᖅᐸᓯᖓᒥ. ᐆᒪᔪᓂᒃ ᑕᑯᔭᐅᒐᓚᖏᑦᑐᓂᒃ ᐱᑕᖃᖃᑦᑕᓕᓇᐊᖓ, ᑎᓇᐊᕐᔪᐊᑦ, ᑕᑯᔭᐅᓯᒪᓕᖅᑐᑦ ᐊᒻᒪᓗ ᑎᖕᒥᐊᑦ ᐊᔾᔩᖏᑦᑐᑦ, ᓇᓗᓇᐃᖅᑕᐅᓪᓗᑎᒃ ᕿᕐᓂᐅᓪᓗᑎᒃ ᖁᔾᔫᓂᖃᖅᑐᑎᒎᖅ ᒥᑏᑦ ᐊᒻᒪᓗ ᖁᖅᓱᖅᑐᑦ ᐊᒻᒪᓗ ᕿᕐᓃᖅ ᖁᐸᓄᐊᑦ. ᐊᒪᑕᐃᓐᓇᖅ, ᐊᒃᕵᑕᖅᑲᒐᔪᒃᓯᔪᐊᓗᒃ.

ᑕᒪᒃᑭᑦ ᐊᐱᖅᓱᖅᑕᐅᓚᐅᖅᑐᑦ ᐅᖃᓚᐅᕐᒥᔪᑦ ᓯᓚᓗᒐᔪᐃᑦᑐᐊᓘᓕᕐᓂᕋᖅᑕᐅᓪᓗᓂ. ᑕᐃᒪᐃᓐᓂᖓᓄᑦ ᐃᒪᐃᑦ

Kugluktuk is located at the mouth of the Coppermine River on Coronation Gulf, which is part of the Northwest Passage (N 67.83°, W 115.08°; elevation 23 m). It is part of the Low Arctic, near the treeline, and is characterized by low shrub tundra vegetation and continuous permafrost. The community was established in the 1930s and its population was estimated at 1,450 in the 2011 Statistics Canada Census.

Kugluktukmiut did not come to much consensus (<50%) about vegetation change specifically, but they did agree on a number of environmental changes occurring in their community. The majority of people interviewed noticed an increase in shrub growth. Other observed changes include later plant blooming times, increased grass abundance, and decreases in Arctic cottongrass abundance.

Participants also noted fewer black flies, mosquitoes, and bumblebees in the Kugluktuk area. New animal species, including marten, have been observed, and some people have also seen new bird species, described as black-and-white ducks and little yellow or black birds. Finally, black bears

ᐊᒃᑎᖕᓂᖅᓴᐅᓕᕆᐊᖕᒐ ᑕᓯᕐᓂᒃ ᑰᖕᓂᑦᑐ ᓄᓇᑐ ᐸᓂᖅᑐᐊᓘᓕᕐᓂᕋᖅᑕᐅᑦᑐᓂ. ᐃᓄᐃᑦ ᐅᖃᖃᑦᑕᓚᐅᖅᑐᑦ ᓂᕐᓇᐃᑦ ᐱᔪᖕᒋᓐᓂᖅᓴᐅᓕᕆᐊᖕᒋᑕ, ᑭᒍᑕᖕᒋᕐᓇᐃᑦ ᐸᐅᕐᖕᒐᐃᑦᑐ ᒥᑭᓐᓂᖅᓴᐅᖃᑦᑕᓕᖅᑐᑦ ᐸᓂᖅᓴᕋᐃᕐᑐᑯᓘᓕᖅᑐᑎᒡᒍ. ᐊᐱᖅᓱᖅᑕᐅᓚᐅᖅᑐᑦ ᐅᖃᓚᐅᕐᒥᔪᑦ ᐅᔾᔨᕆᓱᒃᓯᒪᓂᕋᖅᑐᑎᒃ ᐊᓄᓂ ᓴᙱᓂᖅᓵᓘᓕᕆᐊᖕᒐ, ᑎᑭᑦᓚᑎᑲᑦᓚᖃᑦᑕᕐᓂᖅᓴᐅᓕᕆᐊᖕᒐ ᐊᒻᒪᓗ ᐊᓄᕌᒐᔪᖕᓂᖕᒐ ᐊᓯᔾᔨᖅᓯᒪᓕᕆᐊᖕᒐ. ᐊᐱᓇᓵᖃᑦᑕᓕᕆᐊᖕᒐᓗ ᐊᓄᓂᑐᔪᓄᑦ ᓱᐱᔅᐅᖃᑦᑕᑐᐃᓐᓇᓕᕆᐊᖕᒐ. ᐃᓄᐃᑦᑐᒎᖅ ᐅᖃᐅᑎᖃᖃᑦᑕᓚᐅᕐᒥᔪᑦ ᓄᓇᐅᑉ ᐃᑭᐊᖕᒐᓃᑦᑐᒥᒃ ᓱᑯᐊᓗᖕᒥᒃ ᐊᐅᓕᕆᐊᖕᒐ ᐊᒪᑦᑐ ᖁᕐᓗᖅᑑᑉ ᑰᖕᒐᑕ ᓯᓈ ᓄᖑᓕᕆᐊᖕᒐ. ᖁᕐᓗᖅᑐᕐᓗ ᓯᑯᓇᓵᖃᑦᑕᓕᕆᐊᖕᒐ ;ᐅᖃᐅᑕᐅᓚᐅᕐᒥᔪᖅ, ᐊᓄᓂ ᐱᓗᖅᑯᓯᐅᑎᑦᑐᓂ. ᐃᓄᐃᑦ ᐅᖃᓚᐅᖅᑐᑦ ᑕᐃᔅᓯᒪᓂ, ᓯᑯᐃᓕᕌᖕᒐᑦ ᓂᐱᒃᑭᐊᓘᕙᓚᐅᖅᓯᒪᒋᐊᖕᒐ ᐃᒃᓯᓕᖅᑎᑦᑐᒍ, ᑰᖕᒐᓂ ᓇᐹᖅᑐᕕᓃᑦ ᑎᐱᔅᐅᕙᓚᐅᖅᓯᒪᖕᒪᑕ. ᒫᓐᓇᒃᑯᑦᑕ, ᓯᑯᐃᕐᓇᓵᖃᑦᑕᓕᖅᑐᖅ, ᐃᖕᒋᕐᕋᓂᖕᒐᓗ ᓴᙲᓐᓂᖅᓴᐅᓕᖅᑐᖅ ᐊᒻᒪᓗ ᓇᐹᖅᑐᕕᓃᑦ ᑎᐱᔅᐅᖃᑦᑕᖕᒋᓐᓂᖅᓴᐅᓕᖅᑐᑦ. ᐊᕐᕌᒎᑉ ᐃᓗᐊᓂ ᖃᓄᐃᓐᓂᐅᕙᒃᑐᑦ ᐊᓯᔾᔨᖅᑐᐃᓐᓇᐅᔮᖅᑐᑦ. ᓰᕐᓗ, ᐅᐱᕐᖔᖅ ᐊᐅᔭᓗ ᓂᓪᓚᔪᖕᓂᖅᓴᐅᕙᓕᖅᑑᒃ, ᐅᑭᐊᒃᓵᖅ ᐅᑭᐅᕐᓗ ᐊᓄᓂᒐᔪᒃᑐᐊᓘᑦᑐᑎᒃ ᓂᓪᓚᔪᖕᒋᓐᓂᖅᓴᐅᓕᖅᑐᑎᒡᒍ, ᓇᐃᓐᓂᖅᓴᐅᓕᖅᑐᑎᒃ ᐱᖅᓯᒐᔪᖕᓂᖕᒐᓄᑦ.

are becoming more common around Kugluktuk.

All interviewees noted less precipitation. As a consequence, they have observed lower water levels in lakes and rivers and drying land. People reported that lichens are growing less and blueberries and blackberries are getting smaller and drying out more easily. Participants observed that nowadays the wind appears to be stronger and more frequent, and its prevailing direction has changed. They found that snow does not stay on the ground because it is blown away by the strong winds. People have also noted the thawing of permafrost and more signs of erosion along the coast and the Coppermine River. Participants described that sea ice freeze-up occurs later in Kugluktuk, partly because of the wind. People described that in the past, sea ice breakup was very strong and loud, with the Coppermine River carrying many pieces of driftwood and chunks of ice. Nowadays it arrives later, the current is not as strong, and it no longer carries driftwood. All seasons seem to be affected. For example, spring and summer feel colder, fall is windier, and winter seems warmer and shorter, with fewer blizzards.

ᐃᓐᓇᐃᑦ ᐊᐱᖅᓱᖅᑕᐅᔪᑦ
Elders Interviewed

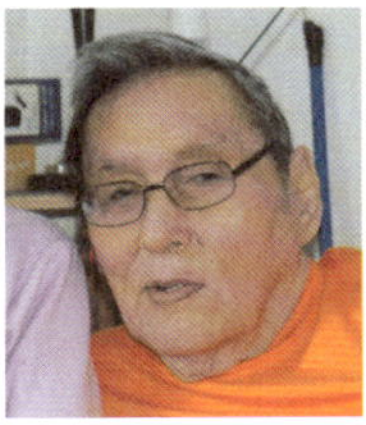

ᔫᓐ ᐊᓗᒃᐱᒃ
John
Allukpik

ᑲᐃᑦ
ᐃᓄᒃᑕᓕᒃ
Kate
Inuktalik

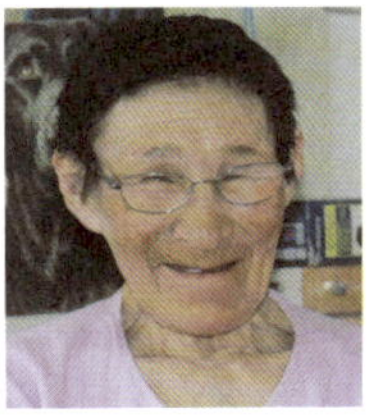

ᓖᓇ ᐊᓗᒃᐱᒃ
Lena
Allukpik

ᕈᐃ
ᐃᓄᒃᑕᓕᒃ
Roy
Inuktalik

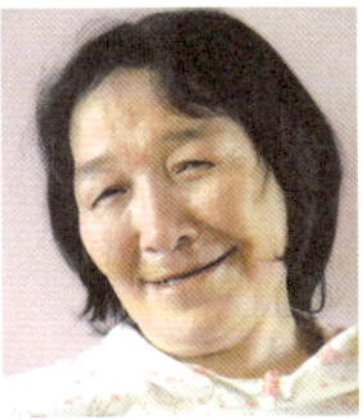

ᒫᑕ ᐊᔭᓕᒐᖅ
Martha
Ayaligak

ᒥᐊᓕ ᕿᓗᒃ
Mary
Kellogok

ᐋᓚᓯ ᐊᔭᓕᒃ
Alice
Ayalik

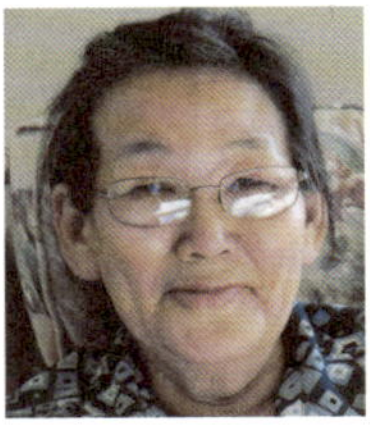

ᐋᓂ ᑭᒍᐃᓇ
Annie
Kigiuna

ᓂᐊᓕ
ᖃᐃᔪᕋᓇ
Nellie
Kaiyogana

ᓚᐅᕋ
ᖁᖅᓱᖅᑕᖅ
Laura
Kohoktak

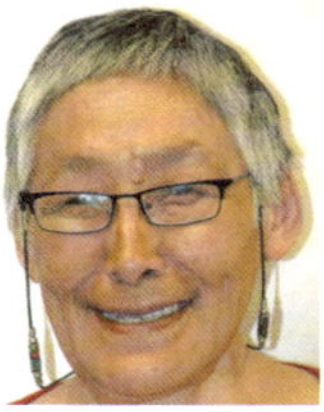

ᐊᓂᐊᓯ
ᖁᖅᑳᖅ
Agnes
Kokak

ᒪᒥ ᐆᓐᓂᖅ
Mamie
Oniak

ᔫᓯᐱ
ᓂᑉᑕᓈᑦᑎᐊᖅ
Joseph
Niptanatiak

ᒫᒃ ᑕᓛᑦᑐᖅ
Mark
Taletok

ᓛᓇ
ᓂᑉᑕᓈᑦᑎᐊᖅ
Lena
Niptanatiak

ᒫᑕ ᑕᓛᑦᑐᖅ
Martha
Taletok

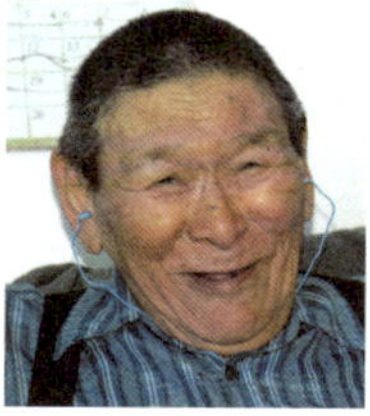

ᔫᓐ ᐅᕼᐅᖅᑲᖅ
John
Ohokak

ᒨᓇ ᑎᒃᑖᓚᒃ
Mona
Tiktalek

ᐸᐅᖕᒐᐃᑦ

“ᑕᒫᓂ ᐊᖅᑯᑎᑕᕐᑲᐅᕐᒪᑦ ᐊᒃᓱᐊᓗᒃ ᐸᑦᑑᖅᑲᑦᑕᖅᑐᒍᑦ. ᓇᓂᑕᒦᖅ ᒪᓂᖅᑲᒥ ᐸᑦᑑᖅᑲᑦᑕᖅᑐᖅ. […] ᐊᖅᑯᑏᑦ ᒪᕐᖂᓗᐃᑦ. […] ᑕᒫᓂ ᐸᐅᖕᒐᖅᑕᖅᐸᓚᐅᖅᓯᒪᔪᖅ. ᑕᒫᓂ ᓯᒡᔭᖅᐸᓯᖕᒥ ᐃᒡᓗᑕᕐᑲᐅᓚᐅᖅᓯᒪᖖᒋᒃᑲᓗᐊᖅᑐᖅ. ᑕᒃᐸᐅᖕᒐ ᐸᐅᖕᒐᖅᑕᕆᐊᖅᐸᓚᐅᖅᓯᒪᒐᓗᐊᖅᑐᒍᑦ ᑭᓯᐊᓂᑕ ᑕᒃᐸᐅᖖᒐᖅᐸᒍᓐᓅᖅᑐᒍᑦ ᐳᔪᓗᐊᕐᖂᓘᑕᕐᒪᑦ.”

—ᐊᓂᐊᓯ ᕐᑯᖅᑳᖅ

“ᐸᐅᖕᒐᖅᑕᕆᐊᕐᕕᒋᕙᒃᑕᓐᓂ ᐊᕚᓚᕐᑭᐊᖅᑲᐅᖅᑲᑦᑕᓕᖅᑐᖅ ᓇᓂᑕᒦᖅ. ᑕᒃᐹᓂ ᐊᖅᐱᒃᑕᕐᑲᕐᑲᑦᑕᓚᐅᖃᓗᐊᕐᒥᔪᖅ, ᑭᓯᐊᓂᓗ ᐸᐅᖕᒐᖅᑕᕆᐊᕐᕕᒋᕙᒃᑕᓐᓂ ᐊᕚᓕᕐᑭᐊᖅᑲᖅᑲᑦᑕᓕᖅᑐᖅ. ᐊᕚᓚᕐᑭᐊᑦ ᑕᒃᓵᕚᑦᑕᐊᑐᐃᓐᓇᓕᖅᑐᑦ. […] ᕐᑯᓕᓂᒃ ᐊᕐᖂᒍᕐᑲᖅᑎᓪᓗᖕᒐ ᐃᕕᐊᓗᖕᓂᒃ ᑕᑯᕙᓚᐅᖅᓯᒪᔪᖕᒐ, ᑕᒫᓂᓗ ᐃᒃᓯᕙᑐᐃᓐᓇᖅᓗᑕ ᐃᔨᖅᓯᒪᕙᓚᐅᖅᓯᒪᔪᒍᑦ. ᐃᕕᖕᓂᒃ ᑕᑯᕙᒍᓐᓃᖅᑐᒍᑦ. ᓇᓃᕐᒪᑕᑭᐊᖅ. ᑕᑭᔪᐊᓗᖕᓂᒃ ᐅᓪᓗᒥ ᑕᑯᕙᒍᓐᓃᖅᑐᖕᒐ. […] ᓯᒡᔭᖅᐸᓯᖕᒥ ᑕᑭᔪᐊᓗᖕᓂᒃ ᑕᑯᕙᓚᐅᖅᓯᒪᒐᓗᐊᖅᑐᖕᒐ.”

—ᐊᓂᐊᓯ ᕐᑯᖅᑳᖅ

Berries

“There are roads here and the sand just flies. There is dust all over the ground. […] The roads are muddy. […] We used to pick berries around here. There used to be no houses along the shore. We would just go up there to pick berries, but now we can’t go because [it’s] too dusty.”

—Agnes Kokak

“If I go where I used to pick, there will be *uqpiit* [willows] everywhere now. There are *aqpiit* [cloudberries/bakeapples] growing up there that way, but if I go where I picked berries there would be *uqpiit*. The branches are growing taller. […] When I was 10 years old I would see big *iviit* [grasses] dry and we would just sit and hide from the other kids. Now I don’t see *iviit* anymore. They might be somewhere, but I don’t see them tall anymore. […] I used to see really tall ones near the beach.”

—Agnes Kokak

“ᓄᓂᕙᖕᓇᐅᑎᓪᓗᒍ
ᐊᖅᐱᓕᐊᓘᓚᐅᖅᑐᖅ.
ᐊᖅᐱᐊᓗᐃᑦ. ᐃᓚᖕᒐᓂ ᓄᓇ
ᐊᐅᐸᖅᑐᑐᐃᓐᓇᐅᕙᓚᐅᖅᓯᒪᔪᖅ
ᐊᖅᐱᓕᖁᓗᐊᐱᒃ. ᑕᒪᑐᒪᓂ
ᐊᖅᖀᒍᒥ ᑭᓯᑕᖅᑲᖕᖏᑦᑎᐊᓚᐅᖅᑐᖅ.
[...] ᐅᑭᐅᖅ. ᐊᐱᓯᒪᔪᐊᓘᒐᑦᑕ
ᐸᐅᖕᒐᖅᑲᑦᑎᐊᖅᓂᐊᖅᑐᒃᓴᐅᕗᖅ.”

—ᐋᓚᓯ ᐊᔭᓕᒃ

“ᐅᓪᓗᒥᐅᓕᖅᑐᖅ ᐸᐅᖕᒐᐃᑦ
ᓈᒻᒪᒃᓯᖅᑲᑦᑕᕈᓯᖕᖏᑦ ᐊᓯᔾᔨᖅᑐᑦ ᐊᒻᒪ
ᑎᐱᖕᖏᑦ. ᒪᒪᖅᐸᓚᐅᖅᓯᒪᔫᒐᓗᐊᑦ
ᒪᒪᓗᐊᕈᓐᓃᖅᑐᑦ. [...]
ᐃᓛᓐᓂᒃᑯᓪᓗ ᓈᒻᒪᒃᓯᒐᓗᐊᖅᑐᑎᒃ
ᐸᓂᑲᐅᑎᕆᖅᑲᑦᑕᓕᖅᑐᑎᒃ.”

—ᒨᓇ ᑎᒃᑖᓕᒃ

“ᐱᕈᖅᑐᑦ ᐊᓯᔾᔨᖅᓯᒪᓕᖅᓂᖕᖏᓐᓂᒃ
ᐅᔾᔨᕆᓯᒪᔪᖕᒐ, ᐃᓛᓐᓂᒃᑯᑦ
ᒪᖅᑯᖅᑲᑦᑕᖕᖏᓗᐊᖅᖔᖕᒐᑦ
ᐸᐅᖕᒐᐃᑦ ᐱᕈᑦᑎᐊᖅᑲᑦᑕᖕᖏᑦᑐᑦ.
ᒪᖅᑯᖅᑲᑦᑕᖕᖏᑉᐸᑦ ᐸᐅᖕᒐᐃᑦ
ᐸᓂᑐᐃᓐᓇᖅᓂᐊᖅᑐᑦ. ᓯᓚᐅᑉ
ᐊᓯᔾᔨᖅᐸᓪᓕᐊᓂᖕᒐᓅᔪᒃᓴᐅᔪᖅ.
ᐃᒻᒪᑲᓪᓚᒃᑎᑑᕈᓐᓃᖅᑐᖅ.
ᐃᒻᒪᒃᑲᓐᓂᖅᑕ
ᐸᐅᖕᒐᓕᒻᒪᕆᐊᓘᖅᑲᑦᑕᓚᐅᖅᓯᒪᔪᖅ.
[...] ᑕᐃᒪᐃᑦᑐᓐᓃᖅᑐᖅ.
ᐸᐅᖕᒐᑭᓐᓂᖅᓴᑯᓘᓕᖅᑐᖅ.
ᑭᒍᑕᖕᖏᖅᓇᐃᑦ ᑕᐃᒪᐃᑦᑐᓐᓃᖅᒥᔪᑦ,
ᐊᖕᖏᔩᑎᐊᓘᕙᓚᐅᖅᓯᒪᒐᓗᐊᖅᒪᑕ.
ᒫᓐᓇ ᒪᖅᑯᖅᑲᑦᑕᖕᖏᓗᐊᖅᓂᖕᒐᓄᑦ
ᑭᒍᑕᖕᖏᖅᓇᐃᑦ ᒥᑭᑦᑐᑯᓘᖅᑲᑦᑕᓕᖅᑐᑦ.
ᑎᐱᖕᖏᑦᑕᐅᖅ ᐊᔾᔨᒋᔭᓐᓃᖅᒥᔭᖕᖏᑦ,
ᔫᖅᓇᖕᖏᓐᓂᖅᓴᐅᓕᖅᒪᑕ.”

—ᓚᐅᕋ ᖁᖅᓱᖅᑕᖅ

“Last season was wonderful for *aqpiit*. Lots of *aqpiit*. The land was just pink in some areas. But the year before that there was next to nothing. [...] This winter there was lots of snow. Going to be lots of berries.”

—Alice Ayalik

“Nowadays [...] the ripeness of the berries is changing, and the taste. They don’t taste so good as they used to [...] and sometimes when they ripen some of them dry right away.”

—Mona Tiktalek

“I’ve noticed some changes [in vegetation], like there is hardly any rain and not very much growth in berries sometimes. No rain and they dry up. It’s something to do with the climate change, I guess. It’s not the same. [...] Not like long ago, there used to be lots of berries. [...] It’s not like that anymore. Less berries. [...] The blueberries are not like long ago; we used to get really big blueberries. Now we don’t get very much rain and we just get small little blueberries. The taste is different. They are not as sweet.”

—Laura Kohoktak

“ᐊᖑᓇᓱᒋᐊᖄᖕᒐᒥᒃ ᐸᐅᖕᒐᖅᑕᖅᐸᓚᐅᖅᓯᒪᖕᒥᔪᑦ. ᑕᒫᓂ ᐸᐅᖕᒐᖅᑕᖅᐸᓚᐅᕐᒥᔪᑦ ᐃᒻᒪᑲᓪᓚᒃ, ᒫᓐᓇᓂ ᐸᐅᖕᒐᖅᑕᖅᐸᑐᓐᓃᖅᑐᑦ ᐃᒡᓗᕐᑲᐅᓗᐊᓕᕐᒪᑦ. ᐊᒃᑕᕐᕕᐅᑉ ᒥᒃᓵᓂ ᐊᖅᐱᒃᑕᕐᑲᑦᑕᓚᐅᖅᓯᒪᒐᓗᐊᖅᑐᒍ, ᑭᓯᐊᓂᓕ ᐅᑦᓗᒥᐅᓕᖅᑐᖅ ᑕᐅᕙᓂ ᑯᕕᖅᑕᕐᕕᕐᑲᓕᖅᑐᖅ ᐊᒻᒪ ᐊᒃᑕᕐᕕᕐᑲᓕᖅᖢᓂ. ᑕᐃᒪᐃᒃᑲᓗᐊᖅᑎᑦᓗᒍ ᓱᓕ ᐸᐅᖕᒐᓕᐊᓗᒃ.”

—ᒨᓇ ᑎᒃᑖᓕᒃ

“When they go hunting they end up picking berries. We used to pick around here, too, a long time ago, but you can’t pick anymore. Too many houses. And toward the dump they used to pick *aqpiit*, but the sewage lagoon and garbage dump [are there] now. There are lots of berries anyway.”

—Mona Tiktalek

ᐱᕈᖅᑐᑦ ᐊᓯᖏᑦ

Other Plants

“ᓄᓇᖃᕐᓂᒃ ᑕᑯᓚᐅᑦᑎᐊᓐᖏᑦᑐᖕᒐ ᒪᕐᑯᓚᐅᑦᑎᐊᓐᖏᒻᒪᑦ.”

—ᐊᓂᐊᓯ ᕐᑯᖅᑳᖅ

“Hardly any flowers I’ve been seeing, because hardly any rain.”

—Agnes Kokak

“ᐊᕚᓚᕐᑭᐊᑦ ᔫᓂᐅᓕᖄᖕᒐᑦ ᐱᕈᖅᐸᑦᓕᐊᓕᓲᖑᔪᑦ. ᐃᓛᓐᓂᒃᑯᑦ ᔪᓚᐃᒥ ᑭᓯᐊᓂ ᐱᕈᕐᑲᑦᑕᓚᐅᖅᓯᒪᔪᑦ. ᒫᓐᓇᓕ ᐆᔭᐅᔭᖕᒍᑲᐅᑎᒋᕐᑲᑦᑕᓕᖅᑐᑦ ᐊᒻᒪ ᐅᑭᐊᒃᓴᕈᖅᐸᑦᓕᐊᓂᐊᓕᖅᑎᑦᓗᒍ ᑐᕐᑯᓕᕐᑲᑦᑕᓕᖅᖢᑎᒃ ᐊᐅᔭᕈᓚᐅᖅᑎᓐᓇᒍᓘᓐᓃᑦ. [...] ᐱᕈᓯᔪᑦ ᐸᐅᖕᒐᕐᑲᕐᕕᕕᓂᐅᒐᓗᐊᓂᑦ.”

—ᐋᓚᓯ ᐊᔭᓕᒃ

“[Shrubs] grow faster in June. Sometimes they used to grow in July. Now they get green right away and just before the fall they’re starting to fall off. Even before the summer starts. [...] They are growing over where the *paurngait* [blackberries/crowberries] are growing.”

—Alice Ayalik

“ᐃᖅᑲᐅᒪᔪᖓ ᐊᓈᓇᒐ ᐆᑯᔅᓯᕈᑐᖕᒥᒃ ᓇᒃᓴᖅᑲᑦᑕᓚᐅᕐᓂᖕᒐᓂᒃ ᑲᖕᒎᔭᖅᑕᕆᐊᕐᓂᐊᓕᕌᖕᒐᒥ. ᐊᓈᓇᒐ ᒪᓕᒃᐸᓚᐅᖅᓯᒪᔭᕋ. ᑖᓐᓇ ᐆᑯᔅᓯᕈᑐᒃ ᑕᑕᑉᐸᓚᐅᖅᓯᒪᔭᖕᒐ ᓵᖅᑲᐃ ᑲᖕᒎᔭᐃᑦ ᖁᓪᓕᕐᒧᑦ ᐃᐸᕋᒃᓴᕆᔭᐅᓂᐊᕋᒥᒃ ᐊᒻᒪ ᑖᓐᓇ ᐅᑭᐅᓕᒫᕐᒧᑦ ᓈᒻᒪᒃᐸᓚᐅᖅᓯᒪᓪᓗᓂ. ᐃᓛᓐᓂᒃᑯᑦ ᐊᖏᖕᓛᓗᒃ ᐱᓱᒋᐊᖅᑲᖅᐸᓚᐅᖅᓯᒪᔪᒧᑦ. [...] ᑕᐃᒪᐃᓐᓂᖕᒐᓄᑦ ᐆᑯᔅᓯᕈᑐᖕᒥᒃ ᓇᒃᓴᖅᐸᓚᐅᖅᑐᑦ ᑕᑕᓐᓂᐊᕋᒥᔅᔪᒃ. ᓯᓕ ᑕᒫᓂ ᑲᖕᒎᔭᕐᓂᒃ ᑕᑯᕙᒃᑐᖓ. ᐃᓛᓐᓂ ᐅᓄᓐᖏᑦᑐᑯᓘᖅᑲᑦᑕᖅᑐᑦ ᐃᓛᓐᓂᒃᑯᑦᑐ ᐅᓄᕐᓂᖅᓴᐅᕙᒃᓱᑎᒃ. ᒪᓕᑐᐃᓐᓇᕐᒥᔪᑦ ᓯᓚᒧᑦ.”

—ᔫᓐ ᐅᕼᐅᖅᑲᖅ

“ᑕᒫᓂ ᐆᓇᔪᐃᓐᓇᐅᔭᖅᑐᖅ. ᓂᒡᓚᓪᓚᑎᒍᓐᓇᖕᖏᑦᑐᖅ; ᑕᐃᒪᐃᓐᓂᖕᒐᓄᑦ ᐱᕈᖅᑐᑦ ᐊᒥᓱᐊᓘᖅᑲᑦᑕᖅᑐᖅ ᐊᐅᔭᐅᓕᕌᖕᒐᑦ. ᑕᐃᒪᐃᓐᓂᖕᒐᓄᑦᑐ ᐊᖅᐱᓕᐊᓘᖅᑲᑦᑕᖅᓱᓂ. ᑕᐅᕙᓂ ᐅᖅᑰᖏᓐᓇᐅᔭᖅᑐᖅ, ᓂᒡᓚᓚᐅᖅᓯᒪᓐᖏᓱᓂ. ᒪᑯᐊ ᐊᕘᓚᖅᑭᐊᓪᓘᓐᓃᑦ. ᓄᓇ ᓯᒡᔭᐅᑉ ᖃᓂᒋᔮᓃᑦᑐᖅ ᐊᔾᔨᒋᓐᖏᒻᒥᔭᖓ, ᑕᒃᐸᐅᖓ ᓄᓇᕕᖕᒧᐊᖅᐸᑦᓕᐊᓃᑦ ᒪᓕᒃᓱᒍ, ᐅᖅᑰᓯᕙᑦᓕᐊᓲᖅ, ᐱᕈᖅᑐᖅᑲᕐᓂᖅᓴᐅᑦᑐᓂ ᐊᒻᒪ ᑭᓱᓕᒫᑦ ᐱᕈᖅᐸᑦᓕᐊᓲᖑᑦᑐᑎᒃ.”

—ᔫᓯᐱ ᓂᑉᑕᓈᖅᑎᐊᖅ

“I remember my mother would carry a big bag when she was going to pick [cottongrass]. I would follow my mom. She would fill up the whole bag so that they could use that with the *qulliq* [stone lamp], and it would last the whole year. Sometimes she would have to go a long way. [...] That’s why they carried a big bag to fill it up. I still see [cottongrass] around this area. Sometimes there is quite few, sometimes there is lots.
It depends.”

—John Ohokak

“Here it’s always warm. It doesn’t get cold; that’s why a lot of plants grow in the summertime. That’s why the cloudberries grow lots. There is always a lot of heat there. It’s never cold. Even shrubs. It’s not like along the coast, but when you go farther up, there is more heat, vegetation, everything grows.”

—Joseph Niptanatiak

“ᐃᓚᖏᓐᓂ ᐱᕈᖅᑐᑦ
ᐱᕈᑦᑎᐊᖅᑐᐊᓘᕐᑲᑦᑕᖅᑐᑦ,
ᐃᓚᖏᓐᓂᓗ ᐱᕈᖅᐸᓐᖏᓗᑎᒃ.
ᓯᓚᒧᑦ ᒪᓕᑐᐃᓐᓇᖅᑐᑦ.
ᓂᒡᓚᓱᓗᐊᖅᐸᑦ ᐱᕈᖅᑐᑦ
ᐱᕈᑦᑎᐊᕐᔫᓐᖏᑦᑐᑦ. ᐊᑯᓐᓂᖅᓱᖅᐸᑦ
ᐱᕈᖅᑐᑦ ᐱᕈᑦᑎᐊᖅᑐᐊᓘᕐᑲᑦᑕᖅᓗᑎᒃ.”

—ᔫᓯᐱ ᓂᑉᑕᓈᑦᑎᐊᖅ

“ᐅᔾᔨᓇᓚᐅᖅᑐᖓ ᒫᓐᓇ ᐊᒥᓱᐊᓗᐃᑦ
ᐱᕈᖅᑐᑦ ᐱᕈᖅᐸᒡᓕᐊᓂᖏᓐᓂᒃ,
ᑭᓯᐊᓂᓗ ᐊᐅᔭᐅᓕᕚᖓᑦ
ᐱᕈᖅᑐᑦ ᐱᕈᑦᑎᐊᕐᑲᑦᑕᓐᖏᑦᑐᑦ.
[...] ᐃᖅᑲᐅᒪᔪᖓ ᐃᒻᒪᑲᒡᓚᒃ
ᐱᖓᓱᓄᑦ ᐅᓪᓗᕐᓄᑦ
ᑲᓪᓕᖅᑐᐊᓘᓚᐅᖅᓯᒪᓂᖓᓂᒃ.
ᒪᕐᕈᓚᐅᖅᓯᒪᔪᖅ ᐅᓪᓗᓄᑦ
ᐱᖓᓱᓄᑦ ᐊᒻᒪ ᐱᕈᖅᑐᑦ
ᐱᕈᖅᑐᐊᓘᓕᖅᓗᑎᒃ. ᒫᓐᓇᓕ ᐱᕈᖅᑐᑦ
ᐱᕈᑦᑎᐊᕐᑲᑦᑕᕈᓐᓃᖅᑐᑦ.”

—ᑲᐃᑦ ᐃᓄᒃᑕᓕᒃ

“ᐃᒻᒪᑲᒡᓚᒃ ᐊᒥᓱᐊᓗᖕᓂᒃ
ᑲᖕᒎᔭᖅᑕᖅᐸᓚᐅᖅᓯᒪᔪᖓ.
ᐊᒥᓱᐊᓘᓚᐅᖅᓯᒪᔪᑦ. ᐃᓖᓐᓂ
ᑲᖕᒎᔭᖅᑕᖅᐸᓚᐅᖅᓯᒪᔪᖓ
ᐊᑭᓯᓕᐊᕆᓂᐊᖅᑕᓐᓂᒃ ᐊᓯᖏᓐᓄᓪᓗ
ᐊᑐᖅᑐᒃᓴᐅᓂᐊᖅᑐᓂᒃ. ᐊᑯᓂᐊᓗᒃ
ᑕᐃᒪᐃᑦᑐᓂᒃ ᑕᑯᓚᐅᖅᓯᒪᔪᓐᓃᖅᑐᖓ.”

—ᓚᐅᕋ ᕐᑯᖅᓱᖅᑕᖅ

“ᐊᓯᔾᔨᖅᑐᒻᒪᕆᐊᓗᒃ. ᐊᕐᕌᓂ
ᑲᖕᒎᔭᕐᑲᑦᑎᐊᓚᐅᖅᓯᒪᓐᖏᑦᑐᖅ
ᓂᒡᓚᓱᓗᐊᖅᑐᐊᓘᓚᐅᕐᓂᖓᓄᑦ.
ᐃᓖᓐᓂ ᐊᒥᓱᐊᓗᐃᑦ
ᐱᕈᖅᐸᓚᐅᖅᑐᑦ, ᑭᓯᐊᓂᓕ ᐊᕐᕌᓂ

“In some seasons [the plants] really grow; in some other seasons they don’t grow. It depends on the weather. If it’s too cold the plants don’t grow very much. If we have medium heat, we get lots of growth.”

—Joseph Niptanatiak

“Right now I notice that there are a lot of plants growing, but in summers the plants don’t really grow. [...] I remember a long time ago, for about three days it was thundering. It rained for three days and it got the plants to grow. But now the plants are not growing much.”

—Kate Inuktalik

“A long time ago I used to pick lots [of cottongrass]. There used to be really a lot. Sometimes I would pick them up to make pillows and other things to fill. I haven’t seen that in years.”

—Laura Kohoktak

“It’s a big change. Last year we didn’t get much [cottongrass] because it’s been so cold. Sometimes they would really grow [...] but last year we didn’t see anything. They didn’t grow.”

—Lena Niptanatiak

ᑕᑯᓚᐅᓐᖏᑦᑎᐊᒻᕆᐊᒃᑐᖓ.
ᐱᕈᓚᐅᓐᖏᑦᑐᑦ."

—ᓖᓇ ᓂᑉᑕᓈᑦᑎᐊᖅ

"ᐱᕈᖅᑐᖅᑲᑦᑎᐊᕈᓐᓃᖅᑐᖅ.
ᐃᒪᕐᑲᓐᖏᓗᐊᕐᒧᑦ ᐱᕈᖅᑐᑦ
ᐸᓂᑐᐃᓐᓇᕐᑲᑦᑕᖅᑐᑦ. [...]
ᐃᕕᕐᑲᑦᑎᐊᓐᖏᒻᒪᕆᓕᖅᑐᖅ.
ᐃᒡᓗᓕᐅᕐᕕᐅᕙᓕᖅᑐᑦ ᐱᕈᖅᑐᓄᑦ
ᐱᕈᕐᕕᐅᕐᑲᑦᑕᓚᐅᖅᓯᒪᔫᒐᓗᐊᑦ."

—ᒪᒥ ᐆᓂᖅ

"ᐱᕈᕐᑲᑦᑕᓕᕋᓗᐊᕐᒪᖕᒑᖏᑦ
ᓇᓗᓇᖅᑐᖅ ᑕᑯᒐᔪᒍᓐᓃᕋᑦᑎᒍᑦ.
ᐱᕈᖅᑐᓂᒃ ᓇᓂᓯᒡᒐᕐᓇᖅᓯᔪᖅ
ᐱᕈᖅᐸᒍᓐᓃᕐᓂᖏᓐᓄᑦ."

—ᒪᒥ ᐆᓂᖅ

"ᐊᕚᓚᕐᑭᐊᓕᐊᓘᓕᖅᑐᖅ ᐅᑦᓗᒥ. [...]
ᑕᐅᕙᓂ ᑐᐱᕐᕕᒋᕙᓚᐅᖅᓯᒪᔭᑦᑎᓐᓂ
ᐱᕈᖅᑐᐊᓗᐃᑦ. ᐅᔾᔨᕆᓚᐅᖅᑐᖓ
ᑐᐱᕐᕕᐅᕙᒃᑐᓂ ᐊᒃᓱᐊᓗᒃ
ᐃᔾᔪᕐᑲᓕᕐᓂᖕᒐᓂᒃ. ᑭᖕᒍᑦᓕᕐᒥ
ᑕᒃᐸᐅᓐᖓᕋᒪ ᑐᐱᕐᕕᒋᕙᒃᑕᑦᑎᓐᓄᑦ
ᐊᕚᓚᕐᑭᐊᖏᓐᓇᕐᓂᒃ ᑕᑯᓚᐅᖅᓯᒪᔪᖓ.
ᕐᑭᒻᒥᖅᑲᕐᕕᒋᖅᑲᑦᑕᓚᐅᖅᑕᕗᑦ ᓘᓐᓃᑦ
ᐃᕖᓐᓇᐅᓕᖅᑐᖅ.
ᐱᕈᕐᓂᖅᓴᕐᔪᐊᒻᒪᕆᐊᓗᐃᑦ
ᑕᐃᒪᖓᓂᑦ ᑕᒃᐹᓂ
ᓄᓇᕐᑲᖅᐸᓚᐅᖅᓯᒪᒐᑦᑕ."

—ᒫᒃ ᑕᓖᑦᑐᖅ

"There are hardly any plants left. There is hardly any water left for the plants to be fed with. [...] There is hardly any growth in the grass, too. It's like where you build houses, where the plants used to grow."

—Mamie Oniak

"Not very much [lichen], neither the cottongrass. We used to see lots of those; we don't anymore. [...] It's really hard to see if they are growing anymore. It's harder to find stuff. They are not growing anymore."

—Mamie Oniak

"[Shrubs] grow so much now. [...] Where we used to have our tents a long time ago, they've been growing so much. I noticed it's from so much dirt around the campsites. I noticed that last trip I went up I saw all the *avaalaqiat* [birches] growing now where we used to have our camps. Even where we used to have our dogs, the *iviit* [grasses] really grow. Where I used to live inland, I noticed they have really grown now."

—Mark Taletok

ᐆᒪᔪᑦ

“ᐅᒃᐱᒡᔪᐊᖃᐅᕐᓂᖅᓴᐅᓕᖅᑐᖅ. ᐅᓪᓗᒥ ᐅᒃᐱᒡᔪᐊᕐᓂᒃ ᑕᑯᒐᔪᖕᓂᖅᓴᐅᓕᖅᑐᒍᑦ.”

—ᐊᓂᐊᓯ ᖁᖅᑳᖅ

“ᑎᖕᒥᐊᑦ ᐃᒻᒪᑲᓪᓚᒃ ᑕᒫᓃᑉᐸᓪᓚᐅᖅᓯᒪᔪᓂᒃ ᑕᑯᒐᔪᒍᓐᓃᖅᑕᕋᑦ. ᖁᐸᓄᐊᑦ.”

—ᐋᓂ ᑭᒍᐃᓇ

“ᒫᓂ ᑐᓗᒐᖅᑲᑦᑎᐊᓚᐅᖅᓯᒪᙱᒃᑲᓗᐊᖅᑐᖅ. ᖃᓂᙱᑦᑐᒦᖅᑲᑦᑕᓚᐅᖅᓯᒪᔪᑦ ᑐᒃᑐᓂᒃ ᐃᖅᓯᒐᒥᒃ. ᐅᓪᓗᒥᓕ ᐃᖅᓯᔪᓐᓃᖅᑐᑦ. ᐃᓄᖕᓂᒡᓗ ᐃᖅᓯᔪᓐᓃᖅᓱᑎᒃ.”

—ᓖᓇ ᐊᓗᖅᐱᒃ

“ᐃᒻᒪᑲᓪᓚᑦ ᐃᓄᐃᑦ ᓄᒃᑕᖅᑎᐅᓯᖕᒐᖅᑎᓪᓗᒋᑦ ᕿᒻᒥᒥᓂᒃ ᕿᒥᓇᕈᓚᐅᖅᓯᒪᔪᑦ ᑐᒃᑑᑉ ᓂᕿᐊᓂᒃ. ᐅᓪᓗᒥᐅᓕᖅᑐᖅ ᐃᓄᐃᑦ ᐅᖃᐅᔾᔭᐅᕙᓕᖅᑐᑦ ᑐᒃᑐᒥᒃ ᕿᒥᓇᖅᑲᑦᑕᖁᔭᐅᔪᓐᓃᖅᓱᑎᒃ. [...] ᑐᓴᐅᒪᓕᕐᒥᔪᒍᑦ ᐊᕐᕌᒍᑕᒫᖅ ᑐᒃᑐᑭᒡᓕᕙᓪᓕᐊᓂᕋᖅᑕᐅᓂᖓᓂᒃ, ᑭᓯᐊᓂᓕ ᐃᒻᒪᑲᓪᓚᒃ ᕿᒻᒥᕗᑦ ᕿᒥᓇᕈᓚᐅᖅᓯᒪᔭᕗᑦ ᑐᒃᑑᑉ ᓂᕿᐊᓂᒃ.”

—ᓖᓇ ᐊᓗᖅᐱᒃ

“ᑐᒃᑐᐃᑦ ᐃᓱᒫᓘᖕᓇᖅᓯᔪᑦ. [...] ᕿᑭᖅᑖᓗᖕᒥᐅᓄᑦ ᔭᓗᓇᐃᒥᐅᓄᓪᓗ ᑐᒃᑐᑦ ᐃᓱᒫᓘᑎᒋᔭᐅᓕᖅᑐᑦ.”

—ᔫᓯᐱ ᓂᑉᑕᓈᑦᑎᐊᖅ

Animals

“There are more snowy owls. People are seeing them more often now.”

—Agnes Kokak

“I hardly see any birds that used to be here a long time ago. The snow buntings.”

—Annie Kigiuna

“There used to be hardly any crows around this area. They used to be really far away because they were afraid of the caribou. Now they are not afraid. They are not scared of us anymore.”

—Lena Allukpik

“There used to be lots of people travelling, and the only thing we would feed our dogs was caribou meat. Nowadays they are telling people not to feed their dogs with caribou meat. [...] You hear that the caribou are becoming fewer and fewer every year, but that was the sort of meat we used to feed our dogs.”

—Lena Allukpik

“I am worried about the caribou. [...] It’s a concern with the Baffin herd and in the Yellowknife area.”

—Joseph Niptanatiak

“ᐃᓛᓐᓂᒃᑯᑦ ᑖᒪᓂ
ᐊᒃᓴᒃᑕᖅᑲᖅᑲᑦᑕᓯᔪᖅ. ᑖᒪᓂ
ᐊᒃᓴᒃᑕᖅᑲᓚᐅᖅᓯᒪᓐᖏᒃᑲᓗᐊᖅᑎᓪᓗᒍ
ᓇᓂᓕᒫᑦ ᑕᑯᔭᐅᖅᑲᑦᑕᓕᖅᑐᑦ.
ᓇᓄᐃᓐᓇᐅᕙᓚᐅᖅᓯᒪᔫᒐᓗᐊᑦ.”

—ᑲᐃᑦ ᐃᓄᒃᑕᓕᒃ

“ᑕᐅᕙᓂ ᖁᒃᑐᕆᐊᓕᒻᒪᕆᐊᓗᒃ,
ᐊᒃᓱᐊᓗᒡᓗ ᐅᖅᑰᓪᓗᓂ.
ᑕᐅᕙᓃᒋᐊᒃᓴᖅ ᓄᓇᖅᑲᕆᐊᒃᓴᖅ
ᐊᔪᖅᓇᒻᒪᕆᓚᐅᖅᓯᒪᔪᖅ.
ᑕᐅᕙᓐᖔᖅᑐᒥᒃ ᐱᓱᒃᑐᒥᒃ ᑕᑯᓐᓂᕈᕕᑦ,
ᑐᓄᑦᑎᐊᖕᒐᓂ ᖁᖅᓂᖅᑖᓗᖕᒥᒃ
ᓄᕗᔮᓘᔮᖅᑐᒥᒃ ᑕᑯᓇᔭᖅᑐᑎᑦ, ᓰᖅᓗ
ᐃᓯᐊᖅᑭᔪᑎᑐᑦ. ᓄᕗᔭᐅᔮᖅᑐᓄᑯᐊ
ᖁᒃᑐᕆᐊᑦ ᐊᒥᓱᑯᓗᐊᐲᑦ ᑕᐅᕙᓂ
ᓄᓇᖅᑲᕈᓐᓇᕋᔭᖅᑐᒃᓴᐅᓐᖏᑦᑐᑎᑦ
ᖁᒃᑐᕆᐊᖅᑲᓗᐊᖅᓂᖕᒐᓄᑦ. ᑕᒃᒪᑮᖕᓂᒃ
ᐱᑕᓕᒃ, ᒥᓗᒋᐊᓂᒃ ᐊᒻᒪ ᖁᒃᑐᕆᐊᓂᒃ.
ᖁᒃᑐᕆᐊᖅᑲᕈᓐᓃᖅᐸᓪᓕᐊᓕᖅᑎᓪᓗᒍ
ᒥᓗᒋᐊᑦ ᓄᐃᖕᒐᓰᑦ. ᓄᕗᔭᓕᕌᖕᒐᑦ
ᒥᓗᒋᐊᑦ ᓯᓂᔾᔮᓐᖏᑦᑐᑦ
ᐊᐅᓪᓚᔾᔮᕋᑎᒡᓗ. ᑐᒃᑐᓯᐅᕐᕕᖕᒥ
ᑕᒃᐸᐅᖕᒐ ᐱᓱᒃᐸᓪᓕᐊᓕᖅᑎᓪᓗᒋᑦ
ᖁᒃᑐᕆᐊᓄᑦ ᒪᓕᒃᑕᐅᕙᓚᐅᖅᑐᑦ.
ᑐᒃᑐᑦ ᐱᓱᒃᑎᓪᓗᒋᑦ ᐊᒻᒪ
ᐊᓂᒍᖅᐸᓪᓕᐊᓕᖅᑎᓪᓗᒋᑦ
ᖁᒃᑐᕆᐊᖅᑲᕈᓐᓃᓲᖅ, ᑭᓯᐊᓂᓗ
ᑕᐃᒪᐃᒃᑲᓗᐊᖅᑎᓪᓗᒍ ᑐᒃᑐᑦ
ᐊᐅᓪᓚᕌᖕᒐᑕ ᒥᓗᒋᐊᓕᒻᒪᕆᐊᓘᓕᓲᖅ.
ᐃᓛᓐᓂᒃᑯᑦ ᐱᓱᒃᒐᓂ
ᒥᓗᒋᐊᖅᑲᓗᐊᒧᑦ ᖁᖕᒐᖅᓄᑦ
ᖃᓂᖅᓄᓪᓗ ᐳᓚᕙᓚᐅᖅᑐᑦ,
ᐅᕆᐊᕆᐊᖅᑲᓕᖅᓱᒋᓪᓗ.”

—ᔫᓯᐱ ᓂᑉᑕᓈᕐᑎᐊᖅ

“Sometimes there are black bears around this area. We never used to get black bears in this area, but now they are everywhere. It used to be only polar bears.”

—Kate Inuktalik

“There are so many *kitturiat* [mosquitoes] in that area. It’s so hot. Can’t stand that area. If you see a person coming from that way, from behind you see big black clouds, just like smoke. Big clouds of *kitturiat*. You probably can’t live in those places where there are so many mosquitoes. Both *milugiat* [black flies] and *kitturiat*. Just after the *kitturiat* are getting fewer, they get a lot of *milugiat*. When it gets cloudy the *milugiat* can stay out all the time; they don’t go away. Around Bathurst area, when the caribou move upwards they go with the herd, the mosquitoes. After the herd has passed on, there are no more *kitturiat*, but when they leave there are lots of *milugiat*. [Sometimes when you’re walking] they just go into your nose and mouth and you have to spit them out.”

—Joseph Niptanatiak

“ᐃᓛᓐᓂ ᑎᖕᒥᐊᓂᒃ ᑕᕆᐅᕐᒥᐅᑕᓂᒡᔫᓐᓃᑦ ᑎᖕᒥᐊᓂᒃ ᑕᑯᕙᒃᑐᒍᑦ. ᑎᖕᒥᐊᕋᓛᓂᒃ ᐊᕐᕌᒍᒐᓴᖕᓄᑦ ᑕᑯᓚᐅᖅᓯᒪᙱᑕᑦᑎᓐᓂᒃ ᑕᑯᖃᑦᑕᖅᑐᒍᑦ. ᑭᓰᓂᒡᔫᓐᓃᑦ ᑕᐃᔭᐅᕙᖕᒪᖕᒑᖕᒋᓐᓂᒃ ᓇᑐᔪᒍᑦ. ᐃᒪᕐᒥᐅᑕᓂᒃ ᑎᖕᒥᐊᓂᒃ ᑕᑯᕙᓕᕐᒥᔪᒍᑦ, ᐃᒪᕐᒧᑦ ᒥᓲᓂᒃ. ᒥᑎᕐᓂᑦ ᐊᖕᑎᓂᖅᓴᒐᓛᑦ ᐊᒻᒪ ᐊᔾᔨᒋᙱᒃᑭᑦᑐᓂᒋᑦ.”

—ᔫᓯᐱ ᓂᑉᑕᓈᑦᑎᐊᖅ

“ᒫᓐᓇ ᐃᒍᑦᑕᖃᑦᑎᐊᕈᓐᓃᖅᑐᖅ. ᐊᒥᓲᐊᓗᕙᓚᐅᖅᓯᒪᔫᒐᓗᐊᑦ.”

—ᑲᐃᑦ ᐃᓄᒃᑕᓕᒃ

“ᐅᓪᓗᒥ ᒥᓗᒋᐊᖅᑕᖃᑦᑎᐊᕈᓐᓃᖅᑐᖅ. ᐅᓪᓗᒥ ᑕᑯᒐᔪᓐᓃᖅᑕᕗᑦ. ᐃᒻᒪᑲᓪᓚᒃ ᐱᑕᖃᐅᓗᐊᓘᑦ ᓂᖃᐅᔪᐊᓘᕙᓚᐅᖅᓯᒪᔪᒍᑦ, ᐅᓪᓗᒥᓕ ᕿᑐᕐᖓᒐᕗᑦ ᒥᓗᒋᐊᖅᑕᐅᕙᒍᓐᓃᖅᑐᑦ.”

—ᒪᒥ ᐆᓂᖅ

“ᑕᓯᑦ ᑰᓪᓗ ᐸᓂᖅᐸᓪᓕᐊᓯᔪᑦ. ᓄᓇᔫᓐᓃᑦ ᐊᒃᓯᓗᐊᕌᓗᒃ ᐸᓂᖃᑦᑕᖅᑐᖅ. ᐅᓇ ᑰᒃ ᐱᓗᐊᓐᖑᐊᖅᑐᖅ, ᒪᖃᒃᐸᓪᓕᐊᓕᖅᑐᕐᔫᓐᓃᑦ. ᐃᒻᒪᒃᑲᓐᓂᖅ ᑕᒪᐅᖓ ᓅᓱᖅᑐᑕ ᐃᓄᐃᑦ ᐃᒥᖅᑕᖅᐸᓚᐅᖅᓯᒪᔪᑦ ᓵᑦᑎᐊᑦᑎᓐᓂ ᐊᒻᒪ ᑏᓕᐅᕈᑎᒋᕙᒃᑕᕗᑎᒃᑯᑦ. ᐅᓪᓗᒥᓕ ᑏᓕᐊᒐᒃᓴᐅᔭᓐᓃᖅᑐᖅ ᒪᕐᕋᐅᓗᐊᓕᕐᓂᖕᒐ ᐱᔾᔪᑎᒋᓪᓗᒍ. ᐅᑭᐊᒃᓵᖅ ᓄᓘᔾᔭᕋᓗᐊᕋᑦᑕ

“Sometimes we see birds or ducks. Little birds we see now that we haven’t seen in many years. We don’t even know what they are called. [...] [We see] different kinds of ducks that can land on water. They are a little bit bigger. They are a little different from eider ducks.”

—Joseph Niptanatiak

“Not very many [bumblebees] now. There used to be a lot.”

—Kate Inuktalik

“There are hardly any *milugiat* left. There used to be so many long ago. Nowadays I don’t see very much of those. Long time ago we used to get really bitten, but now we never see the kids get bit by the *milugiat* anymore.”

—Mamie Oniak

“The lakes and rivers are just about drying up. Even the land, it’s just been so dry. This river, I think, is getting worse. It’s getting muddier. When we first moved here people used to get pails of water from the front here and we used it for tea, but now it’s getting hard to use it for tea because it’s getting muddy. We noticed when we were

ᖅᑲᔅᓰᓐᓇᑯᓗᖕᓂᒃ ᐃᖅᑲᓗᓚᐅᖅᓯᒪᔪᒍᑦ,
ᐊᖕᒋᔩᑎᓂᓪᓗ ᐱᑕᖅᑲᑦᑎᐊᓐᖏᖢᓂ.
ᐃᖅᑲᓗᖅᑲᑦᑎᐊᓚᐅᓐᖏᑦᑐᖅ.”

—ᑲᐃᑦ ᐃᓄᒃᑕᓕᒃ

“ᑐᒃᑐᑦ
ᐊᒃᑐᖅᑕᐅᖅᑲᓯᐅᑎᓂᐅᓴᑐᐃᓐᓇᖅᑐᑦ
ᓄᓇ ᐸᓂᓗᐊᖅᑎᓪᓗᒍ. [...]
ᓄᓇ ᐸᓂᓗᐊᖅᑎᓪᓗᒍ
ᐱᕈᖅᑐᖅᑲᑦᑎᐊᕈᓐᓃᓲᖅ ᐊᒻᒪ
ᑐᒃᑐᑦ ᓂᖅᑭᒃᓴᖅᑲᑦᑎᐊᕈᓐᓃᖅᖢᑎᒃ.
ᐃᓛᓐᓂ ᖅᑯᐃᒃᑎᑦᑎᐊᖅᑲᑦᑕᓐᖏᑦᑐᑦ,
ᓂᖅᑭᒃᓴᖅᑲᑦᑎᐊᓐᖏᓐᓇᒥᓪᓗ
ᐊᓯᐊᓄᓐᖓᐅᔭᕗᐊᖅᑲᓕᖅᖢᑎᒃ
ᓂᖅᑭᒃᓴᖅᑲᖅᓂᖅᓴᒧᑦ.
ᓂᖅᑭᒃᓴᖅᓯᐊᓚᐅᓐᖏᓐᓂᓕᒫᒥᓂᒃ
ᐱᓱᐃᓐᓇᕗᐊᖅᑲᖅᑐᑦ
ᓂᖅᑭᒃᓴᓕᒃᓯᐅᖅᖢᑎᒃ.”

—ᓚᐅᕋ ᖅᑯᖅᓱᖅᑕᖅ

“ᐃᓚᖕᒋᓐᓂ ᐊᖅᑳᒍᓂ
ᑐᒃᑐᓯᓐᓇᑦᑎᐊᓐᖏᑦᑐᖅ; ᑐᒃᑐᓯᐅᖅᑐᑦ
ᐅᖕᒐᓯᒃᑐᓕᐊᕆᐊᖅᑲᓕᖅᐸᒃᑐᑦ.
ᑐᒃᑐᓯᐅᑐᐃᓐᓇᓐᖏᓐᓇᒥᒃ ᐃᓛᓐᓂᒃᑯᑦ
ᐊᒪᕈᖅᓯᐅᖅᐸᒃᑐᑦ ᐅᕝᕙᓘᓐᓃᑦ
ᖅᑲᕝᕕᒑᖅᔭᒃᓯᐅᖅᖢᑎᒃ. ᐃᓛᓐᓂ
ᐅᑭᐅᒃᑯᑦ ᐱᔭᕐᓂᓐᖏᑦᑐᐊᓘᖅᑲᑦᑕᖅᑐᑦ
ᑐᒃᑐᑦ ᐅᖕᒐᓯᒃᓗᐊᖅᑎᓪᓗᒋᑦ.
ᐅᑎᖅᐸᓪᓕᐊᕋᔮᑦᑐᑦ ᐊᖅᑳᒍᑦ ᐱᖕᒐᓱᑦ
ᑎᓴᒪᓪᓘᓐᓃᑦ ᐊᓂᒍᖅᑐᓂ.”

—ᒨᓇ ᑎᒃᑖᓕᒃ

putting fishnets out in the fall we were hardly getting fish, and there were hardly any big fish. There were hardly any fish.”

—Kate Inuktalik

“It affects the caribou, too, when the vegetation is too dry. [...] That’s when the plants are not growing enough for caribou to eat grass. Sometimes they don’t get very fat, so they have to walk to another place to see if there is more grass or shrubs. [...] They grow more fat when they find good vegetation. But when they can’t find a place to graze, they have to move on.”

—Laura Kohoktak

“Some years it’s really hard to find caribou; the hunters have to go really far. [...] It’s not only caribou they like to look for. Sometimes it’s wolf or wolverine. Sometimes it’s really hard in the winter when the caribou go far. It’s three or four years now that they are starting to come closer.”

—Mona Tiktalek

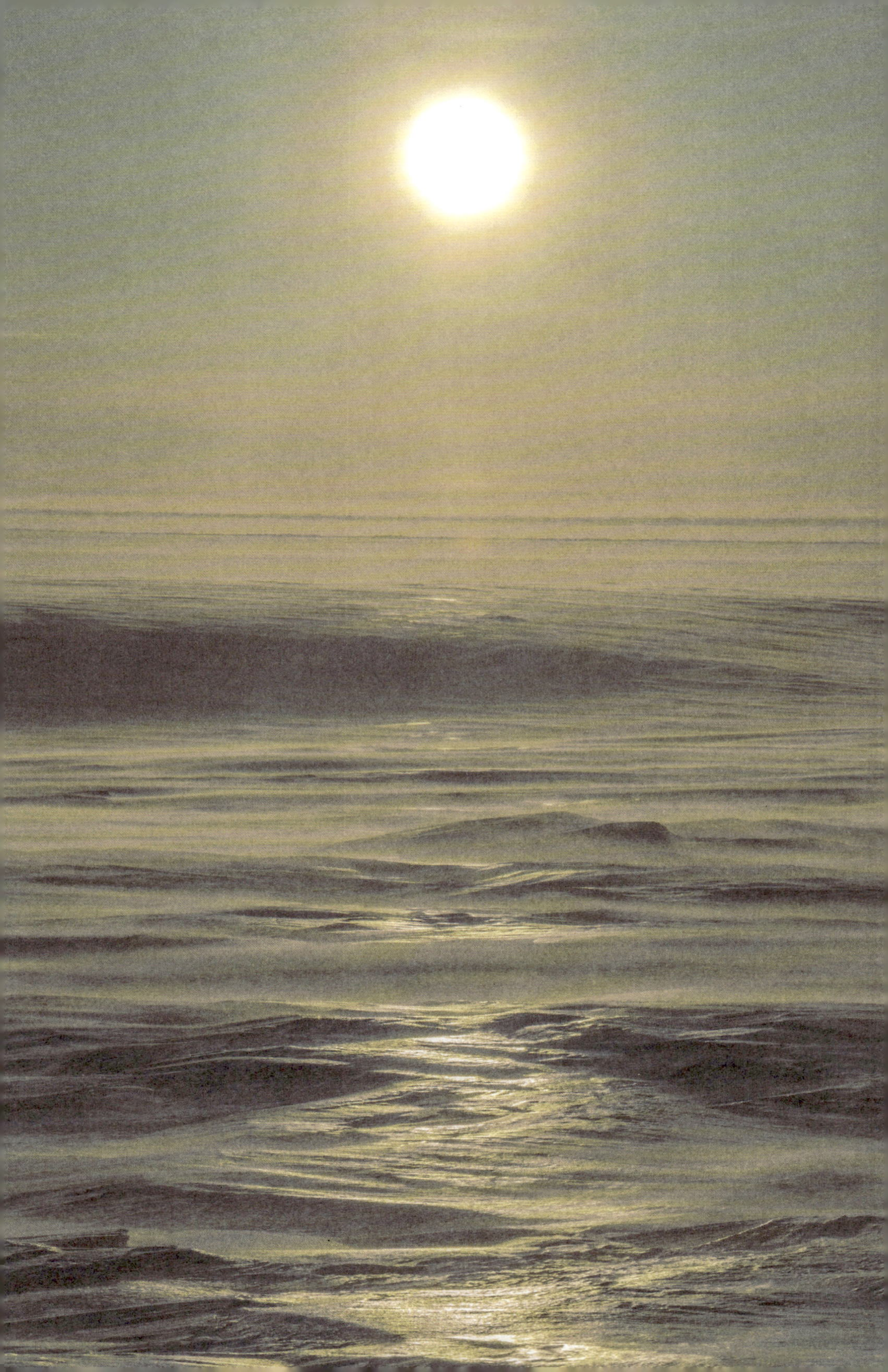

ᐅᑭᐅᒥ ᐊᓱᔾᔨᖅᑕᕐᓂᖅ

“ᑕᖅᑭᐅᓚᐅᖅᑐᒥ ᓰᕐᓗᓕ ᒪᐃ ᑕᖅᑭᖕᒐᓃᑦᑐᒍᑦ ᐊᐃᕐᕆᓕᐅᒐᓗᐊᖅᑎᓪᓗᒍ. ᐊᕐᕿᒍᑕᒫᖅ ᐅᑭᐅᑉ ᐃᓗᐊᓂ ᐊᓱᔾᔨᖅᑕᕐᓂᖕᒐ ᐊᓱᔾᔨᖅᑲᑦᑕᖅᑐᖅ ᓯᒡᔭᒥ. ᐃᒻᒪᑲᓪᓚᒃᑎᑐᑦ ᓯᑯᖅᑲᔾᔮᕆᓐᓃᖅᑐᒍᑦ.”

—ᐊᓂᐊᓯ ᖁᖅᑳᖅ

“ᓄᓇᐅᑉ ᐊᓱᔾᔨᖅᑕᕐᓂᖕᒐ ᐃᓱᒫᓘᑎᒋᕙᕋ. ᓄᓇᓕᕐᑯᑎᕗᑦ ᐃᓱᒫᓘᑎᒋᕙᕋ. ᓯᓚ ᐃᓱᒫᓘᑎᒋᕙᕋ ᐊᒻᒪ ᓄᓇᓕᑦᑎᓐᓂ ᓄᓇᕐᔪᐊᕐᒥᓗ ᖃᓄᐃᑉᐸᓪᓕᐊᓕᕐᓂᐅᔪᑦ ᐃᓱᒫᓘᑎᒋᕙᒃᑲ.”

—ᐋᓚᓯ ᐊᔭᓕᒃ

“ᑕᒪᑐᒪᓂ ᓯᓚ ᐊᔾᔨᒋᓚᐅᙱᑕᖕᒐ. ᐅᖅᑰᓂᖅᓴᐅᓚᐅᖅᑐᖅ. ᑰᑦ ᓰᐱᓯᒪᓕᖅᑐᑦ ᑰᒃᑐᑦ. ᓄᓇ ᕿᓚᒥᑯᓗᒃ ᐊᐅᓚᐅᖅᑐᖅ. ᑕᐃᒪᐃᓐᓂᖕᒐᓄᑦ ᐃᓄᐃᑦ ᓄᓇᓕᕌᕐᒥᖕᓄᙵᐅᓚᐅᙱᑦᑐᑦ ᐊᐅᒃᓴᖅᐃᓗᐊᕐᓂᖕᒐ ᐱᔾᔪᑎᒋᓪᓗᒍ. ᐅᑎᕆᐊᖅᑲᓚᐅᖅᓯᒪᔪᑦ ᑰᑦ ᓰᐱᓯᒪᓕᕇᕐᒪᑕ. ᐊᐃᕐᕆᓕᒥ ᓄᓇ ᐊᐅᓚᔪᙱᑦᑐᖅ. [...] ᐊᑖᑕᒋᓚᐅᖅᓯᒪᔭᕋ ᐅᕙᓐᓄᑦ ᐅᖃᓚᐅᖅᓯᒪᔪᖅ ᐃᑖᓐᓂ ᐅᑭᐅᑉ ᐃᓗᐊᓂ ᐊᓱᔾᔨᖅᑕᕐᓂᖕᒐ ᐊᔾᔨᒋᖅᑲᑦᑕᔾᔮᙱᓐᓂᕋᖅᖢᓂᐅᒃ. ᖃᖕᒐᑭᐊᖅ ᐊᕐᕿᒍᓕᒫᖅ ᐊᐅᔭᖅᑳᖅᑐᒍᑦ. ᓰᕐᓗ ᐅᐱᕐᖔᖅ ᐊᐅᒃᓴᖅᐃᑦᑐᑦᑖᓗᓚᐅᖅᓯᒪᖕᒪᑦ, ᐊᒻᒪ ᑕᑯᖅᑲᑦᑕᓕᖅᖢᑕ ᖃᑦᓗᓃᑦ ᓄᓇᖕᒐᓐᓂ ᐱᖅᓯᖅᑲᑦᑕᓕᖅᑐᖅ. ᐅᑦᓗᕆᐊᑦ ᕿᒥᖅᕈᐊᖅᑲᑦᑕᖅᒥᒐᒃᑭᑦ ᐅᔾᔨᕆᓯᒪᖕᒥᔪᖕᒐ

Seasons

“It was like we were [in] May in April last month. It’s like every year it changes by the shore. There will be no more ice, like long ago.”

—Agnes Kokak

“I’m worried about the seasons. I’m worried about our community. [...] I’m worried about how the weather is and what is going on in other communities and around the world.”

—Alice Ayalik

“It’s different this year. It got warmer. The rivers were running. It really melted up very fast. That’s why the people who were going to go to their outpost camps couldn’t go. They had to come back because the rivers were flowing. It usually doesn’t melt up early in April. [...] My late father told me that maybe someday our seasons will be different. We will get summer all year round here. Like this spring it melted up really fast, and down south we were watching them have blizzards. I have been looking at the stars and I noticed a change in the stars, too. I remember

ᐃᓂᒥᓃᑦᑐᓐᓃᕐᓂᖕᒋᓐᓂᒃ. ᓂᕙᐊᖅᑳᓴᑯᑑᑦᑐᖕᒐ ᐃᖅᑲᐅᒪᔪᖕᒐ ᑕᖅᑭᕐᒥᒃ ᕿᒥᕐᕈᐊᖅᓗᖕᒐ. ᐃᓛᓐᓂ ᐅᓐᓄᒃᑯᑦ ᑕᒃᐸᐅᖕᒐ ᕿᕙᐊᕌᖕᒐᒪ ᑕᖅᑭᓇᒃᐸᓚᐅᖅᓯᒪᔪᖅ, ᓯᕿᓂᕐᓗ ᓂᐱᓚᐅᖅᑎᓐᓇᒍ ᓴᒐᑉᐸᒃᓗᓂ. ᐅᑦᑐᒥᑕ ᐅᔅᓯᓇᓯᒪᑕᖅᑐᖕᒐ ᓄᓇᕐᔪᐊᖅ ᐅᕙᕐᓂᖕᒐᓂᒃ, ᐅᑦᑐᓇᐊᑦ ᐅᖕᒐᓯᖕᓂᖅᓴᐅᖅᑰᔨᑕᖅᑐᑦ ᐅᖕᒐᓯᑐᐊᖅᑰᔨᑕᖅᓗᑎᒡᑐ. ᓄᓇᕐᔪᐊᒍᑦ ᐅᕙᕐᓂᖕᒐᓅᖅᑰᖅᑐᖅ ᑕᐃᒪᐃᑕᖅᑐᖅ.”

—ᐋᓚᓯ ᐊᔭᓕᒃ

“ᓄᓇ ᐊᐅᒃᐸᓕᖅᑐᖅ ᐊᐅᒻᕙᒃᓴᖕᒐ ᓱᓕ ᑎᑭᓚᐅᖅᑎᓐᓇᒍ. ᐃᒻᒪᑲᑦᓚᒃ ᔪᓚᐃᒥᑑᓐᓃᑦ ᐊᐳᑎᓕᐊᑑᕈᓚᐅᖅᓯᒪᔪᖅ. ᐅᑦᑐᒥᐅᓕᖅᑐᖅ ᐊᐅᒃᓴᕋᐃᑐᐊᓕᖅᑐᖅ. [...] ᓯᖅᑲᐃᒻᒪᑑᓐᓃᑦ ᖁᐊᖅᐸᑦᓕᐊᓂᖕᒐ ᑭᖑᕙᖅᓯᒪᓕᖅᑐᖅ ᐊᑯᓂᐅᓂᖅᓴᕐᓗ ᖁᐊᕋᓯᖅᑲᑦᑕᓕᖅᓗᓂ.”

—ᓖᓇ ᐊᓗᖅᐱᒃ

“ᑕᒪᑐᒪᓂ ᐊᕐᕌᒍᒥ ᐊᐅᒃᓴᕋᐃᑦᑐᒻᒪᕆᐊᓘᓚᐅᖅᑐᖅ. ᑕᐃᒪᐃᓐᓂᖕᒐᓄᑦ ᓯᑯ ᕿᕿᓕᖅᐸᑦᓕᐊᓕᑕᐃᓐᓇᖅᑐᖅ. ᐊᐃᕐᕆᓕᒥ ᐊᐅᒃᓴᕋᐃᑦᑐᐊᓘᓚᐅᖅᓯᒪᔪᖅ.”

—ᔩᓯᐱ ᓂᑉᑖᓇᑦᑎᐊᖅ

when I was a little girl I looked at the moon. Sometimes at night it would be up and before sunset it would be gone, but I noticed that the world tilted, because the stars that used to be a little bit farther that way are up here now. I think our world is tilting; that is why.”

—Alice Ayalik

“It’s thawing really early, before the time for it to melt. A long time ago there used to be a lot of snow in July. Now it melts too early. [...] It’s even taking longer to freeze up.”

—Lena Allukpik

“This year we noticed we had a really early melt. That’s why this ice just started to freeze. [...] In April we had [an] early melt.”

—Joseph Niptanatiak

“ᐅᔾᕗᓇᓯᒪᔪᖓ ᓄᓇᐅᑉ ᐃᑭᐊᖓᓂ
ᖅᑭᖅᑭᕚᒃᑑᒐᓗᐊᑦ ᐊᐅᓚᖅᓂᖏᓐᓂᒃ,
ᐃᒻᒪᖄ ᓂᓚᒃ ᐱᔾᔭᑕᐅᑦᑐᓂ.
ᑕᐃᒪᐃᑦᑐᖅᑕᖅᑲᖅᑰᕿᓐᓃᖅᑐᖅ.
ᐃᒻᒪᑲᓪᓛᓗᒃ ᑕᓯᖅᔭᐊᖅ
ᓱᓕ ᓂᓚᖅᑲᖅᑎᓪᓗᒍ, ᑕᒫᓂ
ᑐᒃᑐᓕᒻᒪᓇᐊᓘᓚᐅᖅᓯᒪᔪᖅ
ᑕᒫᓂ ᓂᒡᓚᖅᓯᖅᑐᓂᒃ.
ᐃᓅᔭᖅᖃᒃ ᓄᓇᓕᕋᓛᖓᓂ
ᐃᖅᑲᓇᐃᔭᖅᑲᑦᑕᓕᓵᖅᑎᓪᓗᖓ
ᐅᖅᑰᔭᐊᓘᕙᓚᐅᖅᓯᒪᔪᖅ,
ᑭᓯᐊᓂᓗ ᑭᖑᓪᓕᖅᒥᒃ ᑕᐃᑯᓐᖓᒐᕋᒪ
ᓂᒡᓚᓯᓕᓚᐅᖅᓯᒪᔪᖅ. ᐃᓛᓐᓂᒃᑯᑦ
ᓱᐳᔩᖅᑲᑦᑕᓇᐊᖅᑲᓚᐅᖅᑐᒍᑦ ᑐᐱᖅᐸᑦ
ᓂᒡᓚᓱᓗᐊᓕᕌᖓᑦ.

ᐃᖅᑲᐅᒪᔪᖓ
ᐅᖅᑰᔭᑯᓘᖅᑲᑦᑕᓚᐅᖅᓯᒪᓂᖓᓂᒃ
ᑲᔾᔫᖅᓇᖅᓴᓂᓗ.
ᑕᐃᒪᐃᓚᐅᕋᓗᐊᖅᑎᓪᓗᒍ ᑕᒪᑐᒪᓂ
ᐊᖅᕌᒍᒥ ᐅᖅᑰᓐᖏᓐᓂᖅᓴᐅᓕᓚᐅᖅᑐᖅ.
ᑕᐃᔅᓱᒪᓂ ᐅᖅᑰᖅᑲᑦᑕᓚᐅᕋᓗᐊᖅᑎᓪᓗᒍ,
ᑕᒪᑐᒪᓂ ᐅᐱᕐᖕᒑᖅ ᐊᓯᖏᓐᓂᓗ
ᐊᖅᕌᒍᓂ ᓂᒡᓚᓱᖕᓂᖅᓴᐅᓕᖅᑐᖅ.

ᐊᓯᔾᕗᖅᔭᐊᖅᓂᐊᓗᖕᓂᒃ
ᑕᑯᖅᑲᑦᑕᖅᓯᒪᓕᖅᑐᖓ. ᑰᑦ
ᓱᐱᕆᖃᐃᓐᓂᖅᓴᐅᓕᖅᑐᑦ. ᐃᒻᒪᑲᓪᓚᒃ
ᐱᖅᓯᖃᔪᒃᑐᒻᒪᓇᐊᓘᓚᐅᖅᓯᒪᔪᖅ
ᐊᐱᓂᖅᓴᐅᕚᒃᓴᓂᓗ.
ᐅᑦᓗᒥᐅᓕᖅᑐᖅ ᐱᖅᓯᖃᔪᖑᓐᓃᖅᑐᖅ
ᐊᒻᒪ ᓯᓚᑦᑎᐊᕚᐅᒐᓗᐊᖅᑎᓪᓗᒍ
ᓂᒡᓚᓱᒃᑐᐊᓘᖅᑲᑦᑕᖅᑐᖅ. ᓲᖅᓗᓕ
ᐅᑭᐅᑯᑖᖑᓂᖅᓴᐅᖅᑲᑦᑕᖅᑐᖅ.
ᓯᓚ ᐅᖅᑰᓯᕚᑦᓕᐊᒥᐊᖑᐊ
ᐊᐸᑕᐃᕈᑎᑲᐅᖅᑐᖅᑲᑦᑕᖅᑐᒍᑦ.”

—ᔩᓐ ᐅᕼᐅᖅᑲᖅ

“I notice the permafrost melting and something to do with the *aniugaq* [late snowbeds]. I don’t think there is any of that around anymore. A long time ago, when there was *aniugaq* out around Contwoyto Lake, I saw a lot of caribou hanging around trying to cool off. Where I worked at the outpost camp at Three Rivers, it used to be really hot, but the next time I went there to work it wasn’t so hot anymore. Sometimes it was so cold we had to use Coleman stoves to keep our tents warm.

I remember it used to be nice and warm and cozy. But this year it’s not as hot. In those days it used to be warm, but this spring and other years, it’s not as warm.

I notice big changes; the river here tends to break up early. A long time ago we used to get so many blizzards and more snow buildup, but these years we don’t have many blizzards and get nice days, but cold. It’s like [winter is] longer now. As soon as it gets hot the snow will be gone all at once.”

—John Ohokak

“ᓯᑯᐃᖅᓵᐃᓕᖅᑐᒍᑦ ᓯᑯ
ᓵᑦᑐᑰᓕᕐᓂᖓᓄᑦ. [...] ᐃᓛᓐᓂ
ᓯᑯᓇᓯᑯᑖᒃᑐᐊᑰᖅᑲᑦᑕᖅᑐᒍᑦ.
ᓯᑯ ᓵᓗᐊᓕᕐᓂᖓᓄᑦ
ᐊᐅᒃᓵᐃᓐᓂᖅᓴᐅᓕᖅᑐᖅ. ᑕᒪᑐᒪᓂ
ᐊᕐᕿᒎᒥ ᓯᑯᐃᖅᓵᐃᑦᑐᑰᔪᒍᑦ. [...]
ᓂᒋᕐᒥᑦ ᐊᓄᓇᓚᔪᖕᓂᖅᓴᐅᓕᖅᑐᖅ
ᐊᒻᒪ ᑕᐃᒪᐃᓐᓂᖓᓄᑦ
ᓯᑯᐃᖅᓵᐃᓐᓂᖅᓴᐅᓕᖅᑐᒍᑦ.
ᓂᒋᕐᒥᑦ ᓂᒋᒥᒃᑐᒥᑦ
ᐊᓄᓇᕕᐅᖅᑲᑦᑕᖅᑐᒍᑦ. ᓯᓚ
ᐅᖅᑰᓂᖅᓴᒻᒪᕆᐊᓘᓕᖅᑐᖅ ᐊᐅᔪᖀᖕᒐᑦ.
ᐅᑭᐅᖅ ᓇᐃᓪᓕᕙᓪᓕᐊᖅᑰᓕᖅᑐᖅ
ᐅᖅᑰᓗᐊᓕᕐᓂᖓᓄᑦ.
ᐅᖅᑰᓯᕙᓪᓕᐊᔪᖅ.”

—ᔫᓯᐱ ᓂᑉᑕᓈᑦᑎᐊᖅ

“ᑕᑭᐅᖅ ᐊᑯᓂᐊᓗᒃ
ᓯᑯᓇᓱᖅᑲᑦᑕᖅᑐᖅ.
ᓯᑯᐃᖅᐸᓪᓕᐊᓕᖅᑎᓪᓗᒍᓗᓐᓃᑦ
ᓈᒻᒪᓈᖅᓯᒪᖅᑲᑦᑕᕈᓐᓃᖅᑐᖅ
ᑭᖑᕙᖅᓯᒪᖅᑲᑦᑕᓕᕐᒪᑦ.
ᐅᑭᐅᖕᒍᖅᑲᑦᑕᖅᑐᖅ
ᐱᖓᓱᐃᓐᓇᕐᓄᑦ ᑕᖅᑭᓄᑦ.
ᓇᐃᓪᓕᕙᓪᓕᐊᑐᐃᓐᓇᖅᐸᕐᖕᒋᓛᒃ.
ᐊᐅᔭᖅ ᐃᓯᓕᓵᐃᑉᐸᓪᓕᐊᓕᕐᒥᔪᖅ.
ᐊᐅᔭᑭᓐᓂᖅᓴᐅᓕᖅᑐᒍᑦ, ᓲᕐᓗ
ᑕᐃᒫᑐᐃᓐᓇᖅ ᐅᓐᓄᒃᑳᖕᒐᑦ
ᑖᕐᓂᖅᓵᓗᕐᑲᑦᑕᖅᑐᖅ.
ᐊᐱᑦᑎᐊᖅᐸᒍᓐᓃᖅᑐᒍᑦ.
ᐊᐱᔪᑯᓗᐊᐱᐅᔪᒍᑦ ᐃᒡᓗᕗᑦ
ᓴᐅᔭᐅᑲᓴᒃᐸᓚᐅᖅᓯᒪᔪᖅ.
ᒫᓐᓇ ᑕᐃᒪᐃᑦᑐᓐᓃᖅᑐᖅ.”

—ᑲᐃᑦ ᐃᓄᒃᑕᓕᒃ

“We’re having early breakup because of the ice. It doesn’t get very [thick] anymore. [...] The ice sometimes takes forever to freeze. The ice is too thin. It’s easier to melt. It’s really early this year. [...] We get more winds from the south; that’s why the breakup is early. We keep getting this warm weather from the south. It’s a lot warmer, too, in the summertime. I think the winters are getting shorter. [...] Too warm. It’s getting warmer.”

—Joseph Niptanatiak

“It takes a long time for the ice to freeze on the ocean. Even the breakup time, it’s not at the right time. [...] It’s later. It’s only three months we have winter. It’s getting shorter. The summer is getting shorter, too. [...] It’s shorter, just like the nights are getting darker, too. There is hardly any snow. [...] It used to be high around the house, but it’s not anymore.”

—Kate Inuktalik

“ᐊᐅᔭᐅᓚᖀᖕᒐᑦ
ᐅᖅᑰᑦᑎᐊᖅᑐᑯᓘᕙᓚᐅᖅᓯᒪᒐᓗᐊᖅ.
ᐅᑦᑐᒥᐅᓕᖅᑐᖅ
ᓂᒡᓚᓱᑭᔪᐃᓐᓇᐅᔭᓕᖅᑐᖅ.
ᐅᖅᑰᔪᑦ ᐅᑦᑐᑦ ᐅᖁᑭᓃᖅᑐᑦ.
ᐊᑯᓂᐅᓂᖅᓴᐅᕙᓕᖅᑐᖅ.
ᐊᐅᔭᖅᓴᕋᐃᓐᓂᖅᓴᐅᓕᕐᒥᔪᖅ.
ᑕᐃᒪᐃᑦᑎᐊᖅᐸᒍᓐᓃᖅᑐᖅ.
ᐊᐅᔭᖅᓴᕋᐃᑉᐸᓚᐅᖅᓯᒪᔪᒍᑦ
ᓂᑭᒥᐅᔭᖅᕈᓂ ᐊᒻᒪ ᐅᖅᑰᑦᑐᓂ.
ᒫᓐᓇ ᐊᐅᔭᖅᓴᕋᐃᓐᖕᒋᓐᓂᖅᓴᐅᓕᖅᑐᖅ
ᐊᒻᒪ ᓂᒡᓚᓱᑭᔪᐃᓐᓇᐅᔭᓕᖅᕈᓂ.
ᐅᑭᐊᒃᓵᒃᑯᑦ ᓂᒡᓚᓱᖕᓂᖅᓴᐅᓕᖅᑐᖅ
ᐊᒻᒪ ᐊᓄᓇᒐᔪᖕᓂᖅᓴᐅᓕᖅᕈᓂ.
ᐊᓄᓇ ᓴᓐᖕᒋᔪᐊᓘᖅᑲᑦᑕᖅᑐᖅ ᐊᒻᒪ
ᓇᒍᓐᖕᒐᐅᓐᖕᒋᓐᓂᖅᓴᐅᓕᖅᕈᑕ
ᐊᓄᓇᓗᐊᕐᓂᖕᒐ
ᐱᑦᑐᒍ. ᐅᒥᐊᖅᑐᖅᕈᑕ
ᑕᐃᒪᓐᓇᑐᐃᓐᓈᓘᖅᑰᕗᔪᖅ
ᐊᓄᓇᓕᖅᑲᑦᑕᓕᖅᑐᖅ.”

—ᓚᐅᕋ ᖁᖅᓱᖅᑕᖅ

“ᐊᐅᔭᖅ ᐅᖅᑰᓯᑦᓚᓇᓚᐅᓐᖕᒋᑦᑐᖅ.
ᐃᓚᖕᒋᓐᓂ ᐅᑦᑐᕐᓂ
ᐅᖅᑰᕐᑲᑦᑕᖅᑑᒐᓗᐊᖅ ᑭᓯᐊᓂᓚ
ᓂᒡᓚᓱᑭᔪᐃᓐᓇᕐᑲᑦᑕᖅᑐᖅ. ᐃᓛᓐᓂ
ᐊᐅᒡᒍᓯᒥ ᐊᐅᔭᑦᑎᐊᕚᓗᕐᑲᕐᑲᑦᑕᖅᑐᒍᑦ
ᐅᑭᐊᒃᓴᖅᐸᑦᓚᐊᓇᔭᖕᒍᐊᓕᕋᓗᐊᕋᒥ.
ᐃᓛᓐᓂ ᐅᑭᐊᒃᓵᒃᑯᑦ
ᓂᒡᓚᓱᖕᓂᖅᓴᐅᓲᖅ.
ᐅᑭᐅᖅᓯᐅᑎᐊᓗᖕᓂᒃ ᑭᓯᐊᓂ
ᐊᓐᓄᖀᖅᓯᒪᓕᖅᑲᑦᑕᖅᑐᒍᑦ
ᓂᒡᓚᑐᐊᖅᑲᑦᑕᕐᒪᑦ ᐃᓛᓐᓂᒃᑯᑦ.”

—ᓖᓇ ᓂᑉᑕᓈᑦᑎᐊᖅ

“We used to have really warm summers. Now it’s always cool. We don’t get many warm days. [...] It’s taking longer. The summer is later. It’s not like it used to be. We used to have really early summers. Really warm and hot. Now we have late summers and it’s always cool. [...] In the fall it gets really cool and we have more wind. The wind gets really strong and it’s hard to get around. It gets windy fast when we are out on the boat.”

—Laura Kohoktak

“The summer, it hasn’t been warm enough. Some days it’s hot but most of the time it’s so cold, and sometimes in August we will have a warm summer when it’s supposed to be falltime. Sometimes even in the falltime it gets colder. We have to dress up like for the winter. Sometimes it gets really cold.”

—Lena Niptanatiak

“ᖁᕐᓗᖅᑐᕐᒥ ᓯᑯᐃᓕᖄᖓᑦᑕ ᐊᒃᓱᐊᓗᒃ ᓴᖏᔫᖃᑦᑕᓚᐅᖅᓯᒪᔪᖅ. ᑕᐃᔅᓱᒪᓂᐅᑎᓪᓗᒍ ᓂᐱᓕᐊᓘᕙᓚᐅᖅᓯᒪᔪᖅ ᐃᕗᓕᖄᖓᑦ. ᕿᔭᕙᓂᕐᓂᒃ ᑰᒃᑕᐅᔪᓂᒃ ᑕᑯᖃᑦᑕᓚᐅᖅᓯᒪᒐᓗᐊᖅᑐᒍᑦ ᑭᓯᐊᓂ ᐅᓪᓗᒥᐅᓕᖅᑐᖅ ᐃᖏᕐᕋᓂᑭᑦᑐᑯᓘᓕᖅᑐᖅ. [...] ᐃᒻᒪᑲᓪᓚᒃ, ᓯᑯ ᐃᕗᕈᓪᓕᐊᓕᖅᑎᓪᓗᒍ ᐊᒃᓱᐊᓗᒃ ᓂᐱᖃᖅᐸᓚᐅᖅᓯᒪᔪᖅ. ᐃᓄᐃᑦ ᑕᑯᔭᖅᑐᖅᑐᐊᓘᕙᓚᐅᖅᓯᒪᔪᑦ. ᓯᑯᐊᓗᐃᑦ ᐃᕗᓕᖅᑎᓪᓗᒋᑦ ᑰᒃᑕᐅᕙᓪᓕᐊᓕᖅᑎᓪᓗᒋᑦ ᓂᐱᓕᐊᓘᕙᓚᐅᖅᓯᒪᔪᖅ, ᒫᓐᓇᓕ ᑕᐃᒪᐃᑦᑐᓐᓃᖅᑐᖅ.”

—ᒪᒥ ᐆᓐᓂᖅ

“When we have breakup here in Kugluktuk it used to be really strong. In those days [...] it would be really loud. And we would see driftwood come down, but the current is really not strong anymore. [...] A long time ago, as soon as we started to hear the ice break up, it would be really loud. People would run up to see. These big chunks of ice—when they were travelling down you would hear rumbling, but you don’t hear that anymore.”

—Mamie Oniak

“ᐅᓪᓗᒥ ᐅᔾᔨᕆᖃᑦᑕᓕᖅᑐᒍᑦ ᐅᐱᕐᖔᖅ, ᐊᐅᔭᖅ, ᐅᑭᐅᖅ ᐊᒻᒪ ᐅᑭᐊᒃᓵᖅ ᑕᐃᒪᐃᑦᑎᐊᕈᓐᓃᕐᓂᖏᓐᓂᒃ. ᐅᓪᓗᕗᑦ ᐊᕐᕌᒍᕗᑦ ᖃᓄᐃᖅᑲᑦᑕᓚᐅᕐᓂᖏᓐᓂᒃ ᐃᖅᑲᐅᒪᔪᒍᑦ. ᐅᓪᓗᕆᐊᑦ, ᑕᖅᑭᖅ, ᓯᕿᓂᖅ. ᐅᓪᓗᒥᐅᓕᖅᑐᖅ ᐅᔾᔨᕆᔭᕗᑦ ᑕᒪᒃᑯᐊ ᐊᓯᔾᔨᖅᓯᒪᓕᕐᓂᖏᓐᓂᒃ.”

—ᒪᒥ ᐆᓐᓂᖅ

“We notice even our spring, summer, winter, and fall, they are not the same anymore. We know how our days and years are. The stars, the moon, the sun. We notice the difference now in the changes.”

—Mamie Oniak

“ᐅᑭᐊᒃᓵᖑᓕᖄᖓᑦ ᐊᔾᔨᒋᔭᓐᓃᖅᑕᖓ. ᒪᐃ ᓯᓕ ᐊᑐᖅᑎᓪᓗᒍ ᔫᓂ 15−ᖑᔮᓕᖅᑐᖅ, ᐅᐱᕐᖓᖅᓴᖃᐃᑦᑐᖅ. ᑕᖅᑭᖅ ᑕᒪᓐᓇ ᓯᕗᕙᓯᒃᑐᐊᓗᒃ ᐅᐱᕐᖓᑲᐅᑎᒋᓚᐅᕐᑦᑕ.”

—ᒫᒃ ᐊᒻᒪ ᒫᑕ ᑕᓖᑦᑐᖅ

“It’s getting a lot different, the fall. It’s just like June 15 now even though May is not finished, because the spring came early. This month is like way [ahead] because we got spring early.”

—Mark and Martha Taletok

“ᔭᓄᐊᕆᒥ ᓯᕐᑭᓂᖅ ᓴᖅᑭᓴᕋᐃᓕᖅᐸᑦᓕᐊᑐᐃᓐᓇᖅᑐᖅ. ᐊᑯᓂᐅᓂᖅᓴᖅ ᑕᐅᕙᒡᔪᐊᖑᕙᓚᐅᖅᓯᒪᒐᓗᐊᖅᑐᖅ. ᑕᖅᑭᑦ ᑎᓴᒪᑦ ᓈᓇᓱᒃᑎᑦᓗᒋᑦ ᐅᑦᓗᖅᑐᓂᖅᓴᐅᓕᖅᑐᖅ ᑕᐅᕙᒡᔪᐊᕐᒥᑦ ᑕᑭᓂᖅᓴᐅᓕᖅᖢᓂ. ᑖᖅᑎᑦᓗᒍ ᐃᖅᑲᓇᐃᔭᓕᐊᖅᐸᓚᐅᖅᓯᒪᔪᒍᑦ ᐊᒻᒪ ᑖᖅᑎᑦᓗᒍ ᐊᖏᕐᕋᐅᕙᒃᖢᑕ.”

—ᒨᓇ ᑎᒃᑖᓕᒃ

“ᐅᑭᐅᖅ ᕿᓚᒻᒥᐅᓂᖅᓴᖅ ᐊᓂᒍᖅᑲᑦᑕᖅᑰᓕᖅᑐᖅ. [...] ᑖᑯᑖᖅᑲᑦᑕᕈᓐᓃᖅᑐᖅ ᐊᒻᒪ ᐅᑦᓗᖅᑐᓂᖅᓴᐅᓕᖅᖢᓂ ᐅᑭᐅᒃᑯᑦ. ᐃᓛᓐᓂ ᓯᕐᑭᓂᖅ ᓴᖅᑭᓐᓇᓵᖅᑲᑦᑕᖅᑐᖅ. ᐊᑯᓂᐊᓗᒃ ᓯᕐᑭᓂᖅ ᓴᖅᑭᓐᓇᓱᒃᐸᓚᐅᖅᓯᒪᔪᖅ. ᓄᑲᑉᐱᐊᑯᓘᑦᓗᖓ ᐅᑦᓗᓕᒫᖅ ᐊᓃᕋᔭᒃᐸᓚᐅᖅᓯᒪᔪᖓ ᓯᕐᑭᓂᖅ ᓂᐱᓚᐅᓐᖏᓐᓂᓕᒫᖓᓂ. ᐅᑭᐅᖑᖅᑲᑦᑕᓕᖅᑐᖅ ᐱᖓᓱᐃᓐᓇᕐᓄᑦ ᑕᖅᑭᓄᑦ. ᐅᑭᐅᖑᕙᓚᐅᖅᓯᒪᔪᖅ ᑕᖅᑭᓄᑦ 7-ᓄᑦ 8-ᓄᑦ ᑕᖅᑭᓄᑦ. ᓂᒡᓚᓱᒃᖢᓂ ᑖᖅᖢᓂᓗ. ᓄᐊᕆᒥ ᓯᕐᑭᓂᖅᑲᓐᖏᑦᑎᐊᓚᐅᖅᓯᒪᔪᖅ. ᔫᓂᐅᓕᖀᖓᑦ ᑭᓯᐊᓂ ᓯᕐᑭᓂᖅ ᓄᐃᑉᐸᓚᐅᖅᓯᒪᔪᖅ.”

—ᒫᒃ ᑕᓖᑦᑐᖅ

“Our sun is starting to come up earlier, too, in January. We used to have longer darkness. About four months now the daylight is getting longer than darkness. We used to go to work in the dark and come home in the dark.”

—Mona Tiktalek

“I think the winters are getting shorter. [...] The darkness gets shorter and we have more daylight in the winter. Sometimes the sun doesn’t come out right away. It used to take forever for the sun to come out. When I was a young boy I would play all day before the sunset. [...] Only three months of winter. We used to have it seven, eight months, the winter. Cold and dark. In January we never used to get any sun. The only time we got sun was in June.”

—Mark Taletok

“ᓯᑯᐃᕐᕕᒃᓴᓐᖑᐊᒐᓗᐊᕐᒥᓂ ᓯᑯᐃᖃᑦᑕᕈᓐᓃᖅᑐᒍᑦ ᑭᖑᕙᖅᓯᒪᓕᕐᒪᑦ. [...] ᓯᑯᖃᑯᑖᖃᑦᑕᓕᖅᑐᒍᑦ.”

—ᒫᒃ ᐊᒻᒪ ᒫᑕ ᑕᓖᑦᑐᖅ

“ᐃᒻᒪᒃᑲᓐᓂᖅ ᓯᑯᐃᖅᓴᒐᐃᓐᓂᖅᓴᐅᓚᐅᖅᓯᒪᔪᖅ. ᒫᓐᓇ ᑭᖑᕙᖅᓯᒪᓕᖅᑐᖅ ᓯᑯᐃᕐᕕᐊ. [...] ᓯᑯ ᐃᕗᕙᑦᓕᐊᓕᖅᑎᑦᓗᒍ ᐃᓄᐃᑦ ᖁᓐᖏᐊᓇᐊᖅᐸᓚᐅᖅᓯᒪᔪᑦ. ᐃᓛᓐᓂ ᐊᒃᓯᐊᓗᒃ ᓂᐱᖅᑯᖅᑐᕙᓚᐅᖅᓯᒪᔪᖅ. ᓯᑯᐃᑦ ᑕᐅᓄᖓ ᓯᑐᕙᑦᓕᐊᔪᑦ ᑕᑯᒃᓴᐅᕙᓚᐅᖅᓯᒪᔪᖅ. ᑰᒃᑖᓘᖃᖅᑕᓚᐅᖅᓯᒪᔪᑦ ᓯᑯᐃᑦ ᐊᒻᒪ ᖀᔪᕕᓃᑦ. ᐃᓛᓐᓂ ᑕᒃᐸᐅᖓ ᖃᖅᑲᓕᐊᖅᐸᓚᐅᖅᓯᒪᔪᒍᑦ ᐃᕗᔪᖅ ᖁᓐᖏᐊᓇᐊᖅᑐᖅᓱᑎᒃᑯᑦ.”

—ᓂᐊᓕ ᖃᐃᔪᒐᓇ

“ᐃᒪᖅ ᓂᓪᓚᓇᓐᖏᓐᓂᖅᓴᐅᓕᖅᑐᖅ ᐊᒻᒪ ᐊᑯᓂᐊᓗᒃ ᓯᑯᓇᓯᖃᑦᑕᓕᖅᓱᓂ. [...] ᓯᑯᕙᒃᑲᓗᐊᖅᓱᓂ ᐃᔾᔪᖅᑎᑦᑐᓐᓇᐃᓪᓕᔪᖅ. ᓵᑦᑐᕈᓗᒃ. [...] ᐃᒻᒪᑲᓪᓚᒃ ᓯᑯᑲᐅᑎᒋᕙᓚᐅᖅᓯᒪᔪᖅ ᐊᒻᒪ ᐃᔾᔪᔪᐊᓘᓪᓗᓂ. ᒫᓐᓇ ᐃᓚᖏᓐᓂ ᓵᑦᑐᑯᓘᖃᑦᑕᖅᑐᖅ.”

—ᕈᐃ ᐃᓄᒃᑕᓕᒃ

“The breakup of the ice is really late. [...] The ice stays longer.”

—Mark and Martha Taletok

“It used to break up early a long time ago. Now it’s really late. [...] When it broke up people would be calling to watch the breakup. Sometimes [it was] really loud. You could watch the ice coming down. There used to be lots of ice and driftwood. Sometimes we would go out on the mountains to watch the ice break up.”

—Nellie Kaiyogana

“The water is not as cold, too, and it’s taking forever to freeze up. [...] Even when we have the ice freeze-up it doesn’t get thick anymore. It’s just not too thick. [...] A long time ago the ice would freeze right away and it would be really thick. Now it’s really thin in some places.”

—Roy Inuktalik

ᓯᓚ/ᐊᓄᕆ

Climate/Weather

“ᓄᓇᐅᑉ ᐃᑭᐊᖕᒐ ᖁᐊᖕᒍᔪᓐᓃᖅᑐᖅ. ᑕᒪᕐᒥᐸᓘᖅᑰᖅᑐᖅ ᐊᐅᒃᓯᒪᓕᖅᑐᖅ.”

—ᒫᒃ ᐊᒻᒪ ᒫᑕ ᑕᓖᑦᑐᖅ

“ᓄᓇᐅᑉ ᐃᑭᐊᖕᒐᓂ ᕿᕿᔪᖅᑕᖅᑲᖅᑰᕈᓐᓃᖅᑐᖅ.”

—ᓖᓇ ᐊᓗᖅᐱᒃ

“ᒫᓐᓇᓕ ᐊᓄᕇᓐᓇᐅᔭᐸᓗᓕᖅᑐᖅ ᐅᒥᐊᖅᑐᖅᓯᓂᓗ ᐅᖅᑲᓯᒃᑐᓕᐊᕐᓇᕈᓐᓃᖅᓯᓂ. ᓯᒡᔭᒃᑰᕐᑲᑦᑕᕆᐊᕐᑲᓕᖅᑐᒍᑦ ᐊᓄᕆᓕᑐᐃᓐᓇᕆᐊᕐᑲᕐᓂᖕᒐ ᐃᓱᒪᒋᓗᒍ. [...] ᑕᐃᒪᐃᑉᐸᓚᐅᖅᓯᒪᙱᑦᑑᒐᓗᐊᖅ. ᐊᓄᕆ ᑕᐃᒪᓐᓇᑲᑦᓚᑐᐃᓐᓈᓗᒃ ᑎᑭᖅᑲᑦᑕᓕᖅᑐᖅ ᐊᒻᒪ ᓴᖅᑭᓴᖅᐃᓕᖅᓯᓂ.”

—ᐊᓂᐊᓯ ᖁᖅᑳᖅ

“ᒫᓇ ᐊᐱᓗᐊᖅᐸᒍᓐᓃᖅᑐᖅ. ᐊᐱᔭᐊᓘᖃᓚᐅᖅᓯᒪᒪᒐᓗᐊᖅᑎᑦᓗᒍ ᐅᑦᓗᒥ ᑕᐃᒪᐃᑦᑐᓐᓃᖅᑐᖅ.”

—ᐋᓚᓯ ᐊᔭᓕᒃ

“We don’t have [permafrost] anymore. Most of it is thawed.”

—Mark and Martha Taletok

“I don’t think [the permafrost] is there anymore.”

—Lena Allukpik

“It’s windy most of the time and you can’t go far out on the boat anymore. You have to stay close to shore in case the wind comes. [...] It never used to be like that. The wind now comes all at once and changes.”

—Agnes Kokak

“Nowadays the snow is not as high. We used to get really high snowbanks. Not anymore.”

—Alice Ayalik

“ᐊᐳᑎᖅᑲᑦᑎᐊᕈᓐᓃᖅᑐᖅ. ᐃᓚᖏᓐᓂ ᐊᕐᖄᒍᓂ ᐊᔾᔨᒋᖅᑲᑦᑕᓐᖏᒻᒥᔭᖕᖑ. ᐊᕐᖄᒍᒥ ᐊᑕᐅᓯᕐᒥ ᐊᐳᑎᓕᐊᓘᕙᖅᐳᖅ, ᐊᕐᖄᒍᒻᒥᖕᒪᑦ ᐊᐱᑦᑎᐊᓐᖏᕙᖅᐳᖅ. ᐊᓄᖄᓯᑲᓪᓚᒃᑐᐊᓘᖅᑲᑦᑕᖅᑐᖅ. ᐊᔪᓇᖅᑐᐊᓗᒃ; ᐊᓄᓇᓯᑲᓪᓚᖅᑲᑦᑕᖅᑐᖅ. ᐃᓄᐃᑦ ᐊᐅᓪᓛᖅᓯᒪᔪᑦ ᐱᖕᒋᒋᕙᒃᑕᕗᑦ. ᑕᐃᒪᓐᓇᑲᓪᓗᐊᖄᓗᒃ ᐊᓄᓇᓕᖅᑲᑦᑕᕐᓂᖑᓄᑦ. ᐅᓗᓇᐊᓇᕐᓂᖅᓴᐅᓕᖅᑐᖅ.”

“There is hardly any snow. Some years are different. In one year there is lots of snow; another year, not so much snow. The wind comes all at once. It’s so hard; the wind comes really fast. [...] We worry about people out camping. It just happens all at once. [...] It’s more dangerous.”

—ᓖᓇ ᐊᓗᖅᐱᒃ “ᑰᑦ ᐊᒻᒪ

—Lena Allukpik

ᑕᓰᑦ ᐸᓂᖅᐸᓪᓕᐊᓕᖅᑐᑦ. ᐃᒪᖅᑕᖅᑲᑦᑎᐊᕈᓐᓃᖅᑐᖅ. ᖁᕐᓗᕐᓘᓐᓃᑦ, ᑰᒡᕕᓪᓗᐊᑕᖑ ᓴᓐᖏᓐᓂᖅᓴᐅᓕᖅᑐᖅ ᓴᕐᕿᐅᕝᕕᓪᓗᐊᑕᖑ. ᐊᕐᖄᒍᑕᓖᖅ ᓴᕐᕿᐅᓂᖑ ᓴᓐᖏᒃᑎᑉᐸᓪᓕᐊᑐᐃᓐᓇᖅᑐᖅ.”

“[The rivers and lakes] are drying up. Hardly any water left. Even Bloody Falls, where the river flows, we noticed the flow getting weaker. It’s getting less and less every year.”

—ᓖᓇ ᐊᓗᖅᐱᒃ

—Lena Allukpik

“ᐃᒻᒪᒃᑲᓐᓂᖅ ᐃᖅᑲᐅᒪᔭᖑ ᐊᐱᖅᔭᐊᖅᑐᐊᓘᖅᑲᑦᑕᓚᐅᖅᓯᒪᓂᖑᓂᒃ. ᐅᓪᓗᒥᐅᓕᖅᑐᖅ ᐊᐳᑎᑭᑦᑐᑯᓘᓕᖅᑐᖅ. ᓯᓚᖅᔭᐊᖅ ᐊᓯᔾᔨᖅᐸᓪᓕᐊᓕᕐᒪᖅᑲᐃ. ᐊᕐᖄᒍᓂ ᖃᓂᑦᑐᓂ ᑕᐃᓛᖄᓘᕙᓚᐅᖅᓯᒪᓐᖏᑦᑑᖑᓗᐊᖅ, ᑭᓯᐊᓂᓕ ᐅᓪᓗᒥᐅᓕᖅᑐᖅ ᐊᓄᓇ ᓴᓐᖏᓂᖅᓵᓗᓕᖅᑐᖅ. ᐊᓄᓇ ᓂᓪᓕᓇᕐᓂᖅᓴᐅᓕᕐᒥᔪᖅ. [...] ᖃᑯᑎᒃᑯᐊᓗᒃ ᑭᓯᐊᓂ ᑲᓪᓕᖅᐸᓕᖅᑐᖅ. ᐃᖅᑲᐅᒪᔭᖑ ᓄᓇᖅᐸᖅᓯᒪᓪᓗᑕ ᑲᓪᓕᖅᑐᐊᓘᕙᓚᐅᖅᓯᒪᓂᖑᓂᒃ ᐊᒻᒪ ᐃᒃᑐᓚᐅᖅᔪᓂ.”

“I remember a long time ago there used to be so much snow. Nowadays, hardly any snow. I think our world is changing, I guess. It wasn’t bad so many years ago but nowadays the wind is stronger. We have more cold winds. [...] We don’t get very many thunderstorms now, either. I remember when I used to be inland, there was lots of lightning and lots of thunder.”

—ᔮᓐ ᐅᕼᐅᖅᑲᖅ

—John Ohokak

“ᓄᓇᐅᑉ ᐃᑭᐊᖕᒐᓃᑦᑐᖅ ᓂᓚᒃ ᐊᐅᓯᔪᖅ. ᑐᒃᑑᔭᖅᑐᕐᒥᐅᑦ ᐃᓄᖕᒋᑦ ᐃᒡᓗᓖᑦ ᑲᑉᐱᐊᓯᕈᔪᓕᖅᑐᑦ ᓄᓇᐅᑉ ᐃᑭᐊᖕᒐ ᐊᐅᓕᖅᒪᑦ. ᑕᐃᒪᐃᑦᑐᒃᓴᐅᒋᕗᖅ ᐊᓯᖕᒋᓐᓂ ᓄᓇᓕᖕᓂ.”

—ᔫᓯᐱ ᓂᑉᑕᓈᑦᑎᐊᖅ

“ᓯᓚ ᖃᓄᐃᓐᓂᐊᓕᖅᒪᖕᒑᖕᒐ ᓇᓚᐅᑦᑖᒐᕐᓂᕈᓐᓃᖅᑐᖅ. ᓯᕐᑭᓂᕐᓘᓐᓃᑦ ᐅᕕᓂᖕᒍᑦ ᐱᐅᔪᓐᓃᖅᑐᖅ; ᑕᐃᒪᐃᑉᐸᓚᐅᖅᓯᒪᓐᖏᑦᑑᒐᓗᐊᖅ.”

—ᑲᐃᑦ ᐃᓄᒃᑕᓕᒃ

“ᐃᓛᓐᓂ ᑰᑦ ᐊᐅᑲᐅᑎᒋᖅᑲᑦᑕᓕᖅᑐᑦ ᐊᒻᒪ ᑕᓰᑦ ᐊᐅᒃᓴᕋᐃᓕᖅᖢᑎᒃ. ᐃᓛᓐᓂ ᓯᑯᐃᑲᐅᖅᑐᖅᑲᑦᑕᓐᖏᑦᑐᑦ ᑭᓯᐊᓂᓕ ᐅᓪᓗᒥᐅᓕᖅᑐᖅ ᐃᒃᑲᓗᐊᓕᕐᓂᕐᒥᓄᑦ ᑰᓐᖏᓐᓂᖅᓴᐅᓕᖅᑐᑦ. ᑕᓰᑦ ᓘᓐᓃᑦ ᐸᓂᓕᖅᑐᑦ.”

—ᒪᒥ ᐆᓐᓂᖅ

“ᐅᓪᓗᒥ ᓂᒡᓚᓯᒍᓐᓃᖅᑐᖅ. ᓂᒡᓚᓇᖅᑐᖅᔭᐊᕌᓗᖕᒥᒃ ᐃᒻᒪᒃᑲᓐᓂᖅ ᐊᓄᕌᖅᐸᓚᐅᖅᓯᒪᔪᖅ. ᐅᐊᖕᓇᒥᐅᓐᖏᑦᑐᖅ, ᑲᓇᖕᓇᕐᒥᑦ ᐊᒻᒪ ᐅᐊᖕᓇᒥᑦ. ᐅᔾᔨᕆᓯᒪᔪᖕᒐ ᐊᓄᕆᒃᑲᐅᓂᖅᓴᐅᓕᕐᓂᖕᒐᓂᒃ. ᐃᓛᓐᓂᓘᓐᓃᑦ ᐅᔾᔨᓐᖏᑦᓚᒃᑯᑦᑕ ᐊᓄᕆᓯᑲᑦᓚᒍᓐᓇᖅᑐᖅ ᐊᒻᒪ ᓴᓐᖏᔫᒃᑲᐅᑎᒋᓪᓗᓂ. ᐃᓛᓐᓂᒃᑯᑦ ᐅᔾᔨᕐᓇᓐᖏᑦᑐᕐᓘᓐᓃᑦ ᐊᓄᕇᓕᕐᓂᐊᕐᓂᖕᒐ. ᐃᓛᓐᓂ ᐊᒡᒍᕋᖅᖢᓂ ᐱᓯᑦᑎᐊᕐᓇᓐᖏᒻᒪᕆᓕᖅᑐᖅ. ᐃᓛᓐᓂ

“The permafrost is thawing. The people in Tuktoyaktuk [in the Northwest Territories] are kind of afraid—the people who have houses—for the permafrost is starting to thaw. It’s probably getting to be like that in other places, too.”

—Joseph Niptanatiak

“It’s getting harder [to predict the weather]. Even the sun now is bad for the skin; it never used to be like that.”

—Kate Inuktalik

“Sometimes the rivers tend to [melt] right away and the lakes don’t have ice very long. Sometimes they don’t melt right away, but nowadays [...] there is hardly any water to get the rivers flowing. They are drying up, even the lakes.”

—Mamie Oniak

“It’s not as cold anymore, too. We used to get cold winds. Sometimes the wind is [...] stronger. But it’s not as cold anymore. [...] Not from the north, just from the east and south. I notice the wind getting stronger. Sometimes without even knowing, all at once it comes and gets strong. Sometimes you don’t even know if it’s going to

ᐱᓱᒡᕕᒃᓴᐅᓐᖏᑦᑎᐊᒻᒪᕆᒃᑐᖅ. ᑕᐃᒪᐃᑉᐸᓚᐅᖅᓯᒪᓐᖏᑦᑐᖅ ᐃᒻᒪᒃᑲᓂᖅ. [...] ᓄᕗᔭᐃᑦ ᕿᒥᖅᕈᐊᖅᓱᑎᒃᑯᑦ ᖃᐅᔨᓇᓚᐅᖅᓯᒪᔪᖅ ᖃᓄᐃᑦᑐᒥᒃ ᐊᓄᕆᒥᒃ ᐊᒡᒋᖅᑐᖅᑲᓕᖅᒪᖕᖔᖕᖓ. ᑕᐃᒪᐃᑦᑐᓐᓃᖅᑐᖅ. ᐃᓛᓐᓂ ᓱᓕ ᓄᕗᔭᐃᑦ ᕿᒥᖅᕈᐊᖅᑲᑦᑕᖅᑕᕗᑦ, ᑭᓯᐊᓂᓗ ᐃᓛᓐᓂ ᐅᔾᔨᕆᓐᖏᓱᑕ ᑲᑉᐱᐊᓇᔪᒃᓯᔪᖅ ᐊᓄᕆᓯᑲᓪᓚᒃᑐᐊᓘᖅᑲᑦᑕᖅᓂᖕᖓᓄᑦ ᐃᓱᒫᓗᖕᓇᖅᓯᔪᑦ ᓱᕈᓰᑦ ᐊᓯᖕᒋᓪᓗ ᓄᓇᓕᖕᓂᑦᑐᑦ. ᐅᑭᐅᒃᑯᑦ ᐊᐅᔭᒃᑯᓪᓘᓐᓃᑦ ᑕᐃᒪᐃᖅᑲᑦᑕᓕᖅᑐᖅ.”

—ᒪᒥ ᐆᓐᓂᖅ

“ᐅᓪᓗᒥ ᐊᐱᑦᑎᐊᖅᐸᒍᓐᓃᖅᑐᒍᑦ; ᑕᐃᒪᐃᓐᓂᖕᖓᓄᑦ ᑰᑯᓗᐃᑦ ᑕᓯᑯᓗᐃᓪᓗ ᐸᓂᖅᐸᓪᓕᐊᓕᖅᑐᑦ. ᐊᐱᔪᐊᓘᕙᓚᐅᖅᓯᒪᓪᓗᑕ ᐊᐱᑦᑎᐊᖅᐸᒍᓐᓃᖅᑐᒍᑦ. [...] ᒫᓐᓇ ᐊᓄᕆᓗᐊᖅᑐᐊᓘᖅᑲᑦᑕᓕᖅᑐᖅ—ᑕᐃᒪᐃᓐᓂᖕᖓᓄᑦ ᐊᐱᓚᐅᕋᓗᐊᖅᓱᑕ ᐊᐳᑕᐃᖅᑕᐅᖅᑲᑦᑕᖅᑐᒍᑦ ᐊᓄᕆᒧᑦ. ᐊᐳᑦ ᑕᕝᕙᓂᔅᓴᐃᓐᓈᖅᑲᑦᑕᕈᓐᓃᖅᑐᖅ.”

—ᒫᒃ ᐊᒻᒪ ᒫᑕ ᑕᓖᑦᑐᖅ

“ᐃᓛᓐᓂᒃᑯᑦ ᐅᑭᐊᒃᓴᖔᖕᖓᑦ ᓯᑯᕙᒃᑐᖅ, ᑭᓯᐊᓂᓗ ᓯᑯᓚᐅᕋᓗᐊᖅᓱᓂ ᐊᓄᖔᓕᖔᖕᖓᑦ ᐊᐅᒃᑲᓂᑐᐃᓐᓇᖅᐸᒃᑐᖅ ᐅᕝᕙᓘᓐᓃᑦ ᓯᖅᑯᑎᒃᑲᓂᖅᓱᓂ. ᑕᐃᒪᐃᑉᐸᓚᐅᖅᓯᒪᓐᖏᑦᑑᒐᓗᐊᖅ. ᓯᑯᒐᒥ ᒪᓂᓵᖅᑲᑦᑕᓚᐅᖅᓯᒪᔪᖅ. ᐅᓪᓗᒥ ᓯᖅᑯᒥᑲᑕᓗᐊᒧᑦ ᐊᐅᒃᑲᓂᖅᑲᑦᑕᓗᐊᖅᓂᖕᖓᓄᑦᑐ ᓯᑯ ᒪᓃᓛᓘᖅᑲᑦᑕᓕᖅᑐᖅ.”

—ᒥᐊᓕ ᕿᓗᒃ

blow. Sometimes it's hard to walk toward the wind. You can't even walk. It's not how it used to be long ago. [...] We used to look at the clouds to see what kind of wind was coming. It's not like that anymore. We look at the clouds, but sometimes without knowing. [...] It's really kind of scary; you have to wonder about these kids and other things that are going on in the community because [...] all of a sudden it blows. Even in the wintertime and summertime."

—Mamie Oniak

"We don't get very much snow anymore; that's why our little rivers and lakes are drying out. Not very much snow. [...] We used to get so much snow. Right now we have too much wind—that's why the snow flies and goes to another place. It doesn't stay in one place."

—Mark and Martha Taletok

"During the fall sometimes it freezes up, but [...] when it gets windy it just melts again or breaks up again. It wasn't like that before; we used to have really smooth ice. [...] Now we have lots of wind. That's why it's so rough."

—Mary Kellogok

“ᐃᒪᖅᑕᖅᑦᑎᐊᕈᓐᓃᖅᑐᖅ;
ᐃᒃᑲᑦᑐᑯᓘᓕᖅᑐᑦ.
ᖃᖅᑲᐃᓪᓘᓐᓃᑦ ᐊᕙᓃᑦᑐᑦ
ᑕᓲᑦ ᑰᓪᓗ ᐃᒪᖅᑦᑎᐊᕈᓐᓃᖅᑐᑦ;
ᑕᓯᑯᓗᐃᓐᓇᐅᓕᖅᑐᑦ ᐊᒻᒪ
ᐃᒪᓕᐅᒐᓗᐊᑦ ᒥᑭᔩᑏᓐᓇᐅᓕᖅᖢᑎᒃ.
ᑕᐃᒪᐃᑦᑐᓂᒃ ᑕᑯᕙᒍᓐᓃᖅᑐᒍᑦ.
ᖃᖅᑲᒐᓛᓪᓘᓐᓃᑦ ᓯᕗᖕᒐᔭᓪᓗ.
[...] ᐃᓛᓐᓂ ᑏᓕᐅᒐᒃᓴᑐᐃᓐᓇᕐᒥᒃ
ᐃᒥᖅᑕᕈᓐᓇᖅᐸᓕᖅᑐᒍᑦ.
ᑕᐃᒪᐃᑦᑐᓂᒃ ᑕᑯᕙᒍᓐᓃᖅᑐᒍᑦ.”

—ᒥᐊᓕ ᕿᓗᒃ

“ᖁᕐᓗᖅᑑᑉ ᑰᖕᒐᓘᓐᓃᑦ
ᐃᒃᑲᑦᑐᑯᓘᓕᖅᑐᖅ
ᐊᐱᖅᑲᖅᑕᓐᖏᓗᐊᕐᓂᖕᒐ
ᐱᔾᔪᑎᒋᓪᓗᒍ. ᑕᐃᒃᑯᐊᓘᓐᓃᑦ
ᐃᒡᓗᑦ ᖃᓂᒋᔭᑦᑎᓐᓃᑦᑐᑦ
ᐊᐱᔪᐊᓘᕙᓚᐅᖅᓯᒪᔪᑦ ᐊᐱᓗᐊᕐᒧᓪᓗ
ᐳᐊᕐᓇᔭᓇᐊᖅᑲᖅᐸᓚᐅᖅᓯᒪᔪᑦ.
ᑕᐃᓛᑎᒋᐊᓗᒃ ᐊᐱᕙᒍᓐᓃᖅᑐᖅ.
ᐃᓚᖏᑦ ᐃᒡᓗᑦ
ᓴᐅᔭᐅᑦᑎᐊᖅᐸᓚᐅᖅᓯᒪᔪᑦ,
ᐳᐊᕐᓇᔭᓇᐊᖅᑲᖅᐸᒃᖢᑕᓗ
ᐃᒐᓛᑦ ᓵᖏᓐᓂᒃ ᐃᒡᓘ ᐃᓗᐊ
ᓯᕿᓐᓂᕐᕕᐅᔭᓐᓇᕐᓂᐊᕐᒪᑦ;
ᐃᓚᖏᓪᓘᓐᓃᑦ ᐃᓄᐃᑦ
ᐃᒡᓗᒥᖕᓂᑦ ᐊᔭᖅᓯᑉᐸᓚᐅᖅᓯᒪᔪᑦ.
ᓯᓚᖅᑲᑎᖕᒋᓐᓄᑦ ᑭᓯᐊᓂ
ᐳᐊᕐᓇᔭᖅᑕᐅᓪᓗᑎᒃ
ᐊᓂᔭᓐᓇᖅᓯᕙᓚᐅᖅᓯᒪᔪᑦ.
ᐃᓚᖏᑦ ᐃᒡᓗᐃᑦ ᐊᐱᔭᐅᓗᐊᒧᑦ
ᒪᔭᕐᕈᓐᓇᖅᓯᕙᓚᐅᖅᓯᒪᔪᒍᑦ ᐃᒡᓗᑦ
ᖁᓕᐊᓄᑦ. ᑕᐃᒪᐃᑉᐸᒍᓐᓃᖅᑐᖅ.”

—ᕈᐃ ᐃᓄᒃᑕᓕᒃ

“Not very much water; the level of the water has gone down. Even on the mountains over here, we just have little ponds and little places where there's going to be water. We don't see that anymore. Even on the little hills and steep places [...] sometimes you could get water for tea. We never see that anymore.”

—Mary Kellogok

“Even the Coppermine River, the water level has gone down because we are not getting very much snow. Even these houses around here—when we used to shovel the snow it used to be really high. It's not as high anymore. Some houses, they used to be really covered with snow; we would have to shovel the snow off to get sunlight. Even here some people would get trapped in their houses. Other people would have to dig them out. Outside the houses sometimes there was so much snow you could climb up the house. No more now.”

—Roy Inuktalik

ᐃᓄᐃᑦ ᐱᖅᑯᓯᑐᖃᖓᑦᑕ ᐊᒃᑐᖅᑕᐅᓂᖓ

“ᐃᒻᒪᑲᓪᓛᓗᒃ ᐊᐳᑎᖃᕐᓂᖅᓴᐅᓚᐅᖅᓯᒪᔪᖅ; ᒪᒃᑯᒃᑐᑦ ᕿᑭᖅᑕᑯᓘᑉ ᐅᖓᑖᓄᑦ ᐃᒡᓗᓕᔭᖅᑐᖅᐸᓚᐅᖅᓯᒪᔪᑦ ᐊᐳᑎᑦᑎᐊᕆᖕᒪᑦ. ᐅᓪᓗᒥ ᑕᐃᒪᐃᑦᑐᓐᓃᖅᑐᒍᑦ ᐊᐳᑎᖃᙱᓗᐊᕐᓂᖓ ᐱᔾᔪᑎᒋᓪᓗᒍ.”

—ᐋᓕᓯ ᐊᔭᓕᒃ

“ᐅᓪᓗᒥ ᐊᓄᕆᖃᕐᓂᖅᓵᓘᓕᖅᑐᖅ. ᐃᒻᒪᑲᓪᓚᒃ ᐊᓄᕆᒥᒡᓘᓐᓃᑦ ᐃᓱᒫᓘᑎᖃᖅᐸᓚᐅᖅᓯᒪᙱᑦᑐᒍᑦ ᑕᐃᒪᓐᓇᑲᓪᓛᓗᒃ ᑎᑭᑉᐸᓚᐅᖅᓯᒪᙱᒻᒪᑦ. [...] ᐅᓪᓗᒥᐅᓕᖅᑐᖅ ᐊᓄᕆᓯᑲᓪᓚᒃᑐᐊᓘᖅᑲᑦᑕᓕᖅᑐᖅ. ᐃᖅᑲᐅᒪᔪᖓ ᐊᑖᑕᒐ ᐊᒻᒪ ᑐᓵᔨ ᐊᐅᓪᓚᖅᑲᑦᑕᖅᑎᓪᓗᒋᒃ. ᑕᐃᔅᓯᒪᓂ ᐊᐅᓚᐅᑎᑕᖃᓚᐅᖅᓯᒪᙱᑦᑐᖅ. ᐸᐅᑎᑐᐃᓐᓇᕐᓂᒃ ᐊᑐᖅᐸᓚᐅᖅᓯᒪᔪᑦ. ᐊᓄᕆᒥᒃ ᐃᓱᒫᓘᑎᖃᖅᐸᓚᐅᖅᓯᒪᙱᑦᑑᒃ ᐅᖅᓱᐊᕐᒪᑦ. ᐅᑎᓕᕋᓗᐊᕐᒥᒃ ᐃᓱᒫᓗᒃᐸᓚᐅᖅᓯᒪᙱᑦᑑᒃ ᐱᔭᕐᓂᙱᓐᓂᖅᓴᐅᓕᕋᓗᐊᖅᑎᓪᓗᒍ. ᐅᓪᓗᒥᐅᓕᖅᑐᖅ ᐊᓄᓇᒍᑦ ᑎᑭᑕᐅᑐᐃᓐᓇᖅᑲᑦᑕᓕᖅᑐᒍᑦ ᐊᓄᕆᓯᑲᓪᓚᖅᑲᑦᑕᓗᐊᓕᕐᓂᖓᓄᑦ. ᑕᐃᒪᐃᑉᐸᓚᐅᖅᓯᒪᙱᑦᑑᒐᓗᐊᖅ; ᐃᒻᒪᑲᓪᓚᒃ ᐃᓱᒫᓗᖕᓇᓚᐅᖅᓯᒪᙱᑦᑐᖅ ᐊᓄᕆ. [...] ᒫᓐᓇ ᐊᔪᕐᓇᖅᓯᔪᖅ. [...] ᐊᓄᕆᓕᓚᐅᙱᒃᑲᓗᐊᖅᑎᓪᓗᒍ

Impacts on Traditional Ways of Life

“So many years ago there was a lot of snow; [the young people] used to go behind the island to build *igluit* [igloos] with the snow, but not anymore. Not enough snow.”

—Alice Ayalik

“There is more wind. A long time ago we didn’t worry about the wind because it didn’t come right away. [...] Nowadays it gets windy all at once. I remember my dad and [the interpreter’s] dad used to go out. They didn’t have any motors. They just used paddles. They never worried about the wind. It was calm. Even coming back they didn’t worry, but it gets harder. We are in the modern living now. [...] We can’t beat the wind. It gets windy all at once. It’s not how it used to be; a long time ago you didn’t worry about the weather. [...] Now it’s really hard. [...] Even before it gets windy people rush to come home, or you would be stranded out there for days. [...] It’s really hard to go long distances now. We still worry if we go farther down to where the men used to paddle.

Nunavut / Kugluktuk

ᐃᓄᐃᑦ ᐊᖏᕐᕋᐦᕋᓱᕐᑲᑦᑕᓕᖅᑐᑦ,
ᐊᖏᕐᕋᐦᕋᓱᓐᖏᒃᑯᑎᒃ
ᐅᓪᓗᕙᖕᓄᑦ ᐊᔪᖅᓯᓯᒪᓂᐊᕋᒥᒃ. [...]
ᐅᖓᓯᒃᑐᓕᐊᓇᐊᒃᓴᖅ ᒫᓐᓇ
ᐊᔪᕐᓇᖅᓯᔪᖅ. ᐊᖑᑎᓄᑦ ᐸᐅᖅᖁᑎᒃ
ᐅᐸᒃᐸᓚᐅᖅᓯᒪᔭᖏᓐᓄᑦ
ᐊᐅᓪᓚᖅᓯᒪᔪᕐᑲᓕᕌᖓᑦ
ᐃᓱᒫᓗᕆᕐᑲᑦᑕᓕᖅᑕᕗᑦ.
ᐊᑯᓂᐊᓗᒃ ᑕᐅᕙᓂ
ᐊᔪᖅᓯᓯᒪᑐᐃᓐᓇᓇᐊᕐᑲᓕᖅᑐᒍᑦ."

—ᐋᓚᓯ ᐊᔭᓕᒃ

"ᐊᐳᑎᓯᐊᕿᓐᓇᐃᓪᓕᔪᒍᑦ
ᐃᒡᓗᓕᐅᕐᕕᒃᓴᑦᑎᐊᕙᖕᓂᒃ.
ᑕᐃᒪᐃᓚᐅᖅᓯᒪᓐᖏᑦᑐᖅ ᐃᒻᒪᑲᓪᓚᒃ.
ᐅᓪᓗᒥ ᐊᐳᑎᕐᑲᑦᑎᐊᕿᓐᓃᖅᑐᖅ."

—ᔫᓯᐱ ᓂᑉᑕᓈᑦᑎᐊᖅ

"ᐃᖅᑲᐅᒪᔪᖓ ᐃᒻᒪᑲᓪᓚᒃ
ᓄᓇᑐᐃᓐᓇᕐᒦᖁᑕ ᐊᐳᒻᒥᒃ
ᐊᑐᖅᐸᓚᐅᖅᓯᒪᓂᑦᑎᓐᓂᒃ. ᐊᐳᑦ
ᑕᑯᓪᓗᒍ ᓴᓗᒻᒪᑑᖅᐸᓚᐅᖅᓯᒪᔪᖅ ᐊᒻᒪ
ᑭᓱᑐᐃᓐᓇᕐᓄᑦ ᐊᑐᖅᑕᐅᔪᓐᓇᖅᖁᓂ.
ᐊᐳᑦ ᖃᒃᑯᑑᖅᐸᓚᐅᖅᓯᒪᔪᖅ
ᓱᑲᕐᓇᖅᑐᖅᑲᐅᕋᓂᓗ.
ᐊᐳᑦ ᑭᓱᑯᓘᔭᕐᓄᑦ
ᐊᑐᖅᑕᐅᔪᓐᓇᓚᐅᖅᓯᒪᔪᖅ.
ᐅᓪᓗᒥᐅᓕᖅᑐᖅ ᐊᐳᑎᕐᑲᒡᑲᓕᖅᑐᒍᑦ."

—ᓖᓇ ᓂᑉᑕᓈᑦᑎᐊᖅ

It's kind of scary. You can get stranded for a long time."

—Alice Ayalik

"There are no places to find snow to make *igluit*. It's not like long ago. Not very much snow."

—Joseph Niptanatiak

"I remember a long time ago when we were inland, we used to use the snow. It didn't have any dirt when I used to look at the snow. The snow had so many uses. It was pure and white with no contaminants. There used to be lots of snow [...] but these years there is hardly any snow."

—Lena Niptanatiak

“ᐃᒪᖅᑕᖃᑦᑎᐊᕈᓐᓃᖅᑐᖅ ᓇᓂᓕᒫᖅ. [...] ᑕᒪᒃᑯᐊᓗ ᐃᕚᑦ ᐸᓂᓗᐊᓕᖅᓱᑎᒃ ᐃᒪᖅᑕᖃᙱᓗᐊᕐᓂᖓᓗᒃ. ᐃᒥᖅᑎᑕᐅᒐᓗᐊᕈᑎᒃ ᐃᒥᖅ ᑕᕆᐅᕐᒥᙶᕐᒥ ᐱᐊᓕᖅᑐᖅ; ᑕᐃᒫᖑᔾᔪᓯᓇᓚᐅᖅᓯᒪᔭᒥᓂᒃ ᐊᑐᕈᓐᓃᖅᑐᖅ. ᐃᒥᑦᑎᐊᕙᐅᕙᓚᐅᖅᓯᒪᔫᒐᓗᐊᖅ ᐱᐅᔪᓐᓃᖅᑐᖅ. ᐹᓂ ᓄᓇᑐᐃᓐᓇᕐᒥᓘᓐᓃᑦ ᐃᒥᖅ ᐱᐅᔪᓐᓃᖅᑐᖅ ᑰᒃ ᑕᒪᓐᓇ ᐱᐊᒃᑐᖅ, ᒪᒃᑯᒃᑐᓂᒃ, ᐃᓄᑐᖃᕐᓂᒃ ᐊᒻᒪ ᐊᓯᖕᒋᓐᓂᒃ ᐃᖅᑲᓇᐃᔭᖃᑎᖃᖃᑦᑕᖅᑐᖓ ᐊᐅᓪᓚᖅᓯᒪᖃᑎᒋᓪᓗᒋᑦ. ᐃᒻᒥᐅᕐᓂᐊᓕᕌᖓᑦᑕ ᑏᒃᓴᒥᒃ ᐅᕝᕙᓘᓐᓃᑦ ᓂᕿᓕᐅᕆᑎᒃᓴᑦᑎᓐᓂᒃ ᐃᒥᑦᑎᐊᕙᒃᓯᐅᓚᐅᕐᓗᑕ ᑭᓯᐊᓂᐅᓕᖅᑐᖅ ᐱᐊᓕᕐᓂᖓ ᐱᔾᔪᑎᒋᓪᓗᒍ.”

—ᒪᒥ ᐆᓐᓂᖅ

“ᑕᐃᒫᑎᒋᐊᓗᒃ ᓂᓪᓚᓱᒍᓐᓃᖅᑐᖅ. ᓯᓚ ᐊᒃᓱᐊᓗᒃ ᐊᓯᔾᔨᖅᑐᖅ. ᑕᒪᑐᒪᓂᓘᓐᓃᑦ ᐅᑭᐅᖑᓚᐅᖅᑐᒥ ᓂᓪᓚᓱᓗᐊᓚᐅᙱᑦᑐᖅ. ᐃᓪᓗᒥᐅᑕᐅᓕᕐᓂᕐᒧᑦ ᓯᓚ ᓱᖕᒋᐅᑎᑐᐃᓐᓇᖅᑕᕗᑦ. ᓯᓚ ᓂᓪᓚᓱᒃᑎᓪᓗᒍ ᓱᖕᒋᐅᑎᓯᒪᓚᐅᖅᓯᒪᒐᓗᐊᖅᑕᕗᑦ. [...] ᓄᓇᐅᑉ ᐃᑭᐊᖓ ᐃᓗᐊᓘᓐᓃᖅ ᐅᖅᑰᓂᖅᓴᐅᖅᑰᓕᕐᒪᑦ.”

—ᒪᒥ ᐆᓐᓂᖅ

“There is hardly any water [...] and the grass is too dry. They are not getting enough water. [...] Even if they get some water, the water comes from the ocean and it tastes like salt; it's not as natural as it used to be. It used to be pure water. Now it's not. Even high up on the land [...] the water is not pure. You can taste salt from the river. I work with youth and Elder groups and other people who are camping. When we are going to get water for tea or cooking [...] we need to look around for better water because it tends to taste like salt.”

—Mamie Oniak

“It's not that cold anymore. The weather really changed. Even this winter hasn't been too bad. We are just used to the weather from being in houses. We used to be used to the cold weather. [...] Under the earth I think it's warmer.”

—Mamie Oniak

“The seasons changed. [...] It blows so much that you can't get very much snow. Before, there used to be lots of snow and it was good for travelling [by dog team].”

—Mark and Martha Taletok

“ᓯᓚ ᐊᓯᔾᔨᖅᓯᒪᓕᖅᑐᖅ. [...] ᐊᓄᕌᕋᔪᓗᐊᒧᑦ ᐊᐱᑦᑎᐊᖅᐸᒍᓐᓃᖅᑐᖅ. ᐃᒻᒪᒃᑲᓐᓂᖅ ᐊᐱᕐᔪᐊᖅᐸᓚᐅᖅᓯᒪᔪᒍᑦ ᐊᒻᒪ ᐊᐳᑎᖃᑦᑎᐊᖅᑎᓪᓗᒍ ᕿᒧᒃᓯᖅᖢᓂ ᐱᐅᕙᓚᐅᖅᓯᒪᔪᖅ.”

—ᒫᒃ ᐊᒻᒪ ᒫᑕ ᑕᓖᑦᑐᖅ

“ᐃᓄᐃᑦ ᐃᒡᓗᓕᐅᕋᔪᒍᓐᓃᖅᑐᑦ. [...] ᐃᑖᓐᓂ ᐃᖏᕐᕋᓪᓗᑕ ᐃᒡᓗᓕᐅᕆᐊᖃᕌᖕᒐᑦᑕ ᐊᐳᑦ ᓯᑎᒃᓯᓯᒪᓐᖏᓗᐊᕌᖕᒐᑦ ᓯᑎᒃᑎᓴᖅᑕᐅᓚᐅᖅᑳᖅᖢᑎᒃᑯᑦ ᑭᓯᐊᓂ ᐃᒡᓗᓕᐅᖅᐸᓕᖅᑐᒍᑦ ᓯᓂᒃᑕᕐᕕᐅᑐᐃᓐᓇᕐᓂᐊᖅᑐᒥᒃ.”

—ᒫᒃ ᐊᒻᒪ ᒫᑕ ᑕᓖᑦᑐᖅ

“ᑕᕆᐅᖅ ᐊᑯᓂᐊᓗᒃ ᓯᑯᓇᓱᖃᑦᑕᓕᖅᑐᖅ. ᓯᑯᓴᕋᐃᓐᓂᖅᓴᐅᓚᐅᖅᓯᒪᔫᒐᓗᐊᖅ. ᐃᒻᒪᑲᓪᓚᒃ, ᒫᓐᖓᑦ ᖃᓂᑦᑐᕈᓗᖕᒥ, ᐃᒻᒪᖄ 40 ᒪᐃᓕᓂᒃ ᐅᖓᓯᖕᓂᓕᒃ ᓂᒋᕐᒧᑦ, ᑎᓯᐱᕆᐅᓕᕋᓗᐊᖅᑎᓪᓗᒍ ᓯᓕ ᓯᑯᓯᒪᓐᖏᑦᑐᖅᑕᖅᑲᐅᓚᐅᖅᑐᖅ. ᑕᒪᑐᒪᓂ ᐊᕐᕌᒍᒥ ᐊᖑᓇᓱᒃᑏᑦ ᐅᔾᔨᖅᓱᑦᑎᐊᕆᐊᖃᓚᐅᖅᓯᒪᔪᑦ ᑕᐅᕗᓐᖓᐅᒐᒥᒃ ᑐᒃᑐᓕᐊᖅᖢᑎᒃ. ᑕᐅᓇᓂ ᓘᓐᓃᑦ ᓯᑎᐱᕆᐅᓕᖅᑎᓪᓗᒍ ᓯᑯᓯᒪᓐᖏᑦᑐᖅᑕᖅᑲᐅᓚᐅᖅᑐᖅ. ᑕᓯᕐᓂ ᓘᓐᓃᑦ ᐅᔾᔨᖅᓱᑦᑎᐊᕆᐊᖃᖃᑦᑕᓕᖅᑐᑦ ᐃᖃᓪᓚᐊᕌᖕᒐᒥᒃ. ᓯᑎᐱᕆᒥ ᑕᒪᒃᑯᐊ ᓯᑯᓯᒪᐃᓐᓇᓕᓲᖑᒐᓗᐊᖅᒪᑕ.”

—ᒫᒃ ᑕᓖᑦᑐᖅ

“People hardly make snow houses now. [...] Sometimes the snow doesn’t get really hard, so they have to stomp on it to make *igluit* overnight where they are travelling.”

—Mark and Martha Taletok

“[The waters] take forever to freeze. They used to freeze early. A long time ago, not too far from here—about 40 miles south—in December there was still open water. This year the hunters had to be really careful when they went that way to hunt for caribou. Even down here it was still open water in September. They had to be careful on the lakes to go fishing. Usually in September it was all frozen.”

—Mark Taletok

“ᐊᐱᑦᑎᐊᖅᑲᑦᑕᕈᓐᓃᖅᑐᒍᑦ. ᐳᐊᕐᕆᔭᖅᖢᖓ ᑕᒪᓐᓇ ᐊᐳᑦ ᐱᔭᕐᓂᓐᖏᑦᑐᒻᒪᕆᐊᓘᓚᐅᖅᑐᖅ. [...] ᑕᐃᒪᐃᑦᑐᒥᒃ ᐊᐳᒻᒥᒃ ᑕᑯᓚᐅᖅᓯᒪᓐᖏᑦᑐᖓ. ᐅᐱᕐᖔᑕᒫᖅ ᐊᐳᒻᒥᒃ ᐃᒻᒥᐅᖅᑲᑦᑕᓚᐅᖅᓯᒪᔪᖓ, ᓴᓗᒪᖕᒪᑦ, ᑏᓕᐅᕈᑎᒋᓂᐊᖅᖢᒍ. ᐊᐳᑕᐃᓱᕐᒪ ᐳᐊᕐᕆᔭᐅᒻᒥᒃ ᐊᑐᖅᖢᖓ ᐱᔭᕐᓂᓐᖏᓈᖅᓯᓚᐅᖅᑐᖓ ᑕᒪᓐᓇ ᐊᐳᑦ ᓯᑎᓗᐊᕐᒪᑦ. ᑕᐃᒪᐃᓐᓂᖓᓄᑦ ᐊᐳᑕᐃᓱᖅᐸᒍᓐᓃᖅᑐᖓ.”

—ᒥᐊᓕ ᕿᓗᒃ

“ᐃᒻᒪᑲᓪᓚᒃᑲᓐᓂᖅ ᕿᒧᒃᓯᖅᖢᑕ ᖃᓐᓂᐅᔭᖅᐸᓚᐅᖅᓯᒪᔪᖅ; ᖃᓐᓂᐅᔭᐃᓐᓇᖅᑲᑦᑕᓚᐅᖅᓯᒪᔪᖅ. ᓲᕐᓗ ᐃᒡᒑᒃᑭᒃ ᐳᑦᑎᖅᑎᑦᓗᒋᒃ. ᐊᐳᑎ ᖄᖓᓃᖏᓐᓇᐅᔭᓚᐅᕋᓗᐊᖅᑎᑦᓗᒍ ᐅᓪᓗᒥ ᑕᐃᒪᐃᑦᑐᓐᓃᖅᑐᖅ. ᐸᑐᒃᑐᖅ. ᐃᒻᒪᑲᓪᓚᒃ ᐃᖅᑲᐅᒪᔪᖓ ᐃᓄᐃᑦ ᐅᒥᐊᖅᑐᕆᐊᖅᑐᖅᑲᑦᑕᖅᑎᑦᓗᒋᑦ. ᐊᐅᓚᐅᑎᖃᓐᖏᓐᓇᒥᒃ ᐸᐅᑎᓂᒃ ᐊᑐᑐᐃᓐᓇᖅᐸᓚᐅᖅᓯᒪᔪᑦ. ᐅᓪᓗᓕᒫᖅ ᐅᒥᐊᖅᑐᖅᐸᓚᐅᖅᓯᒪᔪᑦ. ᐅᓪᓗᓕᒫᖅ ᓯᓚᒥᒃ ᐃᓱᒫᓘᑎᖃᖅᐸᓚᐅᖅᓯᒪᓐᖏᑦᑐᑦ ᐊᖑᓇᓱᒃᑲᒥᒃ. ᑕᐃᒪᐃᓚᐅᖅᓯᒪᒐᓗᐊᖅᑎᑦᓗᒍ ᐅᓪᓗᒥ ᓯᓚ ᖃᓄᐃᓕᓂᐊᕐᒪᖔᖓ ᓇᓗᓇᖅᓯᔪᖅ. ᑕᕝᕙᓐᖓᑦᓴᐃᓐᓇᖅ ᐊᓄᕆ ᑎᑭᑉᐸᓕᖅᑐᖅ ᐅᔾᔨᕆᓚᐅᖅᑎᓐᓇᑕᓘᓐᓃᑦ.”

—ᕈᐃ ᐃᓄᒃᑕᓕᒃ

“We don’t get very much snow. When I am shovelling, the snow is hard to shovel. [...] I’ve never seen snow like that. I never go get snow [anymore]. I was getting water from snow every spring. Every spring I used snow for water, for clean water, for drinking tea. I went to go get snow with a shovel. It was too hard for me, so I never bother going to get snow [anymore].”

—Mary Kellogok

“A long time ago, even when we travelled by dog the snow would fall; it would always be there. It was like your glasses get foggy. The snow was always on top, but now it’s not like that anymore. Frosty. A long time ago I remember people used to go out on their boats. They didn’t have kickers or anything. Just paddles. They would be out all day. They didn’t worry about the weather all day long; they would be out hunting. But nowadays [...] you never know what it’s going to be like. All at once the wind comes without [you] even knowing.”

—Roy Inuktalik

ᓄᓇᕗᑦ / ᖁᕐᓗᖅᑐᖅ

“ᐅᓪᓗᒥᐅᓕᖅᑐᖅ ᐅᔾᐱᕐᓇᖅᓯᔪᑦ
ᐃᖅᑲᓗᐃᑦ ᑕᓇᐅᕐᒦᓕᖅᑎᓪᓗᒋᑦ
ᑕᐃᒫᑦᑎᐊᖑᔪᓐᓈᕐᓂᖏᓐᓂᒃ,
ᑎᐱᖏᑦ ᐊᔾᐱᒋᔪᓐᓈᖅᑕᖏᑦ. [...]
ᐊᕐᕌᓂ ᐃᖅᑲᓗᑭᑦᑐᑯᓘᓚᐅᖅᑐᖅ.
ᐃᖅᑲᓗᓕᓱᓪᓕᓚᐅᖅᑐᒍᑦ.”

—ᒥᐊᓕ ᕿᓗᒃ

“ᑭᓱᕈᓘᔮᓗᐃᑦ ᐊᓯᔾᐱᖅᓯᒪᓕᖅᑐᑦ.
ᐅᑭᐅᖑᑎᓪᓗᒍ ᕿᒧᒃᓯᖅᓗᑕ
ᓂᓪᓚᓱᒃᑐᓪᒪᓇᐊᓘᖅᑲᑦᑕᓚᐅᖅᓯᒪᔪᖅ,
ᐊᓪᒪ ᐃᖏᕐᕋᓪᓗᑕ
ᐸᑐᒃᐸᓚᐅᖅᓯᒪᓪᓗᓂ. [...]
ᑕᐃᒪᐃᑦᑐᓂᒃ ᑕᑯᕆᒍᓐᓈᖅᑐᒍᑦ.
ᓂᓪᓚᓪᓚᓇᒃᑳᖓᑦ ᓯᓚ
ᐸᑐᒃᐸᓚᐅᖅᓯᒪᔪᖅ, ᕕᕗᐊᓇᒥᓘᓐᓂᑦ
ᐃᖏᕐᕋᓕᕌᖓᑕ. ᑕᐃᒫᑐᐃᓐᓇᖅ
ᑕᐃᒪᐃᑦᑐᓂᒃ ᑕᑯᕆᒍᓐᓈᖅᑐᒍᑦ.
ᓲᖅᓗ ᐃᓯᐊᖅᑭᔪᓂᒃ ᑕᑯᓪᓗᓂ,
ᐳᔪᖅ ᑕᒃᐸᐅᓐᖓᒍᐅᕆᖕᒪᑦ.
ᑕᐃᒪᐃᑉᐸᒍᓐᓈᖅᑐᖅ.
ᐊᓯᔾᐱᖅᔪᐊᖅᑐᓪᒪᓇᐊᓗᒃ
ᐱᐅᓯᓇᓐᖏᑕᒥᓂᒃ ᐊᑐᓕᖅᒪᑦ.
ᖃᐅᔨᒪᔪᒃᓴᐅᕗᑎᑦ ᑕᓇᐅᖅ
ᐊᓪᒪᓘᓐᓈᓂᑦ ᑕᓰᑦ ᖁᐊᖅᓴᓗᐊᕌᖓᒥᒃ
ᓵᓐᓂᖅᓴᐅᖅᑲᑦᑕᖅᓂᖏᓐᓂᒃ.
ᐅᔾᐱᖅᓱᑦᑎᐊᓇᐊᖅᑲᖅᑲᑦᑕᓕᖅᑐᒍᑦ
ᓇᒧᓐᖓᐅᓂᐊᕌᖓᑦᑕ.”

—ᕈᐃ ᐃᓄᒃᑕᓕᒃ

“Nowadays we find in our ocean the colour of the fish is not the same anymore. The taste is not the same. [...] Last year it was really hard. We had a hard time getting fish.”

—Mary Kellogok

“Lots of change. During the winter when we would travel by dog team it used to be really cold, and when you travelled [you could see] all the snow. [...] You don’t see that anymore. Like the snow flying when it gets really cold. Even when we went out in February on the land, the same thing happened; there was not that snow flying anymore. You know when you see chimneys, the way the smoke is. It’s not like that anymore.

It’s a really big change. It’s not like that anymore. You know the ocean here, even the lakes, when they freeze up the ice is not as thick anymore. You have to be really careful to travel.”

—Roy Inuktalik

“ᐃᓱᒪᔪᖓᓕ ᑕᒪᓐᓇ ᐃᖃᓗᖕᓂᒃ ᐊᒃᑐᐃᓂᖃᕿᓯᒥᓪᓗᒍ. [...] ᑕᒪᑐᒪᓂ ᐅᑭᐅᖑᓚᐅᖅᑐᒥ ᐃᖃᓗᐃᑦ ᐊᐅᐸᕐᓂᖅᓴᐅᓚᐅᖅᑐᑦ. ᐃᓚᖏᓐᓂ ᐊᕐᖀᒍᓂ ᒪᒪᓗᐊᖀᓗᖃᑦᑕᓐᖏᑦᑐᑦ ᐊᒻᒪ ᑎᐱᖏᑦ ᐊᔾᔨᒋᓐᖏᔪᓂᐅᒃ. ᑎᓴᒪᑦ ᑕᓪᓕᒪᐅᖅᑰᓕᖅᑐᑦ ᐊᕐᖀᒍᑦ ᐃᖃᓗᐃᑦ ᑕᐃᒫᑦᑎᐊᖑᓐᖏᑦᑐᖅ ᑎᐱᖃᓚᐅᖅᓯᒪᔪᑦ. ᖃᓄᐃᒻᒪᑦ ᑕᐃᒪᐃᓚᐅᕐᒪᖔᖏᑦ ᓇᓗᔪᖓ. ᐃᓚᖏᑦ ᐃᓄᐃᑦ ᐅᖃᖅᐸᒃᑐᑦ ᐅᔭᕋᖕᓂᐊᕐᕕᖕᓄᑦ ᒪᑯᓄᖓᓗ ᓱᕈᕐᓇᐅᑎᓄᑦ ᓇᑭᓐᖔᖅᑐᓂᒃᑭᐊᖅ ᑕᐃᒪᐃᓐᓇᓱᒋᔭᐅᓪᓗᑎᒃ. ᑏᓕᐊᕋᑦ ᒪᒪᖅᐸᓚᐅᖅᓯᒪᔪᖅ ᑲᑭᐊᒃᔪᓂᓗ ᑏᑦᑎᐊᕆᐅᓪᓗᓂ. ᒫᓐᓇᓕ ᑕᐃᒪᐃᑦᑐᓐᓃᖅᑐᖅ, ᑏᓯᓐᓂᑦᑐᓐᓃᖅᑐᖅ. ᑕᐃᒪᐃᓐᓂᖓᓄᑦ ᓂᓚᒃᑕᖔᖅᑲᑦᑕᓕᖅᑐᒍᑦ.”

—ᕈᐃ ᐃᓄᒃᑕᓕᒃ

“I think there is an effect on the fish. [...] This winter the fish—the colour looked a lot better. [Some] years ago the fish wasn’t too good and [...] the taste wasn’t right. Must be four or five years ago the taste wasn’t right. Don’t know why. Some people were saying from the mine sites and chemicals flowing somewhere. It’s hard to find out the flow of the water. When we used to make tea out of it, it used to be good and clear and good to have tea. Now it’s not the same; it doesn’t taste like tea anymore. That’s why we have to get ice.”

—Roy Inuktalik

ᖃᒪᓂᑦᑐᐊᖅ
Baker Lake

ᖃᒪᓂ'ᑐᐊᖅ

ᖃᒪᓂ'ᑐᐊᖅ ᓄᓇᕗᒻᒥ ᓯᓚᔾᒦᖖᒋᑦᑐᑐᐊᖅ, ᑏᓛᓐ ᑰᒃ ᖃᒪᓂᐊᓅᑦᑐᖅ ᑲᓇᑕᐅᓪᓗ ᕿᑎᓪᓗᐊᑦᑎᐊᙵᓄᑦ ᖃᓂᕐᕗᓂ (N 64.15°, W 95.5°; ᖁᑦᑎᖕᓂᖕᒐ 19 m). ᐅᑭᐅᖅᑕᖅᑐᖅᒥ ᐊᖅᐸᓯᒃᑐᓄᑦ ᐃᓚᒋᔭᖅ, ᓄᓇᖕᒐ ᐊᕚᓚᕿᐊᓛᓂᒃ ᐱᕈᖅᑐᖅᑲᖅᕗᓂ ᓄᓇᐅᓪᓗ ᐃᑭᐊᖕᒐ ᖁᐊᖕᒍᐃᓐᓇᐅᔭᖅᕗᓂ. ᓄᓇᓕᖖᒍᓚᐅᖅᓯᒪᔪᖅ 1950-ᖕᒋᓐᓂ 1,872-ᓂᒃ ᐃᓄᒋᐊᖕᓂᖅᑲᖅᓂᕋᖅᑕᐅᓪᓗᓂ 2011-ᒥ ᑲᓇᑕᒥ ᑭᓪᓕᓯᓂᐊᖅᑕᐅᔪᓂᑦ.

ᖃᒪᓂ'ᑐᐊᕐᒥ, ᖃᔅᓲᓐᓇᒻᒪᙱᑯᓗᐃᑦ ᐊᐱᖅᓱᖅᑕᐅᔪᑦ ᓄᓇᒥ ᐱᕈᖅᑐᓂᒃ ᐊᓯᔾᔨᖅᓯᒪᔪᓂᒃ ᓄᓇᒥᖕᓂ ᐅᖃᐅᓯᖃᓚᐅᖅᑐᑦ. ᑕᐃᒪᐃᔾᔪᑎᖃᖅᑐᒃᓴᐅᔪᖅ ᐃᓅᖕᒋᑦ ᐃᓕᖅᑯᓯᖕᒋᑦᑐ ᓄᓇᓂᑦ ᐊᔾᔨᒌᙱᑦᑐᓂᙵᖅᓯᒪᓂᖕᒋᓐᓄᑦ. ᑭᓯᐊᓂ ᐊᔾᔨᒌᙱᑦᑐᓂᒃ ᐅᔾᔨᕆᔭᖃᖅᓯᒪᓂᕋᖅᑐᑦ, ᓲᖅᓗ ᐱᕈᖅᑐᐃᑦ (ᐱᖃᓯᐅᕐᕗᒋᑦ ᑎᖕᒐᐅᔭᑦ) ᐱᕈᙵᓐᓂᖅᓴᐅᕙᓕᖅᓂᖕᒋᓐᓂᒃ, ᐅᖅᐲᑦ ᓇᐃᓐᓂᖅᓴᐅᕙᓕᖅᓂᖕᒋᓐᓂᒃ, ᐃᕙᖃᖅᓂᖅᓴᐅᕙᓕᖅᓂᖕᒐᓂᒃ, ᐊᒻᒪ ᐱᕈᖅᑐᑦ (ᐱᓗᐊᖅᑐᒥᒃ ᐊᖅᐲᑦ) ᐱᕈᙵᓐᓂᖅᓴᐅᕙᓕᖅᓂᖕᒋᓐᓂᒃ ᒥᑭᓐᓂᖅᓴᐅᕙᓕᖅᕗᑎᓪᓗ. ᐃᓚᖕᒋᑦ ᐱᕈᖅᑐᑦ ᓲᖅᓗ ᐃᔅᓯᒥ ᐊᒥᐊᓛᓕᑦ ᓇᑲᖅᔭᐊᕌᓘᔫᖅᑐᑦ ᐸᐅᓐᓇ ᐱᑕᖅᑲᖅᓂᖅᓴᐅᕙᓕᖅᓂᖕᒋᓐᓂᒃ, ᐊᒻᒪᓗ ᓄᑖᓂᒃ ᐱᕈᖅᑐᓂᒃ ᐱᕈᖅᑐᖃᖅᐸᓕᖅᓂᖕᒐᓂᒃ.

ᐃᓄᑐᖃᐃᑦ ᐅᖃᓚᐅᖅᒥᔪᑦ ᑑᑐᐃᑦ ᓴᓗᖕᓂᖅᓴᐅᕙᓕᖅᓂᖕᒋᑦ,

Baker Lake

Baker Lake (Qamani'tuaq) is Nunavut's only inland community, situated at the mouth of the Thelon River and close to the geographic centre of Canada (N 64.15°, W 95.5°; elevation 19 m). It is part of the Low Arctic, characterized by erect dwarf-shrub tundra vegetation and continuous permafrost. The community was established in the 1950s and its population was estimated at 1,872 people in the 2011 Statistics Canada Census.

In Baker Lake, very few interviewees observed vegetation change in the areas around their community. This may be explained by the presence of individuals from many cultural groups coming from different areas as their reference point for how things were in the past. However, several observations have been made, such as plants (including lichens) growing less, willows growing shorter, grasses growing more, and berries (especially cloudberry/bakeapple) becoming less abundant and smaller. Some plant species, such as fireweed, are becoming more abundant, and new species are appearing.

ᓄᓇᖓ ᐸᓂᕐᓂᖅᓴᐅᓕᕐᓂᖓᓄᑦ ᐱᕈᖅᑐᖃᙱᓐᓂᖅᓴᐅᓕᕐᓂᖓᓄᑦᑐ. ᐃᓚᖏᑦᑐ ᐆᒪᔪᑦ, ᓲᕐᓗ ᐅᒥᖕᒪᑦ, ᐱᑕᕐᑲᕐᓂᖅᓴᐅᕙᓕᕐᓂᖏᑦ, ᐊᓯᖏᑦᑐ, ᓲᕐᓗ ᑲᖑᐃᑦ, ᐅᓄᙱᓐᓂᖅᓴᐅᓕᕐᓂᖏᓐᓂᒃ. ᐊᐱᖅᓱᖅᑕᐅᔪᑦ ᐅᕐᑲᕐᑲᑦᑕᓚᐅᖅᑐᑦ ᐊᓇᙱᕐᑲᕐᓂᖅᓴᐅᓕᕐᓂᖓᓂᒃ, ᕿᒃᑐᕆᐊᑭᓐᓂᖅᓴᐅᓕᖅᔪᓂ ᐊᒪ ᑕᕐᕋᓕᑭᓵᑭᓐᓂᖅᓴᐅᓕᕐᓂᖓᓂᒃ. ᓄᑖᓂᒃ ᑎᖕᒥᐊᓂᒃ ᑕᑯᕙᓕᖅᑐᑦ, ᓲᕐᓗ ᑐᒡᒐᔫᑦ, ᐊᒻᒪ ᓄᑖᑦ ᕐᑯᐱᕐᕈᐃᑦ, ᓲᕐᓗ ᐊᖏᓂᖅᓴᐃᑦ ᑕᖅᓴᓇᖕᓂᖅᓴᐃᑦᑐ ᐃᒍᑦᑕᐃᑦ ᒥᓗᒋᐊᓪᓗ.

ᑰᒐᓛᑦ ᑕᓲᓪᓗ ᐸᓂᖅᐸᓪᓚᐊᓕᖅᑐᑦ, ᐃᒪᑭᒡᓕᕙᓪᓚᐊᓕᖅᔪᑎᒡᓗ ᑕᓯᕐᓂ ᑰᖕᓂᓗ. ᑕᐃᒪᐃᑉᐳᖅ ᖃᒪᓂ'ᑐᐊᕐᒥ ᕿᑭᖅᑕᑯᓗᖕᒥ, ᓄᓇᓕᖓᑦᑕ ᓯᓚᑎᑦᑎᐊᖓᓂ, ᑕᑭᓂᖅᓴᐅᖅᑰᔨᓕᖅᑐᖅ ᑕᓯᖅ ᐃᒪᖅᓂᖅᓴᐅᓕᕐᓂᖓᓄᑦ. ᐅᔭᖅᑲᑦᑐ ᕿᑭᖅᑕᐅᖅᑰᔨᓕᖅᑐᑦ ᑰᖕᒥ, ᐅᒥᐊᖅᑐᖅᑐᓄᑦᑐ ᐊᑦᑕᕐᓇᕐᓂᖅᓴᐅᓕᖅᔪᓂ ᐊᖑᓇᓱᖕᓂᕐᒧᑦ ᐊᒃᑐᐃᓯᒪᓂᕐᑲᓕᖅᔪᓂ. ᐊᐱᖅᓱᖅᑕᐅᔪᑦ ᐅᔾᔨᕈᓱᒃᓯᒪᔪᑦ ᖃᓐᓂᕐᑲᑦᑕᖏᓐᓂᖅᓴᐅᓕᕐᓂᖓ ᒪᕐᑯᕐᑲᑦᑕᖏᓐᓂᖅᓴᐅᓕᕐᓂᖓᓗ, ᐊᒻᒪᓗ ᐊᐳᑎᕐᑲᙱᓐᓂᖅᓴᒻᒪᕆᐅᓕᕐᓂᖓᓂᒃ; ᐃᓚᖏᑦ ᐅᕐᑲᖅᓯᒪᔪᑦ ᑕᐃᒪᐃᓐᓂᖓᓂᒃ ᐊᓄᕆᑐᓂᖅᓴᐅᕙᓕᕐᓂᖓᓄᑦ ᓴᕿᑳᑦᓚᒃᐸᓕᕐᓂᖓᓄᑦᑐ. ᐃᓄᑐᖃᐃᑦ ᑐᑯᕐᑲᑦᑕᕐᓂᖅᓴᐅᓕᕐᓂᕋᖅᑐᑦ "ᐅᓚᓗᔭᐅᔮᖅᑐᓂᒃ" ᐊᓄᕆᓂᒃ ᓴᖅᑭᑳᑦᓚᒃᑑᔮᖅᐸᒃᑐᓂᒃ. ᑕᐃᒪᐃᒻᒪᑦ, ᐊᐳᑦ ᑎᓯᓗᐊᓕᕐᑲᑦᑕᖅᑐᖅ ᐃᒡᓗᓕᕝᕕᒋᒋᐊᖓ. ᑕᐃᔅᓱᒪᓂ, ᐃᓄᐃᑦ ᓯᓚᒥᒃ ᓇᓚᐅᑦᑖᕈᓐᓇᖅᐸᓚᐅᖅᑐᑦ

Elders also described that caribou are thinner nowadays, because the land is getting dryer and the plants are growing less. Some animal species, such as muskoxen, seem to have become more abundant, and others, such as geese, have become less abundant. Participants noted more black flies, but fewer mosquitoes and butterflies. New species of birds have been observed, such as the northern flicker (woodpecker), as well as new insects, such as bigger and darker species of bumblebees and wasps.

Small streams and ponds are drying up, and the water levels are getting lower in lakes and rivers. This is true of the island in Baker Lake, just south of the community, which now appears longer because the lake water depth has decreased. Rocks are also coming up like islands along the Thelon River, making boat travel more hazardous and thus affecting hunting. Participants have also observed less precipitation, as well as a big decline in snow cover; some attributed this to stronger winds that seem to change direction suddenly. Elders have experienced more "hurricane-like" winds that can appear without much warning. Consequently, the

ᓄᕗᔭᑦ ᕿᒥᕿᑦᑐᑎᑦ, ᑭᓯᐊᓂ ᒫᓐᓇ ᓯᓚ ᐊᓯᔾᔨᖅᑕᕐᓂᖅᓴᐅᓕᖅᑐᖅ ᓇᓚᐅᑦᑖᕆᐊᖓᑐ ᐊᔪᕐᓇᕐᓂᖅᓴᐅᓕᖅᖢᓂ. ᐃᓄᑐᖃᐃᑦ ᐅᔾᔨᕆᓯᒪᔭᖏᑎᒍᑦ, ᐊᐅᔭᒃᑯᑦ ᓯᓚ ᐊᓯᔾᔨᖅᓯᒪᓂᖅᐸᐅᓂᖓᓂᒃ. ᐊᑯᓂᒻᒪᕆᐊᓗᒃ ᓯᓚ ᐅᖅᑰᕐᔪᐊᕈᓐᓇᓕᖅᑐᖅ. ᐃᓄᑐᖃᐃᑦ ᓇᑐᓇᐃᖅᓯᓚᐅᖅᑐᑦ ᐅᑭᐊᒃᓵᒃᑯᑦ ᐊᓄᕌᕐᓂᖅᓴᐅᕙᓕᕐᓂᖓᓂᒃ ᐊᒻᒪ ᑕᖅᑭᓕᒫᕐᒥᒃ ᑭᖑᕙᓯᓕᖅᖢᓂ. ᐊᓄᕆᖓ ᓂᒡᓚᓱᖕᓂᖅᓴᐅᖅᑰᕈᓕᖅᑐᖅ, ᑭᓯᐊᓂ ᐱᖅᓯᕋᔪᖕᖏᓐᓂᖅᓴᐅᓕᖅᖢᓂ. ᓄᓇᓕᖕᒥᐅᑦᑐ ᐅᔾᔨᕈᓱᒃᓯᒪᓕᖅᑐᑦ ᐊᓂᐅᕕᖅᑲᖕᒋᓐᓂᖅᓴᐅᓕᕐᓂᖓᓂᒃ, ᓄᓇᐅᑦᑐ ᐃᑭᐊᖓ ᖁᐊᖑᐃᓐᓇᖅᐸᒃᑐᖅ ᐊᐅᒃᐸᑦᑕᐊᕙᓕᕐᓂᖓᓂᒃ ᕿᑎᖅᐸᓯᐊᓂ, ᐊᒻᒪᓗ ᓄᓇᐃᑦ ᑎᓱᐊᖅᓯᒪᕙᓕᖅᖢᑎᒃ ᑰᒥ.

snow becomes too hard to build *igluit* (igloos). In the past, people could predict the weather by observing the clouds, but today the weather seems to be more variable and harder to predict. Based on Elders' observations, it appears that summer is the season that has changed the most. There can be intense, long-lasting heat. Elders identified that fall is windier and can come a month late. The winter appears to be colder, but with fewer blizzards. Community members are also observing fewer permanent ice patches, permafrost thawing resulting in muddier ground, and signs of increased erosion along the Thelon River.

ᐃᓐᓇᐃᑦ ᐊᐱᖅᓱᖅᑕᐅᔪᑦ
Elders Interviewed

ᓴᐃᓚᔅ
ᐊᐃᑦᑕᐅᖅ
Silas
Aittauq

ᕕᐅᕋ
ᐊᒃᕼᐊᒃ
ᐊᕚᓚ
Vera
Avalaa

ᒫᒍᕆᑦ ᑐᓪᓕᒃ
ᓂᕕᐊᑦᓯᐊᖅ
Margaret
Amauruq
Niviatsiaq

ᒦ ᑮᓈᓕᒃ
ᕼᐊᖅᐱ
May
Haqpi

ᓄᐊᒪᓐ
ᐊᑕᖕᒐᓛᖅ
Norman
Attungala

ᔭᐃᑯᑉ
ᐃᑭᓂᓕᒃ
Jacob
Ikinilik

ᕼᐋᑎ ᐊᑐᖕᒑᑦ
ᐊᑐᑐᕚ
Hattie
Attutuvaa

ᕖᓂ ᑕᔾᔭ
ᐃᑭᓂᓕᒃ
Winnie
Ikinilik

ᐹᓪ ᐊᑐᑐᕚ
Paul
Attutuvaa

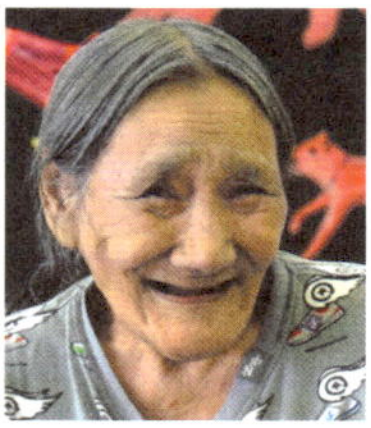

ᔮᓂᑦ ᓂᐱ
ᐃᑰᑕᖅ
Janet
Ikuutaq

ᑑᓇ ᐃᖁᓕᒃ
Toona
Iqulik

ᔫᓐ ᓄᑭᒃ
John
Nukik

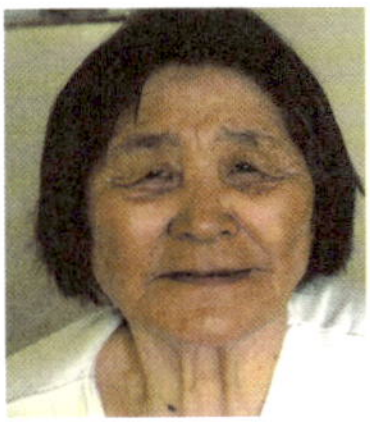

ᐊᐃᕇᓐ
ᑕᕕᓂᖅ
ᑲᓗᕋᖅ
Irene Taviniq
Kaluraq

ᒫᑕ
ᓇᓇᐅᑐᐊᖅ
ᓄᑭᒃ
Martha
Nukik

ᓘᓯ ᖃᐅᓐᓇᖅ
Lucy
Kownak

ᑖᒪᔅ
ᖃᖅᑭᒪᑦ
Thomas
Qaqimat

ᒫᖅᑎᓐ ᖅᑭᓚᒃ
Martin
Kreelak

ᐃᓕᓴᐱ
ᐊᖅᓇᒐᖅᓂᖅ
ᖁᐃᓇᖕᓇᖅ
Elizabeth
Quinangnaq

ᒥᐊᕆ ᖅᑭᓚᒃ
Mary
Kreelak

ᔮᓐ
ᐸᓂᖕᒐᔾ’ᓈᖅ
ᓯᑦᑲᑎ
Joan
Scottie

ᓴᐃᒪᓐ
ᑐᑯᒥᖅ
Simon
Tookoome

ᐹᓯᓪ
ᑐᓗᖅᑐᖅ
Basil
Tuluqtuq

ᔫᓕ ᓴ’ᖑᓵᖅ
ᑐᓗᖅᑐᖅ
Julie
Tuluqtuq

ᓘᓯ ᑎᑭᖅ
ᑐᖑᐊᖅ
Lucy
Tunguaq

ᐸᐅᕐᖓᐃᑦ

“ᐅᓪᓗᒥ ᐸᐅᕐᖓᑦ ᒥᑭᑦᑑᓕᖅᑐᑦ ᓄᓇ ᐸᓂᖅᓯᒪᓕᕐᒪᑦ, ᒪᕐᑯᒐᔭᒍᓐᓃᖅᑐᖅ ᖃᓐᓂᒐᔭᒍᓐᓃᖅᓱᓂᓗ.”

—ᕼᐋᑎ ᐊᑐᖓᑦ ᐊᒻᒪ ᐹᓪ ᐊᑐᑐᕚ

“ᓄᑕᕋᐅᑎᓪᓗᖓ ᓄᓇᓗᒃᑖᖅ ᐊᐅᐸᖅᐸᓚᐅᖅᑐᖅ ᐊᖅᐱᖕᓂᑦ, ᑭᓯᐊᓂ ᑭᖑᓪᓕᖅᐹᕐᒥ ᐊᒡᓕᕐᓇᖅᑐᓕᐊᕋᑦᑕ ᐊᐅᐸᓚᐅᖕᖏᑦᑐᖅ. ᐊᖅᐱᖃᕈᓐᓃᖅᑐᖅ.”

—ᐊᐃᓖᓐ ᑕᕕᓂᖅ ᑲᓗᕋᖅ

“ᓄᓂᕙᒐᑦ ᒥᑭᓐᓂᖅᓴᐅᓕᖅᑐᑦ. ᑕᒪᕐᒥᒃ ᑭᒍᑕᖏᕐᓇᑦ ᐸᐅᕐᖓᓪᓗ.”

—ᔫᓐ ᓄᑭᒃ

“ᐸᐅᕐᖓᕐᑯᑎᑦ ᐱᑕᖃᕈᓐᓃᖅᐸᓪᓕᐊᑐᐃᓐ ᓇᖅᑰᓯᔭᑦ, ᑕᐃᒪᐃᑐᐃᓐᓇᓀᐊᓕᒃ ᓯᕿᓂᐅᑉ ᐆᓇᕐᓂᖓᓄᑦ.”

—ᔫᓐ ᓄᑭᒃ

Berries

“Today the berries are small because the ground is dry, and it hardly rains and hardly snows.”

—Hattie and Paul Attutuvaa

“When I was a kid the whole place would be red from the *aqpiit* [cloudberries/ bakeapples], but the last time we went to Beverly Lake it was not red. No more *aqpiit*.”

—Irene Taviniq Kaluraq

“The berries are smaller. Both blue and black.”

—John Nukik

“[Berry plants] seem to be less and less [abundant], and it could be because of the heat of the sun.”

—John Nukik

Nunavut / Baker Lake

“ᐊᕐᕌᓂ ᐅᖕᒐᓪᓕᕐᒥ
ᐸᐅᕐᖕᒐᖅᑕᕆᐊᓚᐅᖅᓯᒪᔪᖕᒐ, ᑭᓯᐊᓂ
ᐊᓇᖖᒋᓕᒻᒪᕆᐊᓘᓚᐅᖅᓯᒪᔪᖅ.
ᐊᖕᒋᕐᕋᕋᒪ ᖃᑦᑕᕋ ᒪᑐᐃᕋᒃᑯ
ᐊᓇᖖᒋᑦ ᑎᖕᒋᓚᐅᖅᑐᑦ ᖃᑦᑕᕐᒥᑦ
[...] ᐸᐅᕐᖕᒐᓪᓗ ᒥᑭᑦᑐᒻᒪᕆᐅᓪᓗᑎᒃ.
ᑭᓯᐊᓂ ᐊᕐᕌᓂ ᓄᓂᕙᒋᐊᒃᑲᓐᓂᕋᒪ
ᐊᖕᒋᓂᖅᓴᐅᓕᓚᐅᖅᑐᑦ
ᐳᕙᓪᓚᖅᓯᒪᓂᖅᓴᐅᓪᓗᑎᒡᓗ. [...]
ᓄᓂᕙᒋᐊᖅᐸᒃᑐᖕᒐ ᐊᐅᔭᖅᑕᒫᑦ,
ᑭᓯᐊᓂ ᑕᐃᔅᓱᒪᓂ ᐊᔾᔨᐅᓚᐅᖖᒋᑦᑐᖅ.”

“Two years ago I went up to pick blackberries, but there were a lot of black flies. When I got home I opened the can and the flies flew out, and [...] the berries were really small. But last year when I went out again, the berries were bigger and juicier. [...] I go berry picking every year, but that time was different.”

—ᓘᓯ ᑎᑭᖅ ᑐᖕᒍᐊᖅ

—Lucy Tunguaq

“ᑕᐃᔅᓱᒪᓂᑐᑦ ᐱᕈᖅᐸᒍᓐᓃᖅᑐᑦ.
ᐃᒻᒪᖄ ᐱᕈᑦᑎᐊᖅᑐᐊᓘᒐᓗᐊᖅ
ᐃᓄᖃᖖᒋᑦᑐᒥ, ᑭᓯᐊᓂ
ᑕᒫᓂ ᓄᓇᓕᖕᒥ
ᐸᐅᕐᖕᒐᖃᓗᐊᖅᐸᒍᓐᓃᖅᑐᖅ.”

“They grow less than before. Perhaps the berries are growing more abundant and properly where there aren’t people around, but around town there are hardly any berries growing.”

—ᒫᑕ ᓇᓇᐅᑐᐊᖅ ᓄᑭᒃ

—Martha Nukik

“ᑎᑉᔭᓕᒃ ᑕᓯᐊᓂ, ᖁᓕᒥᒎᓕᖕᓄᑦ
ᒥᕝᕕᐅᖃᑦᑕᓕᓚᐅᖅᑎᓐᓇᒍ, ᑭᖕᒥᖕᓇᑦ
ᐱᕈᑦᑎᐊᖅᐸᓚᐅᖅᓯᒪᔪᑦ; ᓄᓇᓕᒫᕌᓗᒃ
ᐊᐅᐸᖅᓯᖅᑰᔨᕙᓚᐅᖅᓯᒪᔪᖅ, ᑭᓯᐊᓂ
ᑕᒫᓂ ᑕᑯᓗᐊᖅᐸᒍᓐᓃᖅᑐᖕᒐ.
[...] ᓯᒡᔭᖅᐸᓯᖕᒥ
ᐊᐅᐸᖅᑐᐃᓐᓇᐅᕙᓚᐅᖅᓯᒪᔪᖅ,
ᑭᓯᐊᓂ ᒫᓐᓇ ᑕᐃᒪᐃᑦᑐᓐᓃᖅᑐᖅ.”

“Around the Ferguson Lake area, before those helicopters started landing, the cloudberries used to grow very well; it almost seemed like all the land would turn red, but around here I can’t see much of that anymore. [...] It used to be all red along the shore, but now you don’t see that anymore.”

—ᒫᑕ ᓇᓇᐅᑐᐊᖅ ᓄᑭᒃ

—Martha Nukik

“ᖃᐅᔨᖃᑦᑕᖅᓯᒪᓕᖅᑕᒃᑲ
ᐸᐅᕐᖓᑦ ᐱᕈᓕᕌᖓᑕ
ᐳᕙᓪᓚᖅᓵᓕᓂᖅᓴᐅᕙᓕᖅᑐᑦ
ᖃᒐᖅᓵᓕᕙᓕᖅᖢᑎᒡᓗ
ᓈᒻᒪᒃᓯᓗᐊᖅᑎᓐᓇᒋᑦ.
ᐸᓂᖅᓴᕋᐃᓕᖅᑐᑦ. [...]
ᒥᑮᓐᓇᖅᑰᔨᕙᓕᖅᑐᑦ ᐱᕈᐊᓂᒃᑑᔮᕋᑎᒃ.
ᖃᓄᐃᑦᑐᑐᐃᓐᓇᑦ ᐱᕈᖅᑐᑦ
ᑐᕐᑯᓕᓵᓕᕙᓕᖅᑐᑦ ᐱᕈᐊᓂᒃᑑᔮᕋᑎᒃ.
ᐱᕈᖅᑐᓂᒃ ᐊᖏᒡᓕᔪᓂᒃ
ᑕᑯᕙᓚᐅᖅᓯᒪᔪᖓ. ᐅᓪᓗᒥ
ᐊᔾᔨᒋᔭᓐᓃᖅᑕᖕᒋᑦ ᑕᐃᔅᓯᒪᓂᑐᑦ.”

—ᒦ ᑮᓈᓕᒃ ᕼᐊᖅᐱ

“ᑕᐃᔅᓱᒪᓂ
ᐸᐅᕐᖓᖅᑲᑦᑎᐊᖅᐸᓚᐅᖅᓯᒪᔪᖅ,
ᐊᒻᒪ ᑕᐃᔅᓱᒪᓂ
ᑭᖕᒥᖕᒪᖅᑲᕐᓂᖅᓴᐅᓚᐅᖅᓯᒪᔪᖅ,
ᐱᑕᖃᙱᓐᓂᖅᓴᐅᓵᓕᖅᑐᑦ
ᓰᕐᓇᙱᓐᓂᖅᓴᐅᓕᖅᖢᑎᒡᓗ.
ᓄᓕᐊᕋᓗ ᖃᑦᑕᕐᒥᒃ
ᑕᑕᑦᑎᑲᐅᑎᒋᕙᓚᐅᖅᓯᒪᔪᒍᑦ
ᐸᐅᕐᖓᖅᑕᓕᕋᓐᓄᒃ. ᐊᕐᕌᒍᑦ
ᑕᓪᓕᒪᑦ ᐊᓂᒍᖅᑎᓪᓗᒋᑦ
ᑕᐃᒪᐃᒃᑲᓐᓂᕆᐊᓚᐅᕋᓗᐊᕋᓐᓄᒃ,
ᒥᑭᓐᓂᖅᓴᐅᓕᖅᑐᑦ
ᑕᑕᑦᑎᑲᐅᖅᑐᕐᓇᕈᓐᓃᖅᖢᓂᓗ.”

—ᓄᐊᒪᓐ ᐊᑕᖓᓛᖅ

“ᑕᐃᔅᓱᒪᓂ, ᓄᓇᒥᐅᑕᐅᑎᓪᓗᑕ,
ᐸᐅᕐᖓᑦ ᒪᒪᖅᑐᐊᓘᕙᓚᐅᖅᓯᒪᔪᑦ.
ᒫᓐᓇᓕ ᒥᑭᓐᓂᖅᓴᐅᓕᖅᑐᑦ
ᒪᒪᓗᐊᕈᓐᓃᖅᖢᑎᒡᓗ.”

—ᑑᓇ ᐃᖁᓕᒃ

“I’ve noticed when the berries start growing, they seem to ripen just like that and get ruined right away and don’t fully ripen at their time. They get dry so easily. [...] They seem to be small all the time and not fully grown. Any plant that I see seems to wither away without being fully grown. I used to see plants that would get quite large. It’s not the same today as I used to see them.”

—May Haqpi

“Years ago there used to be an abundance of berries, and I notice that there used to be more cloudberries around, but there seem to be less and they are not as sweet as before. My wife and I used to carry a kettle around with us and when we started gathering we would fill it up just like that. Within five years we would go to do to the same thing, but they were smaller and it would take more time to pick.”

—Norman Attungala

“Before, when we lived on the land, [berries] would taste really good. Today they seem to be smaller than before and taste less.”

—Toona Iqulik

ᐱᕈᖅᑐᑦ ᐊᓯᖏᑦ

"ᐱᕈᖅᑐᑦ ᐱᕈᓗᐊᖅᐸᒍᓐᓃᖅᑐᑦ. [...] ᒫᓐᓇᓗ ᒪᕐᑯᒐᔭᒍᓐᓃᖅᓱᓂ; ᓄᓇ ᐸᓂᖅᑐᐊᓘᓚᖅᑐᖅ. ᑭᓯᑐᐃᓐᓇᖅ ᐱᕈᖅᐸᒃᑐᑦ ᐱᕈᓗᐊᖅᐸᒍᓐᓃᖅᑐᑦ ᒥᑭᓐᓂᖅᓴᐅᓚᖅᓱᑎᒃᓗ."

—Hᐋᑎ ᐊᑐᖓᑦ ᐊᒻᒪ ᐹᓪ ᐊᑐᑐᕚ

"ᐃᓱᒫᓗᑎᓚᖅᑕᕋ ᓄᓇᕗᑦ ᐸᓂᖅᐸᓪᓚᐊᓚᕐᒪᑦ, ᓱᓇᓗᒃᑖᑦ ᓄᓇᒥ ᐱᕈᖅᐸᒃᑐᑦ ᐱᕈᖅᑐᖅᑲᓗᐊᑦᑎᐊᓚᐅᖅᓯᒪᑦᑎᐊᖏᑦᑐᖅ, ᐊᓯᖏᓐᓂᒃ ᐊᔾᔨᐅᙱᑦᑐᓂᒃ ᐱᕈᖅᑐᖅᑲᓚᑐᐃᓐᓇᕆᐊᖅᑲᖅᑐᖅ ᑕᒪᒃᑯᐊ ᐱᕈᖅᐸᒃᑐᑦ ᓄᖑᑉᐸᑕ."

—Hᐋᑎ ᐊᑐᖓᑦ ᐊᑐᑐᕚ

"ᖃᒪᓂ'ᑐᐊᕐᒥ ᓄᓇ ᐸᓂᖅᑐᐊᓘᖅᑰᔭᓚᖅᑐᖅ. ᐱᕈᖅᑐᓪᓗ ᑕᐃᔅᓯᒪᓂᓂᑦ ᒥᑭᓐᓂᖅᓴᐅᓚᖅᓱᑎᒃ, ᐱᕈᖏᓐᓂᖅᓴᐅᕙᓚᖅᓱᑎᒃᓗ."

—ᐊᐃᕇᓐ ᑕᕕᓂᖅ ᑲᓗᕋᖅ

"ᒥᓂᕙᓚᐅᖅᓯᒪᔭᖅ ᐱᕈᖅᑐᓄᓪᓗ ᐱᕈᑦᑎᐊᕈᑕᐅᕙᒃᓱᓂ, ᑭᓯᐊᓂ ᒫᓐᓇ ᒥᓂᓗᐊᖅᐸᒍᓐᓃᖅᑐᖅ. [...] ᐃᒻᒪᖄ ᒥᓂᕙᙱᑕᕐᓂᖓᓄᑦ ᐱᕈᐊᓚᓴᙱᓐᓂᖅᓴᐅᓚᖅᑐᑦ."

—ᔫᓐ ᓄᑭᒃ

Other Plants

"[Plants] hardly grow now. [...] Now today it hardly rains; the ground is dry. So everything that grows, [grows] less and smaller."

—Hattie and Paul Attutuvaa

"I'm worried about our land that is getting dry, because everything that grows on the land seems to grow less and less, and I'm thinking that something different might start growing if these are gone."

—Hattie Attutuvaa

"It seems like it's so dry now around Baker Lake. The plants seem to be smaller than before, and growing less."

—Irene Taviniq Kaluraq

"We used to get mist that would make the vegetation grow well, but we hardly get mist anymore. [...] That could be one of the things that is slowing everything down."

—John Nukik

“ᐃᒻᒫᕐᔫᓕᖅᑐᖅ
ᑕᐅᓄᖖᒐᐅᓚᐅᖅᓯᒪᔪᒍᒃ ᐸᓂᒐᓗ
ᐅᒥᐊᒃᑯᑦ. ᑕᐃᔅᓱᒪᓂ ᐸᐅᖕᒐᑦ, ᐊᕐᐲᑦ
ᑭᒍᑕᖕᒋᕐᓇᑦᑐ ᐱᑕᖃᑦᑎᐊᓚᐅᖅᓯᒪᔪᑦ.
ᐱᕈᖅᑐᖃᕐᓂᖅᓴᐅᓚᐅᖅᓯᒪᔪᖅ
ᑕᐅᓇᓂ. ᑭᖑᓂᐊᒍᑦ
ᐅᑎᖁᓗᐊᖁᑦᑕ ᑕᐅᓄᖕᒐᔅᓴᐃᓐᓇᖅ,
ᐱᕈᖅᑐᖃᓗᐊᑦᑎᐊᓚᐅᖅᓯᒪᑎᙱᑦᑐᖅ.
ᐸᐅᖕᒐᖃᓚᐅᙱᑦᑐᖅ.”

—ᓘᓯ ᖃᐅᓐᓇᖅ

“ᐅᑦᑐᒥ ᓄᓇᓕᖕᒥ
ᐱᕈᖅᑐᓕᐊᓘᓕᖅᑐᖅ.
ᐸᐅᓐᓇᓕᐊᓘᓕᖅᑐᖅ ᐃᒡᓗᐃᑦ
ᑭᒡᓕᖕᒋᓐᓂ ᐊᖅᑯᑎᑦᑐ ᑭᒡᓕᖕᒋᓐᓂ.”

—ᓘᓯ ᑎᑭᖅ ᑐᖑᐊᖅ

“ᐱᕈᖅᑐᖃᖅᐸᓕᖅᑐᖅ
ᐱᕈᖅᐸᓚᐅᖅᓯᒪᙱᑦᑐᓂᒃ,
ᕿᒻᒥᖃᕐᕕᖕᒥ. ᑕᑭᔪᐊᓘᑦᑐᑎᒃ
ᖁᖅᓱᖅᑐᑯᓗᖃᖅᓱᑎᒃ,
ᐱᕈᖅᓯᐊᑯᓘᔮᖅᑐᑦ ᓄᕗᖕᒋᓐᓂ.”

—ᒫᒍᓇᑦ ᑐᑦᓕᒃ ᓂᕕᐊᑦᓯᐊᖅ

“I went down a few years ago with my daughter by boat. That was the time I found quite a few blackberries, *aqpiit* [cloudberries/bakeapples], and blueberries. There were more plants down there. And a few years later we went back down to the same place, but there was hardly anything growing. No berries.”

—Lucy Kownak

“Today around town we have a lot of flowers. The fireweeds are all over outside the houses. Along the road they are growing.”

—Lucy Tunguaq

“There are some plants that never grew before, where the dogs are. They are really tall, with little yellow things, almost like flowers, at the tips.”

—Margaret Amauruq Niviatsiaq

Nunavut / Baker Lake

“ᐅᖅᐲᑦ ᐅᖅᐱᒐᐃᓪᓗ ᐊᖏᓂᖅᓴᐅᓪᓗᑎᒃ ᐱᕈᖅᐸᓚᐅᖅᓯᒪᔪᑦ ᑕᑯᕙᓚᐅᖅᑕᒃᑲ, ᑭᓯᐊᓂ ᐅᓪᓗᒥ ᐱᕈᓗᐊᖅᑑᔮᖅᐸᒍᓐᓃᖅᑐᑦ. ᐊᖑᓇᓯᒋᐊᖅᓯᒪᓪᓗᑕ ᐸᓂᖅᑐᓂᒃ ᐅᖅᐱᓂᒃ ᐊᕝᕗᖅᐸᓚᐅᖅᓯᒪᔪᒍᑦ ᐃᒐᔾᔪᑎᒃᓴᓂᒃ, ᑭᓯᐊᓂ ᐱᑕᖃᖅᐸᒍᓐᓃᖅᑐᑦ. ᐃᓱᒪᓇᖅᓯᔪᖅ ᐱᕈᖅᑐᑦ ᐅᖅᐲᓪᓗ ᑐᖁᓕᕐᒪᖕᒑᑕ.”

—ᓄᐊᒪᓐ ᐊᑕᖑᒑᖅ

“I used to see these shrubs or willows growing larger in the past, but today it doesn’t seem [like] they grow very much. When we were hunting we would have to look for dried-up leaves or shrubs to make a fire, but we don’t see those anymore. So we would start wondering if the plants or shrubs were dying or what.”

—Norman Attungala

“ᐅᖅᐲᑦ ᒥᑭᓂᖅᓴᐅᕙᓕᖅᑐᑦ ᓇᐃᓐᓂᖅᓴᐅᓕᖅᑐᑎᒡᓗ. ᐱᑕᖃᓗᐊᖅᐸᒍᓐᓃᖅᑐᑦ. ᐸᓂᖅᓯᒪᔪᒻᒪᕆᐅᓕᖅᑐᖅ. ᑕᐃᔅᓱᒪᓂ ᑕᑭᔪᐊᓘᓪᓗᑎᒃ ᐱᕈᖅᐸᓚᐅᖅᓯᒪᔪᑦ. ᒪᖁᖕᒍᑦ ᐱᕈᖅᐸᓚᐅᖅᑐᑦ ᑕᑭᓂᖅᓴᐅᓪᓗᑎᒃ ᐱᕈᖅᓵᓚᓂᖅᓴᐅᓪᓗᑎᒡᓗ. [...] ᐃᕼᖃᕐᓂᖅᓴᐅᓪᓗᓂᓗ.”

—ᑖᒪᔅ ᖃᕿᒪᑦ

“[Shrubs] are smaller and shorter now. There is hardly anything now. It’s really dried up. A long time ago they used to grow really long. With the rain, often they would grow taller and faster. [...] More grass growing.”

—Thomas Qaqimat

ᐆᒪᔪᑦ

Animals

“ᐊᒡᓛᑦ ᑐᒃᑐᑦ ᑐᓐᓄᖃᖅᐸᒍᓐᓃᖅᑐᑦ, ᐱᕈᖅᑐᑦ ᐱᕈᖅᐸᒍᓐᓃᕐᒪᑕ. [...] ᓄᓇ ᖃᐅᓯᕈᔪᐃᓐᓇᐅᔮᖅᑎᓪᓗᒍ [...] ᐱᐅᔪᖅ ᑐᒃᑐᓄᑦ. ᓄᓇ ᐸᓂᖅᑎᓪᓗᒍ ᐱᐅᙱᑦᑐᖅ ᑐᒃᑐᓄᑦ.”

—ᐹᓪ ᐊᑐᑐᕚ

“Even the caribou aren’t fat today, because the plants don’t grow. [...] When the ground is a little bit wet all the time [...] it’s good for the caribou. When the earth is dry it’s no good for the caribou.”

—Paul Attutuvaa

“ᓯᒃᓰᑦ ᓂᕿᑐᖅᐸᓚᐅᖅᓯᒪᙱᑦᑐᑦ;
ᐅᓪᓗᒥ ᓂᒃᑯᓂᒃ ᓂᕆᕙᓕᖅᑐᑦ.
[...] ᑭᓯᐊᓂ ᓇᐅᔭᑦ ᑕᐃᔅᓱᒪᓂ
ᓂᕿᑦᑎᓐᓂᒃ ᐱᕙᓚᐅᖅᓯᒪᔪᑦ,
ᒫᓐᓇ ᐊᕕᙵᑦ ᓯᒃᓯᑦ ᑎᖕᒥᐊᓪᓗ
ᓂᕿᑦᑎᓐᓂᒃ ᓂᕆᕙᓕᖅᑐᑦ.
ᐊᓯᔾᔨᖅᓯᒪᓕᖅᑐᖅ.”

—Hᐋᑎ ᐊᑐᖓᑦ ᐊᑐᑐᕚ

“ᐅᓪᓗᒥ ᐊᔾᔨᙱᑦᑎᓕᖅᑕᖓ,
ᑕᑯᕙᓚᐅᖅᓯᒪᙱᒃᑲᓗᐊᕋᑦᑕ
ᑎᖕᒥᐊᓂᒃ ᓂᕿᑐᖅᑐᓂᒃ. ᒫᓐᓇ ᑎᖕᒥᐊᑦ
ᓂᕿᑐᖅᐸᓕᖅᑐᑦ. ᓂᒃᑯᓕᐅᓕᕌᖓᑦᑕ,
ᑎᖕᒥᐊᓄᑦ ᓂᖄᐅᕙᒃᑐᑦ. ᐊᒡᓛᑦ
ᓯᒃᓰᑦ ᑕᐃᔅᓱᒪᓂ ᐸᐅᕐᖓᐃᓐᓇᕐᓂᒃ
ᓂᕆᕙᓚᐅᖅᓯᒪᔪᑦ. ᐅᓪᓗᒥ
ᓂᕿᑐᖅᐸᓕᖅᑐᑦ—ᓂᒃᑯᓕᐊᑦᑎᓐᓂᒃ.”

—ᐊᐃᕆᓐ ᑕᕕᓂᖅ ᑲᓗᕋᖅ

“ᕿᒃᑐᕆᐊᓕᒻᒪᕆᐊᓘᓕᖅᑐᖅ
ᐊᓇᙱᕐᑲᐅᓕᖅᕗᓂᓗ.
ᐅᒥᓗᐊᖅᑐᒻᒪᕆᐊᓗᐃᑦ—ᑕᐃᔅᓱᒪᓂᓂᑦ
ᐅᒥᕐᓂᖅᓴᒻᒪᓃᑦ. ᑕᐃᔅᓱᒪᓂ ᕿᒃᑐᕆᐊᑦ
ᓴᖅᑭᖅᑳᖅᕗᑎᒃ, ᐱᓇᓱᐊᕈᓯᖅ
ᐊᓂᒍᕐᒪᑦ ᐊᓇᖕᒌᑦ ᓴᖅᑭᓕᕆᓪᓗᑎᒃ,
ᐅᓐᓄᐊᒃᑯᓪᓗ ᐊᓇᙱᕐᑲᖅᐸᒐᓂ.
ᐅᓪᓗᒥᓗ ᖃᖓᓕᒫᖅ ᓴᖅᑮᔭᖅᐸᓕᖅᑐᑦ
ᐅᓐᓄᐊᖑᒐᓗᐊᒃᑯᑦ.”

—ᔮᓂᑦ ᓂᐱ ᐃᑰᑕᖅ

“ᐱᕈᖅᓴᑎᓪᓗᙵ,
ᑕᑯᕙᓚᐅᖅᓯᒪᙱᑦᑐᖑᑦ ᐅᒥᖕᒪᖕᓂᒃ
[...] ᑭᓯᐊᓂ ᑕᒫᓂ ᐅᒥᖕᒪᖕᓂᒃ
ᑕᑯᓕᖅᑐᖑᑦ. ᑕᑯᓚᐅᖅᑐᖓ
ᐊᖑᓴᓪᓗᐊᓗᖕᒥᒃ ᐊᖏᔪᐊᓗᖕᒥᒃ.”

—ᔫᓐ ᓄᑭᒃ

“The *siksiit* [Arctic ground squirrels] never used to eat meat; today, they seem to be eating dried meat. [...] Only seagulls used to eat our meat a long time ago, but now even mice or *siksiit* and birds eat the meat. It’s changed.”

—Hattie Attutuvaa

“Today it’s different, because we never saw birds eating meat before. Today we see birds eating meat. When we dry meat, the birds are eating it. Even the *siksiit*, they seemed to only eat berries before. Today they eat a lot of meat—our dried meat.”

—Irene Taviniq Kaluraq

“I notice that there are a lot of mosquitoes and a lot of flies. Too much—more than before. [Before], the mosquitoes would arrive, and then a week later the flies would come out, and during the night the flies would go to sleep. But today they seem to be up all the time, even at night.”

—Janet Ikuutaq

“When I was growing up, I never used to see muskoxen [...] but we are starting to see muskoxen around. I saw one huge male.”

—John Nukik

“ᑕᐃᔅᓱᒪᓂ [...]
ᕿᑯᑦᑐᓇᐊᓕᐊᓘᕙᓚᐅᖅᓯᒪᔪᖅ, ᑭᓯᐊᓂ
ᒫᓐᓇ ᓯᕿᓂᖅ ᐆᓇᓗᐊᓕᕐᓂᖓᓄᑦ
ᕿᑯᑦᑐᓇᐊᖅᑲᓗᐊᖅᐸᑎᓐᓃᖅᑐᖅ.”

—ᔫᓐ ᓄᑭᒃ

“ᕿᑯᑦᑐᓇᐊᖅᑲᕈᓐᓃᖅᓵᓕᕙᓕᖅᑐᖅ,
ᑭᓯᐊᓂ ᐊᓇᖏᑦ, ᓯᓚ
ᓂᒡᓚᓕᕋᓗᐊᖅᑎᑦᑐᒍ ᓱᓕ
ᐱᑕᖅᑲᖅᐸᓕᖅᑐᑦ.”

—ᔫᓐ ᓄᑭᒃ

“ᖃᑦᑐᓈᑦ ᓄᓇᖕᖓᓐᓄᐊᕌᖕᖓᒪ
ᑕᑯᕙᒃᑐᖓ ᑕᑭᔪᐊᓗᖕᓂᒃ
ᖁᐱᕐᕈᓂᒃ ᓇᐹᖅᑐᓂᑦ. ᑕᐃᒪᐃᑦᑐᓂᒃ
ᐱᑕᖅᑲᖅᐸᓚᐅᖅᓯᒪᙱᑦᑑᒐᓗᐊᖅ
ᑭᓯᐊᓂ ᑕᒫᓂ ᐱᑕᖅᑲᖅᐸᓕᖅᑐᑦ.
ᑕᑯᕙᓚᐅᖅᑕᒃᑲ ᐐᓂᐸᐃᒃᒥ. ᒫᓐᓇ
ᒫᓂ ᐱᑕᖅᑲᓕᖅᑐᑦ. ᑕᑭᔫᒻᒪᕆᐊᓗᐃᑦ.
ᕿᕐᓂᖅᑕᐅᑦᑐᑎᒃ ᑎᖏᔪᓐᓇᖅᑐᑦ.”

—ᒫᑕ ᓇᓇᐅᑐᐊᖅ ᓄᑭᒃ

“ᐅᑦᑐᒥ ᐊᓇᖏᑦ
ᐅᓄᓗᐊᖅᑐᐊᓘᓕᖅᑐᑦ.
ᑕᐃᔅᓱᒪᓂᓂᑦ ᐱᑕᖅᑲᕐᓂᖅᓴᐅᓕᖅᑐᑦ,
ᐱᑕᖅᑲᔅᓲᔭᕐᓂᖅᓴᐅᕙᓕᖅᑑᑎᒡᓗ.
ᐅᑭᐊᒃᓵᖕᒍᓕᕋᓗᐊᖅᑎᑦᑐᒍ
ᓱᓕ ᐱᑕᖅᑲᖅᐸᒃᑐᑦ, ᑕᓰᑦ ᑰᑦᑐ
ᖁᐊᕋᓗᐊᖅᑎᑦᑐᒋᑦ. ᐊᓄᕌᓕᕌᖕᒐᑦ
ᓯᕿᓐᓂᓕᖅᑑᓂᓗ, ᓱᓕ ᐱᑕᖅᑲᖅᐸᒃᑐᑦ.
ᕿᑯᑦᑐᓇᐊᓕᐊᓘᓕᖅᑑᓂᓗ.
ᕿᑯᑦᑐᓇᐊᖅᑲᓗᐊᖅᐸᓚᐅᖅᓯᒪᙱᑦᑑᒐᓗᐊᖅ,
ᒫᓐᓇ ᐊᒥᓱᐊᓘᓕᖅᑐᑦ,
ᐱᑕᖅᑲᑯᑖᖕᓂᖅᓴᐅᕙᓕᖅᑑᑎᒡᓗ.
ᖃᐅᔨᓴᐅᑦ ᐊᓯᔾᔨᙱᒃᑲᓗᐊᖅᑎᑦᑐᒍ,
ᐊᓇᖏᑦ ᓴᖅᑭᓵᓕᕙᓕᖅᑐᑦ,

“In the past [...] there were a lot of mosquitoes, but now because the sun is so hot, there are hardly any mosquitoes anymore.”

—John Nukik

“The mosquitoes are gone earlier, but with the black flies, even after the air gets cold they are still around.”

—John Nukik

“When I go down south I see those really long insects coming from the trees. We never had those kinds before, [but] they are starting to come up here. I used to see them down south in Winnipeg. Now they are up here. They are really long. They are black and they fly.”

—Martha Nukik

“Today there are too many black flies. More than before, and they seem to stay longer. Even when the fall comes, they are still around, even when the pond or river freezes. When the wind comes and the sun comes out, they are still around. There seem to be lots of mosquitoes. There used to be fewer mosquitoes, and now there are so many, and they stay

ᐅᓄᕐᓂᖅᓴᐅᓪᓗᑎᒃ,
ᐱᑕᕐᑲᔅᓰᔭᕐᓂᖅᓴᐅᕙᓪᓕᖅᕈᑎᒡᓗ."

—ᓘᓯ ᖃᐅᓐᓇᖅ

“ᐅᔅᐲᓇᓯᒪᖕᒥᔭᒃᑲ
ᐃᒍᑦᑕᑦ. ᑕᐃᔅᓱᒪᓂᑐᑦ
ᑕᑯᕙᓚᐅᖅᓱᒪᔭᑦᑎᑑᖅᑰᐱᔪᓐᓃᖅᑐᑦ.
ᐊᖏᔪᐊᓘᓕᖅᑐᑦ, ᐊᔅᐱᖖᒋᑕᖕᒋᓐᓂᓪᓗ
ᐃᒍᑦᑕᕐᓂᒃ ᑕᑯᕙᓚᖅᕈᖕᒐ."

—ᒫᑕ ᓇᓇᐅᑐᐊᖅ ᓄᑭᒃ

“[ᕿᒃᑐᕆᐊᑦ ᐊᓇᖖᒋᑦᓗ]
ᓴᖅᑮᓴᓕᕙᓕᖅᑐᑦ, ᑕᐃᔅᓱᒪᓂ
ᖃᐅᔨᒪᕙᓚᐅᖅᑐᒍᑦ ᖃᖓ
ᕿᒃᑐᕆᐊᑦ [ᐊᓇᖖᒋᑦᓗ]
ᓴᖅᑭᓂᐊᕐᒪᖔᑕ [...] ᑭᓯᐊᓂ ᒫᓐᓇ
ᐊᑕᐅᑦᑎᒃᑰᓕᖅᑐᑦ. ᐆᒃᑑᑎᒋᓗᒍ
ᖃᐅᔨᒪᕙᓚᐅᖅᑐᒍᑦ ᖃᖓ ᑐᒃᑐᑦ
[...] ᓴᒡᒐᐅᓕᕌᖓᑕ ᓄᑖᓂᒡᓗ
ᐊᒥᖃᓕᕌᖓᑕ, ᕿᒃᑐᕆᐊᖃᕈᓐᓃᖅᕈᓂ
ᐊᓇᖖᒋᖃᓕᖅᐸᓚᐅᖅᑐᖅ, ᑭᓯᐊᓂ
ᒫᓐᓇ ᑕᐃᒪᐃᑦᑐᓐᓃᖅᑐᖅ."

—ᓄᐊᒪᓐ ᐊᑕᖕᒐᓛᖅ

“ᐃᖅᑲᐅᒪᖖᒋᓚᖕᒐ [ᕐᑯᐱᕐᕈᑦ]
ᐱᑕᕐᑲᓗᐊᖅᑐᐊᓘᓂᖕᒋᓐᓂᒃ,
ᒫᓐᓇ ᕐᑯᐱᕐᕈᓪᓚᕆᐊᓘᓕᖅᑐᖅ.
ᐅᓄᖅᑐᒻᒪᕆᐊᓘᓕᖅᑐᑦ."

—ᕓᐅᕋ ᐊᒃᕼᐊᒃ ᐊᕚᓚ

longer. The time doesn't seem to change, but I notice that the black flies come so soon, too many, and stay long."

—Lucy Kownak

"I also notice the bumblebees. They don't seem to be the original ones that I used to see. They are so big, and I've seen different species of bumblebees."

—Martha Nukik

"[Mosquitoes and black flies] are arriving earlier, and years ago we used to know when the mosquitoes [and black flies] would be coming, [...] but today it's all together, all at once and too early. For example, we knew that when the caribou [...] had shed and the new fur was on them, the mosquitoes would be gone and the black flies would be arriving, but it's not like that anymore."

—Norman Attungala

"I don't remember [insects] being so bad, but now they are bad. Today they are really bad."

—Vera Avalaa

“ᑐᓐᓄᖅᑲᑦᑎᐊᓪᓗᐊᓕᖅᑎᓪᓗᒋᑦ [ᑐᒃᑐᑦ], ᐃᓛᓐᓂᒃᑯᑦ ᑐᓐᓄᖅᑲᖖᒋᑲᓴᑦᑎᐊᖅᐸᒃᑐᑦ. ᐃᓚᖕᒋᑦ ᓴᓗᒃᑐᒻᒪᕆᐊᓘᓪᓗᑎᒃ. ᐃᓛᓐᓂᓗ ᓂᖅᑭᖕᖓᓐᓂ ᖅᑲᐅᓪᓗᖅᑐᓂᒃ ᑕᑯᓐᓇᖅᓯᔪᖅ. ᓇᓂᓕᒫᖅ ᓂᖅᑭᒦᓕᖅᑐᑦ. ᒫᓐᓇ ᑕᐃᔅᓱᒪᓂᓂᑦ ᑕᐃᒪᐃᑦᑐᖅᑲᖅᓂᖅᓴᐅᓕᖅᑐᖅ. ᖅᑲᐅᓪᓗᖅᑐᑦ ᒥᑭᑦᑐᑦ ᑎᓯᔪᑦ— ᒪᖅᑭᐅᔮᖅᑐᑦ, ᑯᒪᐅᖖᒋᑦᑐᑦ. ᑐᒃᑐᑦ ᓂᖅᑭᖕᒋᓐᓂ ᐱᑕᖅᑲᖅᓂᖅᓴᐅᓕᖅᑐᑦ. [...] ᐅᔾᔨᖅᓇᖅᓯᔪᑦ ᑐᒃᑐᑦ— ᐊᖅᑯᐱᓪᓗᐊᓕᖅᑎᓪᓗᒋᑦ, ᓱᒃᑲᓂᖅᓴᒥᒃ ᐱᓱᒃᐸᓕᖅᑐᑦ. ᑕᐃᔅᓱᒪᓂᑐᑦ ᑕᒫᓂᓯᐅᔅᓲᔭᖅᑎᒋᖖᒋᓕᖅᑐᑦ.”

—ᓄᐊᒪᓐ ᐊᑕᖑᓛᖅ

“When there should be a lot of fat [in the caribou meat], there is hardly any sometimes. Some of them are so skinny. Even in the meat, sometimes you see white stuff. They are all over the meat. There are more of those than before. The white little hard ones—like pus or something, not bugs. There is more on the meat of caribou than before. [...] You can't help but notice the caribou—when they should be lingering around the ground, they seem to be moving a bit too fast. Not lingering like they used to.”

—Norman Attungala

ᐅᑭᐅᒥ ᐊᓯᔾᔨᖅᑕᖅᓂᖅ

“ᑰᑉ ᓯᓈᓂ ᓯᑯᖅᑲᖅᐸᓚᐅᖅᓯᒪᔪᖅ. [...] [ᓯᑯᑦ] ᐊᐅᔭᓕᒫᖅ ᐱᑕᖅᑲᖅᐸᓚᐅᖅᑐᑦ. ᐅᓪᓗᒥ ᑕᐃᒪᐃᑦᑐᓐᓃᖅᑐᖅ.”

—ᐹᓯᓪ ᐊᒻᒪ ᔫᓕ ᓴ’ᖕᒍᓵᖅ ᑐᓗᖅᑐᖅ

“ᐊᑯᓂᐅᓂᖅᓴᖅ ᐊᐅᔭᐅᕙᓕᖅᑐᖅ. ᐊᐅᒃᓵᓕᕙᓕᖅᓱᓂ ᖅᑯᐊᖅᓇᓵᖅᓂᖅᓴᐅᕙᓕᖅᓱᓂ. ᐊᐱᓇᓵᖅᐸᓕᖅᑐᖅ. [...] ᒪᑯᐊ ᐃᒪᒐᓛᖑᕙᓚᐅᖅᑐᑦ ᐃᒪᖅᑲᕈᓐᓃᖅᑐᑦ, ᑕᓯᕈᐃᑦ ᓘᓐᓃᑦ. ᑕᐃᒪᐃᓕᖅᑐᖅ ᐊᐳᑎᖅᑲᖖᒋᓐᓂᖅᓴᐅᕙᓕᖅᓂᖕᖓᓄᑦ.”

—Hᐋᑎ ᐊᑐᖕᒐᑦ ᐊᒻᒪ ᐹᓪ ᐊᑐᑐᕚ

Seasons

“There used to be ice along the shore of the Thelon River. [...] [It would] stay there all through summer. Today we don't have [that] anymore.”

—Basil and Julie Tuluqtuq

“The summer seems to be longer now. It melts fast and freezes slow. The snow comes late. [...] We are finding that all these little waters are drying, even the ponds. Because there is hardly any snow today.”

—Hattie and Paul Attutuvaa

“ᒫᓐᓇ ᑕᐃᔅᓱᒪᓂᓂᑦ
ᓂᒡᓚᓱᖕᓂᖅᓴᒻᒪᕆᐅᓕᖅᑐᖅ. ᑕᒪᑐᒪᓂ
ᐅᑭᐅᖅᒥ ᓂᒡᓚᓱᒃᑐᒻᒪᕆᐅᓚᐅᖅᒪᑦ
ᐅᐱᕐᖔᖅ ᐆᓇᖅᓂᖅᓴᐅᓂᐊᖅᑰᖅᑐᖅ.”

—ᐊᐃᕇᓐ ᑕᕕᓂᖅ ᑲᓗᕋᖅ

“ᐊᐅᔭᒃᑯᑦ ᐆᓇᖅᑐᒻᒪᕆᐊᓘᕙᓕᖅᑐᖅ,
ᑭᓯᐊᓂ ᐅᑭᐅᒃᑯᑦ [...]
ᐱᖅᓯᖅᑐᐊᓘᕙᒃᑐᖅ, ᓯᑯ ᐊᐳᑦᑐ
ᑎᓯᓗᐊᓕᖅᕈᑎᒃ.”

—ᔮᓐ ᓄᑭᒃ

“ᐅᓪᓗᐃᑦ ᐃᓚᖓᓐᓂ
ᓂᒡᓚᓱᒃᑐᒻᒪᕆᐊᓘᓪᓗᓂ
ᖃᐅᒻᒪᓪᓗ ᓂᒡᓚᓱᓗᐊᕿᓐᓃᖅᕈᓂ.
ᔮᓄᐊᕆᒥᓘᓐᓃᑦ, ᑕᐃᔅᓱᒪᓂᑐᑦ
ᖁᐊᖀᓘᕙᒃᑐᑦ. ᐊᒻᒪ
ᓯᓚᐅᑉ ᓂᒡᓚᓱᖕᓂᖓ
ᐊᓯᔾᔨᖅᓯᒪᔫᒻᒪᕆᐅᓕᖅᑐᖅ. ᑕᐃᔅᓱᒪᓂ
ᓯᓚ ᓂᒡᓚᓱᒃᑳᙱᑦ, ᐅᓪᓗᓴᓄᑦ
ᑕᐃᒪᐃᑉᐸᓚᐅᖅᑐᖅ. ᔮᓄᐊᕆᒥᓘᓐᓃᑦ,
ᓂᒡᓚᓱᖕᓂᖅᐸᐅᓗᐊᖅᑎᓪᓗᒍ,
ᓯᓚ ᐊᓯᔾᔨᖃᐅᖅᑐᖅᐸᓚᐅᙱᑦᑐᖅ.
ᒫᓐᓇ ᓂᒡᓚᓱᒃᑐᐊᓘᓚᐅᖅᑎᓪᓗᒍ,
ᖃᐅᑐᐃᓐᓇᖅᐸᑦ
ᐊᐅᒃᐸᓪᓕᐊᓕᕈᓐᓇᖅᑐᖅ.”

—ᒫᑕ ᓇᓇᐅᑐᐊᖅ ᓄᑭᒃ

“I’m finding it much colder now than before. I’m just thinking that this winter was too cold, so I think the spring is going to be hot.”

—Irene Taviniq Kaluraq

“In summertime the heat is so intense, but in the wintertime [...] we get very bad blizzards, and the ice and snow are too hard.”

—John Nukik

“One day it’s really cold, and the next day it’s not that cold. Even in January, sometimes things are as frozen as they used to be in the past. Also, temperatures have changed a lot. Years ago when it was cold outside, it would be the same temperature for a few days. Even in January, although it would be the coldest month, the temperatures would not change quickly. Now it could be really cold, and the next day almost melting.”

—Martha Nukik

“ᑕᒫᓂ ᓄᓇᐅᑉ ᐃᓗᐊᓂ
ᓂᕿᒃᑯᕕᓕᐅᖅᐸᓚᐅᖅᓯᒪᔪᒍᑦ.
ᐃᓗᕐᑐᓗᐊᙱᒃᑲᓗᐊᕐᓗᒍ
ᓂᓪᓕᓇᖅᑐᒃᑯᕕᖃᖅᐸᓚᐅᖅᑐᒍᑦ.
[...] ᒫᓐᓇ ᐃᓗᑐᔪᓕᐅᕋᓱᒃᑲᓗᐊᕈᑦᑕ
ᕐᑯᐊᖑᔪᒍᑦ ᑎᑭᐅᑎᕙᒍᓐᓃᖅᑐᒍᑦ.
[...] ᐊᐅᒃᐸᓪᓕᐊᑐᐃᓐᓇᓕᖅᑐᖅ.
ᑕᐃᒪᐃᓚᐅᖅᓯᒪᙱᑦᑑᒐᓗᐊᖅ.
ᐅᓪᓗᒥ ᐃᑎᔪᒻᒪᕆᖕᒥᒃ
ᐃᓗᑐᔪᓕᐅᕆᐊᖃᓕᖅᑐᒍᑦ
ᓂᓪᓕᓇᖅᑐᒃᑯᕕᒃᓴᒥᒃ.”

—ᓄᐊᒪᓐ ᐊᑕᖓᓛᖅ

“ᐅᓪᓗᒥ ᐊᐳᑦ
ᐊᐅᒃᓵᓕᓂᖅᓴᐅᕙᓕᖅᑐᖅ,
ᑕᐃᔅᓱᒪᓂ ᐅᐱᕐᖔᒃᑯᑦ ᐊᐳᑦ
ᐊᐅᒃᐸᓪᓕᐊᔅᓲᔭᖅᐸᓚᐅᖅᓯᒪᔪᖅ,
ᓱᒃᑲᐃᑦᑐᒻᒪᕆᐅᓪᓗᓂ. ᐅᓪᓗᒥ
ᐊᐅᒃᓵᓕᓂᖅᓴᐅᕙᓕᖅᑐᖅ.”

—ᓴᐃᒪᓐ ᑐᑯᒻᒥᖅ

“ᐊᖑᓇᓱᒋᐊᓚᐅᖅᓯᒪᔪᖓ ᐊᕗᖓ
ᐃᒡᓗᐊᓄᖔᖅ [ᑕᓯᐅᑉ]. ᑐᒃᑐ
ᓂᕿᐊ ᓄᓇᒧᐊᕈᒪᓚᐅᖅᓯᒪᔭᕋ
ᐊᐃᓛᕋᒃᑯ. ᐅᕙᓐᓄᑦ ᐃᓱᒪᓪᓗᖓ
‘ᓄᓇᐅᑉ ᕐᑯᐊᖑᓂᖓ ᑕᐃᔅᓱᒪᓂᑐᑦ
ᖃᓂᑦᑐᒃᓴᐅᕗᖅ,’ ᑭᓯᐊᓂ
ᑕᐅᓄᖓᑲᓪᓚᒃ ᐳᐊᕆᔭᓚᐅᖅᓯᒪᔪᖓ
ᕐᑯᐊᖑᓂᖓᓄᑦ ᑎᑭᐅᑎᓇᓱᒃᖢᒍ.”

—ᑑᓇ ᐃᕐᑯᓕᒃ

“Around here we used to dig underground for food storage. We didn’t really have to dig deep to get the cold storage. [...] Today if you try to dig that deep you can’t even reach the frozen area anymore. [...] It’s just thawing. It never used to be like that; we never had to dig very deep to get cold storage. Today you have to dig deeper to get to the cold area.”

—Norman Attungala

“Today it seems like the snow melts faster than before, and in the past in the springtime the snow would be melting for a while, really slowly. Today it seems to be melting faster.”

—Simon Tookoome

“One time I went out hunting on the other side [of the lake].
I was going to put the caribou under the ground to save it.
I was saying to myself, ‘The frozen ground will be really close like always,’ but I was digging really far down to find the frozen ground.”

—Toona lqulik

“ᐅᑭᐊᒃᓵᖅ ᑕᖅᑭᓕᒫᕐᒥᒃ ᑭᖑᕙᖅᓯᒪᓕᖅᑐᖅ. ᓯᑯᔭᕌᖕᖓᑦ, ᐊᐳᑎᖃᙱᓐᓂᖕᖓᓗᑦ ᓯᑯ ᐃᔾᔪᔪᐊᓘᖅᑲᑦᑕᖅᑐᖅ. ᐅᐱᕐᖓᖕᒪᑦᑐ, ᐊᐳᑦ ᐊᐅᑲᐅᑎᒋᑦᑐᓂ, ᓱᒃᑲᔪᒻᒪᕆᐊᓗᖕᒥᒃ. ᑭᓯᐊᓂ ᓯᑯᖅᑯᑖᖕᓂᖅᓴᐅᕙᓕᖅᑐᖅ.”

—ᕖᓂ ᑕᔾᔭ ᐊᒻᒪ ᔭᐃᑲᑉ ᐃᑭᓂᓕᒃ

“The fall is late by about a month. When the ice comes, because there is no snow, it’s really thick. And when the spring comes, the snow melts just like that, fast. But the ice stays here for a long time.”

—Winnie and Jacob Ikinilik

“ᐅᖃᖅᑐᖃᖅᐸᒃᑐᖅ ᓯᓚ ᒫᓂ ᐅᖅᑰᓂᖅᓴᐅᓂᐊᓕᕐᓂᕐᖓᓂᒃ. ᑕᐃᒪᓐᓇ ᐅᖃᖅᑐᖃᖅᐸᒃᑐᖅ, ᑭᓯᐊᓂ ᓂᒡᓚᖅᐸᑦᓕᐊᔪᒻᒪᕆᐅᔪᖅ. ᐊᐅᔭᒃᑯᑦ ᐅᖅᑰᓂᖅᓴᐅᑦᑐᓂ, ᑭᓯᐊᓂ ᐅᑭᐅᒃᑯᑦ 60-ᖏᓐᓂ ᓂᒡᓚᓱᖕᓂᖅᑲᖅᐸᓚᐅᖅᓯᒪᙱᑦᑐᖅ. ᐅᑦᑐᒥ ᓂᒡᓚᓱᖕᓂᖕᖓ 50-ᖏᓐᓂ 60-ᖏᓐᓃᑉᐸᓕᖅᑐᖅ. ᑕᐃᒪᓐᓇ ᓂᒡᓚᓱᒃᑎᒋᕙᓚᐅᖅᓯᒪᙱᑦᑐᖅ. ᐊᐅᔭᒃᑯᑦᑐ ᐊᓄᕌᖕᒋᓐᓇᖅᑐᐊᓘᔮᓕᖅᑑᓂ. ᑕᐃᒪᙵᑦ ᐊᓄᕌᖕᒋᓐᓇᖅᑐᖅ. ᐅᑭᐅᒃᑯᑦᑐ ᓯᓚᑐᐃᓐᓇᐅᔭᓕᖅᑑᓂ, ᑭᓯᐊᓂ ᐊᐳᑎᖅᑲᑐᐊᕋᓂ.”

—ᕖᓂ ᑕᔾᔭ ᐊᒻᒪ ᔭᐃᑲᑉ ᐃᑭᓂᓕᒃ

“They say that the weather is going to get warmer up here. That is what they are saying, but it’s getting really cold up here. It’s hot in summertime, but in the wintertime it never used to get into the minus 60s. Today it comes up to the minus 60s or 50s. It never used to be that cold. And in the summertime it always seems to be windy nowadays. Never calm. Windy all the time. In the winter we always seem to have bad weather nowadays, but hardly any snow.”

—Winnie and Jacob Ikinilik

ᓯᓚ/ᐊᓄᕆ

“ᐃᒪᕐᓂᖅᓴᐅᓕᖅᑐᖅ. ᐃᑎᔪᐊᓘᕙᓚᐅᖅᑐᖅ ᐃᒃᑲᓐᓂᖅᓴᐅᓕᖅᑐᖅ. [...] ᐃᑖᓐᓂᒃᑯᑦᑐ ᐃᒃᑲᖅᑕᓐᓇᖅᐸᓕᖅᑑᓂ.”

—ᐲᓯᑦ ᐊᒻᒪ ᔫᓕ ᓴ’ᖑᓵᖅ ᑐᓗᖅᑐᖅ

Climate/Weather

“The water is lower. Used to be really deep and today it’s really shallow. [...] Sometimes you even hit a rock by boat.”

—Basil and Julie Tuluqtuq

Nunavut / Baker Lake

“ᒪᖅᑯᓕᓚᐅᖅᕗᓂ, ᒪᖅᑯᒍᓐᓃᖅᐸᓕᖅᑐᖅ. [...] ᓯᓚᓗ ᑐᖕᒍᔪᓚᐅᖅᓯᒪᔪᓐᓃᖅᕗᓂ. ᓄᕙᔭᐃᓐᓇᐅᔭᓕᖅᑐᖅ; ᓂᑦᑕᐃᓕᓚᐅᖅᓯᒪᔪᓐᓃᖅᑐᖅ. ᐊᓄᖀᖕᖏᓐᓇᐅᔭᖅᕗᓂᓗ. ᑕᐃᔅᓱᒪᓂ ᓄᕙᔭᓇᓂ, ᓯᓚᑦᑎᐊᕙᐅᓂᖅᓴᐅᕙᓚᐅᖅᑐᖅ. ᒪᖅᑯᔅᓲᔭᖅᐸᒃᕗᓂᓗ. ᒫᓐᓇ ᐅᑦᓗᒐᓴᓄᑦ ᒪᖅᑯᒃᐸᓕᖅᑐᖅ.”

—ᐲᓯᓪ ᐊᒻᒪ ᔫᓕ ᓴ’ᖕᒍᓵᖅ ᑐᓗᖅᑐᖅ

“It seems to rain, but it stops again. [...] [It] looks like the sky is never blue anymore. Always cloudy; no more clear days. The wind is always here. There used to be more clear, nice days. It would rain steadily. Now it will rain for quite a few days.”

—Basil and Julie Tuluqtuq

“ᐊᐱᓗᐊᖅᐸᒍᓐᓃᖅᑐᖅ. ᐃᒪᐃᑦ ᓄᓇᓗ ᐸᓂᓛᕿᓯᒋᓕᖅᑕᒃᑲ.”

—ᐲᓯᓪ ᐊᒻᒪ ᔫᓕ ᓴ’ᖕᒍᓵᖅ ᑐᓗᖅᑐᖅ

“Today we hardly get any snow. [I am] worried that water and the land are going to dry up.”

—Basil and Julie Tuluqtuq

“ᑕᐃᔅᓱᒪᓂ ᒪᖅᑯᓕᖀᖕᒐᑦ ᐊᓯᔾᔨᖅᑕᖅᐸᓚᐅᖅᓯᒪᙱᑦᑐᖅ, ᒪᖅᑯᐃᓐᓇᖅᐸᒃᕗᓂ, ᒪᖅᑯᓗᐊᖀᓗᒐᓂ. [...] ᒫᓐᓇᓕ ᒪᖅᑯᓕᖀᖕᒐᑦ ᓯᓚᓗᕐᔪᐊᖅᑐᐊᓘᕙᒃᑐᖅ. ᓱᒃᑲᔪᐊᓗᖕᒥᒃ ᒪᖅᑯᒃᑐᒻᒪᕆᐊᓘᕙᒃᑐᖅ, ᒪᖅᑯᒍᓐᓃᑳᑦᓚᒃᐸᒃᕗᓂᓗ. ᐊᔾᔨᒋᔪᓐᓃᖅᑕᖕᒐ. ᑲᑉᐱᐊᓇᖅᓯᔪᖅ ᑐᐱᖅᒦᑦᑕᕆᐊᒃᓴᖅ, ᐊᓄᖀᓕᖅᔪᐊᖅᑲᑦᑕᓕᖅᒪᑦ ᒪᖅᑯᖅᔪᐊᖀᓗᒃᐸᓕᖅᕗᓂᓗ.”

—ᐃᓕᓴᐱ ᐊᖅᓇᒐᖅᓂᖅ ᖁᐃᓇᖕᓇᖅ

“Before when it rained it would rain steadily, not changing, just rain, and not too much. [...] But today when it rains it’s a storm. Like rain really fast and very much, and then it stops. It’s not the same as before. It’s scary now to be in a tent, because the wind picks up and the rain is strong.”

—Elizabeth Quinangnaq

“ᑰᒐᓛᑦ ᑰᒍᓐᓃᖅᑐᖅ ᑕᓯᓪᓗ ᒥᑭᑦᑐᖅ ᐃᒪᖅᑲᕈᓐᓃᖅᑐᑦ.”

—ᐃᓕᓴᐱ ᐊᖅᓇᒐᖅᓂᖅ ᖁᐃᓇᖕᓇᖅ

“The small rivers are dry; even the small ponds are dry now.”

—Elizabeth Quinangnaq

“ᐅᓪᓗᒥ ᐊᐳᑎᖅᑲᖅᐸᒍᓐᓃᖅᑐᖅ. ᐊᓄᖓᑐᓗᐊᓕᖅᒪᑦ. ᐊᐳᑦ ᑎᓯᒃᑎᑦᑐᐊᓘᑲᐅᑎᕆᕙᓕᖅᑐᖅ. ᐊᐳᒻᒧᑦ ᐃᒡᓗᓕᕝᕕᒃᓴᐅᔪᓐᓃᖅᑐᖅ. ᐅᓪᓗᒥ ᒪᓂᕋᖅ ᑕᑯᒃᓴᐅᒻᒪᕆᒃᐸᓕᖅᑐᖅ. [...] ᐊᓄᖓᑐᓗᐊᖏᓐᓇᐅᔭᓕᖅᒪᑦ. ᑕᐃᔅᓯᒪᓂᑑᕈᓐᓃᖅᑐᖅ.”

—Hᐋᑎ ᐊᑐᖓᑦ ᐊᒻᒪ ᐹᓪ ᐊᑐᑐᕚ

“ᐊᐅᔭᒃᑯᑦ ᒪᖁᒃᐸᓚᐅᖅᑐᖅ, ᒪᖁᒐᔪᓗᐊᕋᓂ. ᒫᓐᓇ ᒪᖁᖅᔪᐊᖅᑑᐊᓘᓕᑳᓪᓚᒃᐸᓕᖅᑐᖅ ᓄᖅᑲᑳᓪᓚᒃᐸᒃᓱᓂᓗ. ᑐᑭᓯᐊᙳᐱᓪᓕᑎᐊᓕᖅᑕᕋ. ᓯᓚᕗᑦ ᐊᓯᔾᔨᖅᓯᒪᔫᒻᒪᕆᐅᓕᖅᑐᖅ. ᑕᐃᔅᓯᒪᓂᑑᕈᓐᓃᖅᑐᖅ. [...] ᐅᓪᓗᒥ ᒪᖁᖕᓂᖅᓴᐅᕙᓕᖅᑐᖅ, ᐊᓄᖐᖏᓐᓇᐅᔭᓕᖅᓱᓂ, ᐱᓗᐊᖅᑐᒥᒃ ᐅᑭᐊᒃᓵᒃᑯᑦ.”

—ᐊᐃᓖᓐ ᑕᕕᓂᖅ ᑲᓗᕋᖅ

“ᐃᒡᓗᐃᑦ ᐊᐳᑕᐃᓐᓇᐅᕙᓚᐅᖅᓯᒪᔪᑦ. ᐅᓪᓗᒥ ᖃᒪᓂ’ᑐᐊᕐᒥ ᐊᐱᓗᐊᖅᐸᒍᓐᓃᖅᑐᖅ.”

—ᔮᓂᑦ ᓂᐱ ᐃᑰᑕᖅ

“ᐃᒻᒪᖄ ᓯᓚᐅᑉ ᐊᓯᔾᔨᖅᐸᓪᓕᐊᓂᖓᓄᑦ, ᓯᓚ ᐅᖅᑰᓂᖅᓴᐅᓕᖅᓱᓂ ᑭᓯᓕᒫᓪᓗ ᐱᕈᖅᓂᖅᓴᐅᓕᖅᓱᑎᒃ. ᐃᕕᑐᐃᓐᓇᐅᙱᑦᑐᑦ, ᐃᔾᔪᖅᓘᓐᓃᑦ ᐊᓯᔾᔨᖅᓯᒪᓕᖅᑐᖅ ᐱᕈᖅᕕᐅᓂᖅᓴᐅᓕᖅᓱᓂᓗ.”

—ᔫᓐ ᐸᓂᖓᔭ’ᓈᖅ ᓯᑦᑲᑎ

“Today there’s just no snow. The wind is too strong. The snow gets really hard right away. There is no more snow for *igluit* [igloos]. Today you can even see the ground. [...] The wind is too strong all the time now. It’s not like before.”

—Hattie and Paul Attutuvaa

“In the summertime we would have rain, but not too often. Today when the storm comes it hits so fast and it clears up so fast. It doesn’t make sense to me anymore. The weather is so changed. It’s not like before. [...] Today it rains more. It’s always more windy, especially in the fall.”

—Irene Taviniq Kaluraq

“The houses used to be covered in snow. Today, living in Baker Lake, we hardly get the snow anymore.”

—Janet Ikuutaq

“Maybe because of climate change, the weather is warmer now and everything is just growing. Not only the grass and vegetation, but I think the soil itself is emerging and producing growth.”

—Joan Scottie

Nunavut / Baker Lake

“ᐅᓪᓗᒥ ᐃᒪᖅᑲᑦᑎᐊᕈᓐᓂᖅᑐᖅ ᓇᓂᑐᐃᓐᓇᖅ. ᖃᒪᓂ’ᑐᐊᕐᒥᓪ ᓅᓐᓂᑦ ᐃᒪᕆᓐᓂᖅᓴᐅᓕᖅᑐᖅ.”

—ᓘᓯ ᖃᐅᓐᓇᖅ

“ᐅᓪᓗᒥ ᐊᔾᔨᒋᔭᓐᓂᖅᑕᖓ. ᐅᓪᓗᒥ ᒪᖁᓕᖀᖕᒐᑦ, ᐱᖅᓯᖅᑐᐊᓘᔾᔭᖅᐸᒃᑐᖅ, ᑕᕝᕙᙳᒐᑦᑕᐃᓐᓇᖅ. ᑕᐃᒪᐃᑉᐸᓚᐅᖅᓯᒪᙱᑦᑑᒐᓗᐊᖅ. ᐅᓪᓗᒥ ᒪᖁᓕᖀᖕᒐᑦ ᒪᖁᖅᔪᐊᖅᑐᐊᓘᕙᒃᑐᖅ. ᓴᙱᔪᐊᓗᖕᒥᒃ ᐊᓄᕆᑐᔪᐊᓗᖕᒥᒃ.”

—ᓘᓯ ᑎᑭᖅ ᑐᖕᒍᐊᖅ

“ᑕᐃᔅᓯᒪᓂ, ᖁᕐᓗᖅᑐᖅ ᑐᓵᔭᓐᓇᖅᐸᓚᐅᖅᑕᕗᑦ, ᖃᒪᓂ’ᑐᐊᕐᒥᑦ ᐅᖓᓯᒃᑐᐊᓘᓗᐊᖅᑎᓪᓗᒍ. ᑭᓯᐊᓂ ᐃᒪᕆᓐᓂᖅᓴᐅᓕᕐᓂᖓᓄᑦ ᖁᕐᓗᖅᑐᖅ ᑐᓵᕙᒍᓐᓂᖅᑕᕗᑦ. [...] ᐅᕙᓐᓄᑦ, ᐃᓛᓐᓂᒃᑯᑦ ᐃᒃᑐᓚᐅᕐᕙᓗᒃᐸᓚᐅᖅᓯᒪᔪᖅ.”

—ᒫᑕ ᓇᓇᐅᑐᐊᖅ ᓄᑭᒃ

“ᑕᐃᔅᓯᒪᓂ ᑕᒪᐅᖓ ᓅᓵᖅᑕ ᐊᐳᑎᓪᓚᒪᕆᐅᓚᐅᖅᓯᒪᔪᖅ. ᑎᓯᔪᐊᓘᓇᓂ, ᒫᓐᓇᓕ ᖃᓐᓂᓵᖅᑐᕕᓂᐅᒐᓗᐊᖅᑎᓪᓗᒍ ᑎᓯᔪᐊᓘᕙᓕᖅᑐᖅ. [...] ᑕᓯᕐᒥ ᕿᒧᒡᔭᓪᓚᒪᕆᐅᕙᓚᐅᖅᓯᒪᔪᖅ, ᒫᓐᓇ ᐱᑕᖅᑲᓗᐊᖅᐸᒌᓐᓂᖅᑐᑦ, ᐊᐳᑎᖅᑲᙱᓗᐊᓕᕐᒪᑦ. ᐅᔾᔨᕆᔭᕋ ᐊᐳᑎᖃᓗᐊᖅᐸᒌᓐᓂᖅᑐᒋᑦ, ᓄᓇᓕᑦᑕ ᓯᓚᑖᓂᓪ ᓅᓐᓂᑦ.”

—ᒫᑕ ᓇᓇᐅᑐᐊᖅ ᓄᑭᒃ

“Today there is hardly any water anywhere. Even in Baker Lake, the water level is low.”

—Lucy Kownak

“So different today. Today when we get rain, it’s like a storm, just like that. This didn’t happen before. Today when it rains, it rains so much and so hard. So strong and so windy.”

—Lucy Tunguaq

“Years ago, we used to hear the Kazan Falls, [even though they] are very far from Baker. But because of the water level going down, we don’t hear it anymore. [...] To me, sometimes it sounded like thundering.”

—Martha Nukik

“Years ago, just when we first moved here, there was a lot of snow. It wasn’t really hard, but now even after a snowfall the snow is really hard. [...] There used to be a lot of [snowdrifts] on the lake over there, but there is hardly anything anymore, because there is not enough snow. I have noticed we hardly get snow anymore, even outside of town.”

—Martha Nukik

“ᑕᐃᔅᓯᒪᓂ, ᐸᐅᕐᖓᑦ ᐱᕈᖅᑐᑦᑐ
ᐱᕈᑦᑎᐊᖅᐸᒃᑎᑦᑐᒋᑦ,
ᒪᖁᒃᐸᓚᐅᖅᓯᒪᔪᖅ—
ᐊᒃᓱᕈᐊᕌᕉᖖᒋᑦᑐᖅ, ᐃᓛᓐᓂ
ᒥᓂᑎᑦᑐᒍ ᐃᒃᐱᙵᖅᐸᓚᐅᖅᓯᒪᔪᖅ
ᓄᓇᑐ ᒪᓯᐅᑦᑐᓂ. ᑭᓯᐊᓂ ᐅᑦᑐᓂ
ᒪᖁᓕᕌᖓᑦ, ᐊᓄᕌᓯᑳᑦᓚᒃᑐᐊᕉᕗᒃᑐᖅ.
ᑕᒪᔅᓴ ᓯᓚᐅᑉ ᐊᓯᔅᓴᖅᐸᑦᓚᐊᓂᖓ,
ᐃᒃᐱᒋᔭᕋ.”

—ᓄᐊᒪᓐ ᐊᑕᖓᓛᖅ

“ᐃᓯᒫᖑᑎᑐᐊᕋ
ᐊᐳᑎᖃᕈᓐᓃᖅᐸᑦᓚᐊᑐᐃᓐᓇᕈᑦᑕ,
ᐃᒪᑭᓐᓂᖅᓴᐅᓚᕐᓂᐊᖅᑐᒍᑦ. [...]
ᐅᕙᓐᓄᑦ ᐃᓯᒫᑐᑐᐊᖖᒋᑦᑐᖓ ᑭᓯᐊᓂ
ᑭᖑᕚᕆᓂᐊᖅᑕᒃᑲ ᐃᓯᒫᑐᒍᑎᕙᒃᑲ.”

—ᓴᐃᒪᓐ ᑐᑯᒥᖅ

“ᐃᒃᑲᑦᑐᒻᒪᕆᐅᓕᖅᑐᖅ. ᖁᖕᒥ
ᕿᑭᖅᑕᒐᓛᖅᑲᖅᐸᓚᐅᖅᓯᒪᖖᒋᑦᑐᖅ.
ᒫᓐᓇ ᕿᑭᖅᑕᒐᓛᖅᑲᓕᖅᑐᖅ. ᐅᔭᖅᑲᑦ
ᑕᑯᔭᒃᓴᐅᓕᖅᑐᑦ. ᓇᖏᐊᕐᓇᖅᓯᔪᖅ.”

—ᕚᐅᕋ ᐊᒃᕼᐊᒃ ᐊᕚᓚ

“Years ago, when berries or any plants were growing properly, there would be rain—not hard, pouring rain, but sometimes in the misty days you could feel the mist when you were outside and the ground was wet. But today when it rains, suddenly it becomes windy just like that. That’s the changes in the climate, I feel.”

—Norman Attungala

“The only concern I have is if someday we have less and less snow, we will have less water. [...] Although I’m not really concerned for myself, but for my descendants.”

—Simon Tookoome

“It’s getting shallow. There used to be no small islands along the Thelon River. There are some now. You can see rocks today. That’s scary.”

—Vera Avalaa

“ᖁᓕᒥᒨᓕᒃᑯᑦ ᐱᕈᖅᓴᕕᒋᓯᒪᔭᓐᓄᐊᓚᐅᖅᑐᖓ. ᐃᒪᖓ ᑲᑭᐊᑦᑎᐊᖅᐸᓚᐅᖅᑐᖅ, ᑭᓯᐊᓂ ᑕᒪᑐᒪᓂ ᐊᐅᔭᕐᒥ ᐊᔾᔨᒋᓚᐅᙱᑕᖓ; ᖃᐳᐊᖃᖅᑰᔮᓚᐅᖅᑐᖅ. ᑕᒫᓂ ᖃᒪᓂʼᑐᐊᕐᒥ ᐊᓯᔾᔨᖅᓯᒪᓂᖓ ᑕᑯᓯᒪᔭᕋ. ᐸᓂᕐᓂᖅᓴᐅᓕᖅᑐᖅ. ᐅᑭᐅᒃᑯᑦ ᐃᑎᕕᐊᓄᐊᓚᐅᖅᑐᖓ, ᐃᓅᓕᕐᕕᓐᓄᑦ. ᑕᐅᕙᓂ ᐊᐅᒃᑲᕐᓂᖃᐃᓐᓇᖅᐸᓚᐅᖅᑐᖅ. ᓲᕐᓗ ᑰᖑᑦᓗᓂ, ᐅᑭᐅᖑᒐᓗᐊᒃᑯᑦ. ᐃᒻᒪᑦᑎᐊᖅ ᑕᐅᕗᖓ ᐅᑎᓚᐅᖅᑐᖓ ᑕᑯᔭᖅᑐᖅᓱᒍ. ᓱᓕ ᐊᐅᒃᑲᕐᓂᐅᒐᓗᐊᖅ, ᑭᓯᐊᓂ ᐃᒪᑭᓐᓂᖅᓴᒻᒪᕆᐅᓕᖅᑐᖅ.”

—ᑑᓇ ᐃᖁᓕᒃ

“I went up by helicopter to where I come from. The water used to be clear, but this summer it was different; it looked like there were bubbles or something. Here around Baker Lake I've seen the difference. It's more dry. I went to the other side in the wintertime, where I was born. There is open water all year round. It used to run like a river, even in the wintertime. A few years ago I went back there to see it. It was still open, but the water was really low.”

—Toona Iqulik

ᐃᓄᐃᑦ ᐱᖁᓯᑐᖃᖓᑦᑕ ᐊᒃᑐᖅᑕᐅᓂᖓ

“ᐊᐳᑦ ᐃᒡᓗᓕᔾᔪᑎᒋᔭᒃᓴᐅᔪᓐᓃᖅᑐᖅ.”

—ᐃᓕᓴᐱ ᐊᕐᓇᒐᕐᓂᖅ ᖁᐃᓇᖕᓇᖅ

“ᓯᓚᐅᑉ ᐊᓯᔾᔨᓕᕐᓂᖓ ᐅᔾᔨᕆᔭᕋ. ᐃᓄᑐᖃᐃᑦᑕᐅᖅ ᐅᔾᔨᕈᓱᒃᓯᒪᖕᒥᔪᑦ. ᑕᐃᔅᓯᒪᓂ ᓯᓚ ᖃᓄᐃᑦᑑᓂᐊᕐᒪᖔᑦ ᖃᐅᔨᒪᖃᑦᑕᓚᐅᖅᓯᒪᔪᒍᑦ ᓄᕗᔭᑦ ᑕᑯᑐᐃᓐᓇᖅᖢᒋᑦ, ᑭᓯᐊᓂ ᐅᓪᓗᒥ ᓄᕗᔭᑦ ᑕᑯᓪᓗᒋᑦ ᐃᓱᒪᒃᑐᒍᑦ, ‘ᖃᐅᒃᐸᑦ ᓯᓚᑦᑎᐊᕆᐅᓂᐊᖅᑐᖅ,’ ᑭᓯᐊᓂ ᓯᓚᑦᑎᐊᕆᐅᓪᓚᕆᓚᐅᖅᓯᒪᙱᑦᑐᖅ. ᒫᓐᓇ ᐊᔾᔨᒋᔪᓐᓃᖅᑕᖓ;

Impacts on Traditional Ways of Life

“There's no snow for building *igluit* [igloos].”

—Elizabeth Quinangnaq

“I see changes in the weather. It's like the rest of the Elders have seen. A long time ago we could tell what the weather was going to be just by looking at the clouds, but today when we look at the clouds we say, ‘It's going to be nice out tomorrow,’ but it's never really nice out. It's so

ᖃᓄᐃᑦᑑᓂᐊᖅᒪᖕᒑᑦ ᖃᐅᔨᒪᕙᒍᓐᓃᖅᑐᒍᑦ. ᑕᐃᔅᓯᒪᓂ, ᓄᓇᒥᐅᑕᐅᑎᓪᓗᑕ, ᖃᐅᔨᒪᕙᓚᐅᖅᑐᒍᑦ ᑕᖅᑭᓕᒫᖅ ᓯᓚᖕᒐ ᖃᓄᐃᑦᑑᓂᐊᖅᒪᖕᒑᑦ ᕿᒥᕐᕈᓪᓗᒋᑦ ᑕᖅᑭᖅ ᓄᕗᔭᓪᓗ. [...] ᑕᐃᔅᓯᒪᓂ ᐃᓄᑐᖃᐃᑦ ᐅᖃᖅᐸᓚᐅᖅᓯᒪᔪᑦ ᓯᓚᑦᑎᐊᕙᐅᓂᐊᓕᖅᓂᖕᒐᓂᒃ, ᐊᒻᒪ ᑐᒃᑐᖃᑦᑎᐊᖅᓂᐊᖅᓂᖕᒐᓂᒃ ᓂᖅᔪᑎᖃᑦᑎᐊᖅᓂᐊᖅᓂᖕᒐᓂᒃ ᑕᖅᑭᕆᓂᐊᖅᑕᑦᑎᓐᓂ. [...] ᑕᖅᑭᐅᑉ ᐅᓪᓗᖅ ᓯᕗᓪᓕᖅᐹᖕᒐᓂ ᑕᖅᑭᖅ ᑕᑯᓪᓗᒍ ᓯᓚ ᖃᓄᐃᑦᑑᓂᐊᖅᒪᖕᒑᑦ ᖃᐅᔨᒪᕙᓚᐅᖅᑐᒍᑦ. [...] ᑕᐃᔅᓯᒪᓂ ᐊᑐᖅᐸᓚᐅᖅᓯᒪᔭᕗᑦ, ᒫᓐᓇ ᐊᑑᑎᖃᕈᓐᓃᖅᑐᖅ.”

—ᐊᐃᕇᓐ ᑕᕕᓂᖅ ᑲᓗᕋᖅ

“ᐅᒥᐊᖅᑐᖅᐸᓚᐅᖅᓯᒪᔪᒍᑦ ᐊᕙᒻᒧᑦ ᑕᓯᖅᒥ. ᑕᐃᒪᐃᑦᑐᓐᓇᐃᓪᓕᔪᑦ. ᐃᓛᓐᓂ ᐃᒪᖃᖖᒋᑦᑎᐊᒻᒪᕆᒃᐸᓕᖅᑐᖅ!”

—ᔫᓐ ᐸᓂᖕᒐᔭ’ᓈᖅ ᓯᑦᑲᑎ

“ᐊᐳᑦ ᑎᓯᔪᒻᒪᕆᐊᓘᓕᖅᑐᖅ, ᐊᐳᒧᑦ ᐃᒡᓗᓕᒡᕕᒃᓴᐅᔪᓐᓃᖅᑐᖅ. [...] ᐅᓪᓗᒥ ᐊᐳᑦ ᑎᓯᒃᑎᑲᐅᑎᒋᓕᖅᑐᖅ, ᓄᓇᒥᐅᑕᐅᑎᓪᓗᑕ, ᐊᐳᑦ ᐃᒡᓗᓕᔅᔨᑎᑦᑎᐊᕙᐊᕙᓚᐅᖅᑐᖅ. ᐅᓪᓗᒥ ᐱᑕᖃᕈᓐᓃᖅᑐᖅ. ᐅᓪᓗᒥ ᐊᐳᑎᖃᕈᓐᓃᖅᑐᖅ. ᐱᑕᖃᖅᑰᔨᔪᓐᓃᖅᑐᖅ—ᐃᒡᓗᐃᑦ ᓯᓚᑏᓐᓇᖕᒋᓐᓂ ᐊᐳᑎᖃᓕᖅᑐᖅ, ᓄᓇᒥᐅᖖᒋᑦᑐᖅ ᓯᑯᒥᐅᖖᒋᑦᑐᖅ. ᐊᐳᑎᖃᑦᑎᐊᖅᐸᓚᐅᖅᓯᒪᔪᖅ ᐃᖃᓪᓕᐊᖅᕕᑦᑎᐊᕙᐅᓪᓗᓂ; ᐅᓪᓗᒥ ᐱᑕᖃᕈᓐᓃᖅᑐᖅ.”

—ᓘᓯ ᑎᑭᖅ ᑐᖕᒍᐊᖅ

different now; we can’t really tell what it’s going to be like anymore. A long time ago, when we used to live on the land, we could tell how the month would go by looking at the moon or clouds. [...] A long time ago the old people would say that [we will have] really good weather, or that there will be a lot of caribou or animals for the next month. [...] I look at the first day of the month’s moon to tell what the weather will be like. [...] It used to work a long time ago, but it’s not working anymore.”

—Irene Taviniq Kaluraq

“We used to go boating back and forth across the lake. You can’t do that anymore. Sometimes there isn’t even water there!”

—Joan Scottie

“The snow is really hard, snow that is not good for making *igluit*. [...] Today it gets too hard right away, and when we used to be out there, there was good snow for *igluit*. Today there is nothing. Today the snow is gone. It looks like it’s gone—it’s just around the houses, not on the land or ice. There used to be good snow for fishing; today there is nothing.”

—Lucy Tunguaq

“ᐅᔾᔨᕆᓱᖅᑲᑦᑕᖅᑐᖕᒐ ᐃᖅᑲᓪᓚᐊᕌᖕᒐᒪ ᐊᒡᓗᐊᓕᐅᖅᖢᖕᒐ ᑕᓯᕐᒥ, ᐊᕐᒪᕌᖕᒐᑦ ᐃᒪᖅ ᓱᒃᑲᓪᓕᔪᐊᓘᑦᓗᓂ ᐃᒻᒥᖅᐸᓪᓕᐊᕙᓕᖅᑐᖅ, ᐊᒡᓛᑦ ᓯᑯᑉ ᖄᖕᒐᓄᑦ. ᑕᐃᒪᐃᑉᐸᓕᖅᑐᖅ ᒫᓐᓇ.”

—ᒫᒍᓇᑦ ᑐᓪᓕᒃ ᓂᕕᐊᑦᓯᐊᖅ

“ᐊᕙᓂ ᐊᑭᐊᓂ ᕿᑭᖅᑕᖅᑕᖅᑲᖅᑐᖅ. ᐃᒪᓕᐊᓘᕙᓚᐅᖅᓯᒪᔪᖅ ᐅᒥᐊᓄᑦ ᐃᑳᖅᕕᒃᓴᐅᓪᓗᓂ ᒫᖖᒐᑦ ᑕᐅᕗᖕᒐ, ᑭᓯᐊᓂ ᒫᓇ ᐃᒪᖅᑭᓐᓃᖅᑐᖅ, ᒫᓇ ᐅᓄᖕᒐᐅᔮᓗᒃ ᕿᑭᖅᑕᑦ ᐅᖕᒐᑖᒍᑦ ᐃᑳᓇᐊᖅᑲᖅᐸᓕᖅᑐᑦ.”

—ᒫᑕ ᓇᓇᐅᑐᐊᖅ ᓄᑭᒃ

“ᑕᐃᔅᓯᒪᓂ, ᑕᐃᔅᓯᒪᓂᓗᐊᕐᔪᖖᒋᑦᑐᖅ, ᒥᑦᑕᖅᕕᐅᑉ ᓴᓂᐊᓂ ᑎᕐᒐᐅᔭᓂᒃ, ᒪᓂᖅᓂᒃ, ᕿᔪᒃᑖᓂᒡᓗ ᓯᓚᒥ ᐃᒐᔾᔪᑎᒃᓴᓂᒃ ᐊᕗᖅᐸᓚᐅᖅᓯᒪᔪᒍᑦ. ᑕᐅᕗᖖᒐᐅᕆᐊᓚᐅᖅᑐᖕᒐ ᐅᖕᒐᓯᓗᐊᖖᒋᑦᑐᒧᑦ, ᑭᓯᐊᓂ ᑕᐅᕙᓂ ᑭᓱᖅᑲᕐᓃᖅᑐᖅ.”

—ᒫᑕ ᓇᓇᐅᑐᐊᖅ ᓄᑭᒃ

“I notice when I go out fishing and make a hole in the lake, when it’s open the water starts to come up so fast, even over the ice. That’s how it usually is now.”

—Margaret Amauruq Niviatsiaq

“There is an island just across from here. There used to be a lot of water for the boats to cross from here to there, but there is no more water, so you have to go way out there to go behind the islands and across.”

—Martha Nukik

“Years ago, not too long ago, I used to go along just by the airport to collect lichens, moss, or those bushes to gather for a fire if I was cooking outside. I tried to go to where it was not too far, but there is nothing around there anymore.”

—Martha Nukik

“ᖃᐅᔨᒪᔪᖓ ᓂᓚᓕᒃᑕᖃᖅᐸᓚᐅᖅᓯᒪᖕᒪᑦ ᐊᖏᔪᐊᓗᖕᒥᒃ ᐊᐅᓚᐅᖅᓯᒪᙱᑦᑐᒥᒃ. ᐱᑕᖃᕈᓐᓃᖅᑐᖅ. ᑕᐃᔭᐅᔪᖅ ᓂᓚᓕᖕᒥᒃ. ᐊᖑᓇᓱᒃᑎᓄᑦ ᐅᖃᐅᓯᐅᖏᓐᓇᖅᐸᓚᐅᖅᑐᖅ, ᖃᐅᔨᒪᕙᓚᐅᕐᒪᑕ ᓂᓚᖃᕐᓂᖓᓂᒃ, ᐅᖅᑰᓗᐊᕆᖀᖕᒐᓪᓗ ᑐᒃᑐᑦ ᑕᐅᕙᓂᓯᐅᖅᐸᓚᐅᖅᑐᑦ. ᑕᐅᕗᖓ ᑐᒃᑐᓕᐊᖅᐸᓚᐅᖅᓯᒪᔪᑦ. [...] ᐊᖏᔪᐊᓘᓚᐅᖅᓯᒪᔪᖅ ᐊᐅᒍᓐᓇᖏᓐᓂᕋᖅᑕᐅᓪᓗᓂ, ᑭᓯᐊᓂ ᐱᑕᖃᕈᓐᓃᖅᑐᖅ. ᑐᒃᑐᓄᑦ ᓂᒡᓚᖅᓯᕐᕕᐅᕙᓚᐅᖅᓯᒪᔪᖅ.”

—ᓄᐊᒪᓐ ᐊᑕᖕᒑᖅ

“I know about a huge ice that never used to melt. It’s not there anymore. It’s called Nilalik. Hunters talked about that all the time, because they knew there would be ice there, and when it was too hot the caribou would linger around that area. That’s where they would go to catch the caribou. [...] It was so huge they said it would not melt, but it’s not there anymore. The caribou would use it to keep themselves cool.”

—Norman Attungala

“ᓇᓂᑐᐃᓐᓇᖅ ᐃᒡᓗᓕᕐᕕᒃᓴᐅᔪᓐᓃᖅᑐᖅ, ᐱᓗᐊᖅᑐᒥᒃ ᐃᓕᓐᓂᐊᖅᑐᓂᒃ ᐃᓕᓐᓂᐊᖅᑎᑦᑎᓇᓱᒃᕈᓂ ᐃᒡᓗᓕᓂᕐᒥᒃ ᐅᑭᐅᒃᑯᑦ. ᐅᓪᓗᒥ ᕿᓂᒻᒪᕆᒡᓗᓂ ᑭᓯᐊᓂ ᐃᒡᓗᓕᕐᕕᒃᓴᒥᒃ ᐃᒡᓗᓕᓐᓇᖅᓯᔪᖅ ᐊᐳᑦ ᑎᓯᔪᐊᓘᓕᕐᒪᑦ ᐃᒡᓗᓕᔭᓇᐊᒃᓴᖅ ᐃᓚᖏᓪᓗ ᐊᕿᓗᐊᖅᕈᑎᒃ, ᓂᕈᐊᑐᐃᓐᓇᕐᓗᓂ ᐃᒡᓗᓕᕐᕕᒃᓴᐅᔪᓐᓃᖅᑐᖅ.”

—ᓄᐊᒪᓐ ᐊᑕᖕᒑᖅ

“You can’t just go anywhere and make an *iglu* anymore, especially when you’re trying to teach students how to build one in the wintertime. Today you have to look really hard for where to build one because the snow is harder and part of it is too soft, and you can’t just pick any spot and build an *iglu* anymore.”

—Norman Attungala

“ᑕᐃᔅᓯᒪᓂ, ᐃᓚᓖᑦ ᐊᐅᓪᓛᖅᓯᒪᒃᐸᑕ
[...] ᐃᓱᒫᓗᒃᐸᓚᐅᖅᓯᒪᙱᑦᑐᑦ
ᐊᐳᑦ ᐊᐅᑲᐅᖅᑐᕐᓂᐊᕆᐊᒃᓴᖓ,
ᑭᓯᐊᓂ ᐅᓪᓗᒥ, ᐃᒻᒪᖄ
ᐊᐳᑦ ᑎᓯᓗᐊᓕᕐᓂᖓᓄᑦ,
ᐊᐅᑲᐅᖅᑐᖅᐸᓕᖅᑐᖅ. [...] ᑕᒪᓐᓇ
ᐃᓱᒫᓗᓇᖅᓯᔪᖅ ᐊᐅᓪᓛᖅᓯᒪᓪᓗᓂ.
ᑕᖅᑭᕐᒥ ᖁᕕᐊᓱᒃᐸᓚᐅᖅᑐᑦ
ᐊᐅᓪᓛᖅᓯᒪᔪᑦ; ᓯᓚᑦᑎᐊᕙᐅᑎᓪᓗᒍ
ᐊᐅᓪᓛᖅᐸᓚᐅᖅᑐᑦ ᐊᐅᓂᐅᔪᓄᑦ.
ᑐᐱᖅᐸᓚᐅᖅᑐᑦ ᐊᐳᑎᖃᙱᓐᓂᕐᒥ,
ᑭᓯᐊᓂ ᐅᓪᓗᒥ ᐊᐅᓪᓛᖅᓯᒪᔭᕆᐊᒃᓴᖅ
ᐃᓗᐊᕈᓐᓃᖅᑐᖅ [...] ᐊᐳᑦ
ᐊᐅᑲᐅᖅᑐᖅᐸᓕᕐᒪᑦ ᐅᑎᕈᓯᕆᐊᒃᓴᖅ
ᓄᓇᓕᖕᒍᑦ ᐊᑲᐅᖅᑲᑦᑕᕈᓐᓃᖅᑐᖅ.”

—ᓄᐊᒪᓐ ᐊᑕᖕᒐᓛᖅ

“1940-ᖏᓐᓂ ᖃᒪᓂ’ᑐᐊᕐᒧᐊᖅᑎᓪᓗᑕ
ᓯᒡᔭᖅᐸᓯᖕᒥ ᐃᒡᓗᓕᔪᓐᓇᓚᐅᖅᓯᒪᔪᒍᑦ,
ᐃᒡᓗᓕᔭᕆᐊᒃᓴᖅ
ᐊᐳᑎᖃᑦᑎᐊᓚᐅᖅᓯᒪᖕᒪᑦ. ᐅᓪᓗᒥ
ᐱᑕᖃᕈᓐᓃᖅᑐᖅ. ᐃᑉᐸᒃᓵᓂᑦᑎᐊᖅ
ᑕᑉᐸᐅᙵᐅᓚᐅᖅᑐᖓ
ᐃᒡᓗᓕᔪᒪᓪᓗᖓ ᐱᓇᓱᐊᕈᓯᐅᑉ
ᓄᙳᐊᓂ ᐃᖃᓪᓕᐊᖅᑕᕈᒪᓪᓗᖓ,
ᑭᓯᐊᓂ ᐃᒡᓗᓕᐊᓂᒎᓇᓚᐅᙱᑦᑐᖓ
ᐊᐳᑎᖃᙱᓗᐊᕐᒪᑦ.”

—ᓴᐃᓚᔅ ᐊᐃᑦᑕᐅᖅ

“ᐅᖅᑯᐊᓕᐅᖅᐸᓚᐅᖅᓯᒪᔪᖅ ᑕᓯᕐᒧᑦ
ᐃᖅᑲᓪᓕᐊᖄᖕᒐᑦᑕ, ᑭᓯᐊᓂ ᐅᓪᓗᒥ
ᑕᓯᕐᒥ ᐊᐳᑎᖅᐸᒍᓐᓃᖅᑐᖅ. ᐅᓪᓗᒥ
ᐱᖅᓯᓕᖄᖕᒐᑦ, ᐱᖅᓯᖅᑐᒻᒪᕆᐊᓘᕙᒃᑐᖅ,
ᐊᐳᑦ ᑎᒃᑕᐅᕙᒃᖢᓂ. ᓯᓚ
ᐊᓯᔾᔨᖅᓯᒪᔪᒻᒪᕆᐅᓕᖅᑐᖅ. ᐊᒡᓛᑦ
ᑕᖅᑭᖅ ᐊᓯᔾᔨᖅᓯᒪᓕᖅᑐᖅ.”

—ᑖᒪᔅ ᖃᕐᑭᒪᑦ

“Years ago, if a family was out on the land [...] they never used to be worried about the snow melting right away, but today, maybe because the snow is too hard, it melts just like that. [...] You have to be worried about that while you’re out of town. In the same month they used to be happy being out camping; on a nice day they would be travelling and camping where the snow had melted. [They would] pitch the tent where there was no snow, but today it makes you uneasy being out of town [...] because the snow melts just like that and you can have problems getting back to town.”

—Norman Attungala

“In the 1940s when we went to Baker Lake we would make *igluit* along the shore, because there was enough snow to make *igluit*. Today there is nothing. Just a few days ago I went up there on the lake to make an *iglu* so I could go fishing up there on the weekends, but I couldn’t finish the *iglu* because there was hardly any snow.”

—Silas Aittauq

“ᐋᒡᒋᓯᒥ ᒪᕐᑯᖕᓂᖅᓴᐅᕙᓕᖅᑐᖅ, ᓄᓇᒥᐅᑕᐅᑎᑦᑐᑕ ᐊᒥᕐᓂᒃ ᐊᓐᓄᕌᒃᓴᓂᒃ ᐸᓂᖅᓲᕙᓚᐅᖅᓯᒪᔪᒍᑦ ᐸᓂᑦᑎᐊᖅᐸᓚᐅᖅᕈᑎᒡᓗ. ᐅᑦᓗᒥ ᐊᒥᕐᓂᒃ ᐸᓂᖅᓯᑦᑎᐊᕈᓐᓇᐃᑦᓕᔪᒍᑦ ᒪᓴᐅᖏᓐᓇᐅᔭᓕᕐᒪᑦ.”

—ᕕᐅᕋ ᐊᒃᕼᐊᒃ ᐊᕚᓚ

“ᑕᐃᔅᓱᒪᓂ, ᓄᓇᒥᐅᑕᐅᑎᑦᑐᑕ, ᐅᖃᕈᓐᓇᖅᐸᓚᐅᖅᓯᒪᔪᒍᑦ ᖃᐅᒃᐸᑦ ᓯᓚᑦᓴᐅᓂᐊᕐᓂᖓᓂᒃ ᐅᒡᕙᓘᓐᓃᑦ ᐅᑦᓗᒥ ᓯᓚᑦᑎᐊᕙᐅᔾᔫᖏᓐᓂᖓᓂᒃ, ᑭᓯᐊᓂ ᐅᑦᓗᒥ ᖃᐅᔨᒪᔪᓐᓇᐃᑦᓕᔪᒍᑦ ᓯᓚᑦᓴᐅᓂᐊᕋᓱᒋᔭᕋᓗᐊᕗᑦ ᓯᓚᓗᒃᐸᓕᕐᒪᑦ. ᓯᓚᓗᖕᓂᐊᕋᓱᒋᔭᕗᑦᓗ ᓯᓚᑦᓴᐅᕙᓕᖅᕈᓂ. ᑐᑭᓯᓇᕈᓐᓃᖅᑐᖅ. ᓄᕗᔭᑎᒍᑦ ᖃᐅᔨᔪᓐᓇᖅᐸᓚᐅᖅᑐᒍᑦ ᖃᐅᔨᒪᔪᓐᓇᐃᑦᓕᔪᒍᑦ.”

—ᕕᓂ ᑕᔾᔭ ᐊᒻᒪ ᔭᐃᑲᑉ ᐃᑭᓂᓕᒃ

“We used to make shelters when out fishing on the ice, but today there is no snow on the lakes. Today when the storm comes, it hits really hard, so the snow doesn’t stay. The weather has changed a lot. Even the moon has changed.”

—Thomas Qaqimat

“In August we get more rain, because when we used to live on the land we dried skins for clothing and they dried good. Today we can’t even seem to dry the skins good because it’s always wet.”

—Vera Avalaa

“Before, when we lived on the land, we used to say it would be a nice day tomorrow or it’s going to be bad weather today, but today we can’t tell anymore because when we think it’s going to be nice it’s not nice. When we think it’s going to be bad weather, it’s good. It doesn’t make sense. We used to tell by the clouds, but today we can’t tell.”

—Winnie and Jacob Ikinilik

ᒥᑦᑎᒪᑕᓕᒃ
Pond Inlet

ᒥᑦᑎᒪᑕᓕᒃ

ᒥᑦᑎᒪᑕᓕᒃ ᕿᑭᕐᑲᑖᓗᖕᒥ ᖁᑦᑎᒃᑐᒦᑦᑐᖅ, ᓯᕐᒥᓕᐅᑉ ᑕᓪᓗᕈᑎᐅᑉ ᒥᒃᓵᓃᑦᑐᖅ (N 72.70°, W 77.96°; ᖁᑦᑎᖕᓂᖕᖓ 55 m). ᓄᓇᓕᖕᖓ ᖁᑦᑎᒃᑐᕐᒥᐅᑕᐅᓂᕋᕐᑕᐅᔪᖅ, ᐅᕐᐱᐊᒑᓂᒃ ᐱᕈᕐᑐᕐᑲᕐᓱᓂ ᓄᓇᐅᓪᓗ ᐃᑭᐊᖕᖓ ᖁᐊᖑᐃᓐᓇᕐᓱᓂ. 1961-ᒥ ᓄᓇᓕᖖᒍᓚᐅᕐᓯᒪᔪᖅ 1,549-ᓂᓪᓗ ᐃᓄᒋᐊᒃᑎᒋᓚᐅᕐᓱᓂ 2011-ᒥ ᑲᓇᑕᒥ ᑭᓪᓕᓯᓂᐊᕐᑕᐅᔪᓂᑦ.

ᐊᖏᕐᑲᑎᒌᓗᐊᕐᑲᑦᑕᓚᐅᖖᒋᑦᑐᑦ ᐃᓄᑐᕐᑲᐃᑦ ᒥᑦᑎᒪᑕᓕᖕᒥ ᐱᕈᕐᑐᑦ ᐊᓯᔾᔨᕐᐸᓪᓕᐊᓂᖏᓐᓂᒃ; ᑭᓯᐊᓂ ᐃᓚᖏᑦ ᐊᓯᔾᔨᕐᓃᑦ ᐅᔾᔨᕆᔭᐅᓯᒪᔪᑦ, ᓲᕐᓗ ᐱᕈᕐᑐᕐᑲᕐᓂᕐᓴᐅᓕᕐᓂᖕᖓᓂᒃ (ᓲᕐᓗ ᐃᕖᑦ) ᐱᕈᐊᓂᒃᓵᓕᓂᕐᓴᐅᕙᓕᕐᓂᖏᓐᓂᓪᓗ. ᐃᓄᑐᕐᑲᐃᑦ ᐅᕐᑲᓚᐅᕐᑐᑦ ᑭᒍᑕᖏᕐᓇᕐᑲᕐᓂᕐᓴᐅᓕᕐᓂᖕᖓᓂᒃ, ᐸᐅᕐᖓᕐᑲᕐᓂᕐᓴᐅᕙᓕᕐᓂᖕᖓᓂᓪᓗ. ᐃᓚᖏᑦ ᐊᐱᕐᓲᕐᑕᐅᔪᑦ ᑕᐃᒪᐃᔾᔪᑎᕐᑲᕐᓂᕋᐃᓚᐅᕐᐳᑦ ᑐᒃᑐᕐᑲᖏᓐᓂᕐᓴᐅᓕᕐᓂᖕᖓᓄᑦ ᐸᐅᕐᖓᑦ ᓂᕆᔭᐅᖏᓐᓂᕐᓴᐅᓕᕐᓂᖏᓐᓄᑦ, ᐅᕝᕙᓘᓐᓃᑦ ᐊᓂᐅᕙᕐᑲᖏᓐᓂᕐᓴᐅᓕᕐᓂᖕᖓᓄᑦ ᐃᔾᔪᕐᓄᑦ ᓂᓪᓚᕈᑕᐅᕙᒃᑐᓂᒃ. ᑲᖕᒎᔾᔭᐃᑦ ᐅᓄᕐᓂᕐᓴᐅᕙᕐᑰᔨᓕᕐᑐᑦ, ᓄᑖᓂᓪᓗ ᐱᕈᕐᑐᕐᑲᓕᕐᐸᓪᓕᐊᓪᓗᓂ, ᓲᕐᓗ ᖁᕐᓱᕐᑐᓂᒃ ᐊᒻᒪᓗ ᐊᕐᐱᐅᔫᕐᑐᓂᒃ, ᑕᑯᔭᕐᓂᖏᓐᓂᒃ.

ᒥᑦᑎᒪᑕᓕᖕᒥ, ᐱᓪᒪᖓᐅᔪᓂᒃ ᐅᔾᔨᕈᓱᒃᓯᒪᓕᕐᑐᑦ ᓂᕐᔪᑎᑦ ᐅᓄᕐᓂᖏᓐᓂᒃ, ᐊᒻᒪᓗ ᑎᖕᒥᐊᑦ ᐃᕐᑲᓗᐃᓪᓗ

Pond Inlet

Pond Inlet (Mittimatalik) is located on the northern tip of Baffin Island, which looks out on Eclipse Sound and the mountains and glaciers of Bylot Island (N 72.70°, W 77.96°; elevation 55 m). The community is classified as High Arctic, with prostrate dwarf-shrub and herb tundra vegetation and continuous permafrost. It was established in 1961 and its population was estimated at 1,549 people in the 2011 Statistics Canada Census.

There was little consensus among Elders in Pond Inlet regarding vegetation change; however, some changes have been observed, such as increased plant growth (e.g., grasses) and earlier blooming times. Elders indicated that blueberry plants are producing more fruits, and that blackberry plants are becoming more abundant. Some participants attributed these changes in berry plants to fewer caribou digging the soil and eating the plants, or to fewer permanent snow patches that cool off the soil. Arctic cottongrass seems to be more abundant, and new plant species, such as some kind of dandelion and a furry willow, have been seen.

ᐅᓄᕐᓂᖏᓐᓂᒃ, ᐊᒻᒪ ᓄᑖᓂᒃ ᐱᑕᖃᖅᐸᓕᕐᓂᖓᓂᒃ. ᐆᒃᑑᑎᒋᓗᒍ, ᑕᑎᒡᒐᖃᕐᓂᖅᓴᐅᓕᕐᓂᖓᓂᒃ ᓂᕐᓕᖃᕐᓂᖅᓴᐅᕙᓕᕐᓂᖓᓂᓪᓗ, ᑭᓯᐊᓂ ᒥᑭᓐᓂᖅᓴᓂᒃ ᑎᖕᒥᐊᖃᙱᓐᓂᖅᓴᐅᓕᖅᓱᓂ. ᐃᓚᖏᑦ ᐅᖃᖅᓯᒪᖕᒥᔪᑦ ᓇᑦᑎᖃᙱᓐᓂᖅᓴᐅᓕᕐᓂᖓᓂᒃ ᑕᐃᔅᓱᒪᓂᓂᑦ ᐊᒻᒪ ᒥᖅᑯᐃᔭᕐᓂᖅᓴᐅᓕᕐᓂᖏᓐᓂᒃ. ᐊᓇᙱᕐᔪᐊᓪᓗ ᖃᔅᓰᓐᓇᐅᓂᖅᓴᐅᖅᑰᔨᓕᖅᓱᑎᒃ, ᐃᓚᖏᓐᓄᑦ ᑕᐃᒪᐃᓐᓂᕆᖅᑕᐅᔪᖅ ᑐᒃᑐᖃᙱᓐᓂᖅᓴᐅᓕᕐᓂᖓᓄᑦ.

ᐅᑭᐅᒃᑯᑦ ᓯᓚ ᐊᓯᔾᔨᖅᓯᒪᓂᖅᐸᐅᓂᕆᖅᑕᐅᔪᖅ, ᓲᕐᓗ ᓂᓪᓚᓱᙱᓐᓂᖅᓴᐅᓕᕐᓂᖓ ᐊᐳᑎᑭᓐᓂᖅᓴᐅᓕᕐᓂᖓᓗ. ᑕᐃᔅᓱᒪᓂ ᐊᐳᑦ ᓯᑯᒥ ᐃᔾᔪᓂᖅᓵᓘᕙᓚᐅᕐᓂᖓᓄᑦ ᐄᑉᐳᕆᒥ ᐃᓄᐃᑦ ᐃᖅᑲᐅᒪᕗᑦ ᐊᖅᑯᓯᐅᕆᐊᖃᓚᐅᖅᓯᒪᓂᖏᓐᓂᒃ ᐃᒐᕐᔪᐊᕐᒥ ᒥᑦᑎᒪᑕᓕᖕᒧᑦ ᐱᓱᒃᓱᑎᒃ ᕿᒧᒃᓯᕆᐊᖏᓐᓂᕐᒥᓂᒃ. ᓯᑯᐃᖅᓵᓕᓂᖅᓴᐅᕙᓕᖅᑐᖅ, ᓯᑯ ᓵᓐᓂᖅᓴᐅᕙᓕᕐᓂᖓ ᐱᔾᔪᑕᐅᓪᓗᓂ. ᐊᓄᕌᕐᓂᖅᓴᐅᕙᓕᖅᑐᖅ, ᓯᑯᓗ ᐊᐳᑎᑭᓐᓂᖅᓴᐅᕙᓕᖅᓱᓂ ᐊᓄᕆᒧᑦ, ᐊᐳᑎᑭᓐᓂᖓᓄᑦ ᐊᖑᓇᓱᖕᓂᖅ ᐱᒡᒐᓇᕐᓂᖅᓴᐅᕙᓕᕐᓂᖓᓂᒃ. ᐊᐱᖅᓱᖅᑕᐅᔪᑦ ᐅᖃᓚᐅᖅᑐᑦ ᑰᖕᓃᑦ ᐃᒪᖃᕐᓂᖓᓗ ᐊᓯᔾᔨᖅᓯᒪᓕᕐᓂᖏᓐᓂᒃ ᒥᑭᓐᓂᖅᓴᓂ ᑰᖕᓂ ᑕᓯᕐᓂᓗ. ᐅᔾᔨᕆᓱᖃᑦᑕᖅᐸᓪᓕᐊᓕᕐᒥᔪᑦ ᓄᓇᐅᑉ ᐃᑭᐊᖓᓂ ᖁᐊᖑᐃᓐᓇᖅᑐᖅ ᐊᐅᒃᐸᓕᕐᓂᖓᓂᒃ ᓄᓇᒥᓗ ᑎᓱᐊᕆᑕᐅᕙᓕᕐᓂᖓᓂᒃ ᑰᖕᓃᑦᑐᓂ. ᓯᕐᒦᑦ ᒥᑭᓐᓂᖅᓴᐅᓕᖅᑐᑦ ᐃᓚᖏᓪᓗ ᑕᕆᐅᕐᒧᑦ ᑎᑭᐅᖅᑲᔪᓐᓃᖅᓱᑎᒃ.

In Pond Inlet, important changes have been observed in mammal abundance, as well as some changes in bird and fish populations, including new species. For example, there are more cranes and Canada geese, but fewer small birds. Some people also noticed that there are fewer ringed seals than there used to be and that their fur sheds more easily. There also appear to be fewer big black flies (*anangirjuaq*) nowadays, which some attribute to declines in caribou numbers.

Winter is the season in which people have witnessed the most changes, such as warmer temperatures and less snow. In the past, the snow on the sea ice was so deep in April that people remember having to make a trail from Mount Herodier to Pond Inlet by foot before they could go by dog team. Sea ice breakup now arrives earlier in the season, partly because the ice is thinner. The wind is also more prevalent, blowing the snow off the sea ice and making it more difficult to travel. Participants noted that river current patterns and water levels are changing in small rivers and lakes. They have also started to notice permafrost thawing and signs of erosion on riverbanks. The glaciers are getting smaller, and some no longer touch the ocean's edge.

ᐃᓐᓇᐃᑦ ᐊᐱᖅᓱᖅᑕᐅᔪᑦ
Elders Interviewed

ᐅᓘᑕ
ᐊᕐᓇᑲᓪᓚᒃ
Rhoda
Arnakallak

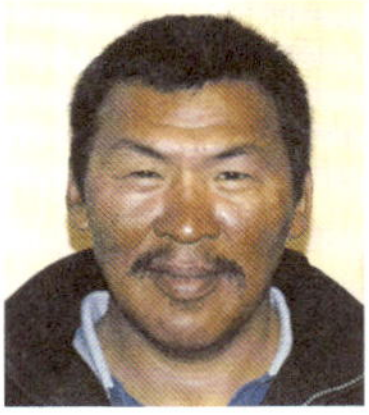

ᐊᐃᕐᕋᓚᕼᐊ
ᑯᓄᒃ
Abraham
Kunnuk

ᕼᐊᒻ ᑲᓪᓗᒃ
Ham
Kadloo

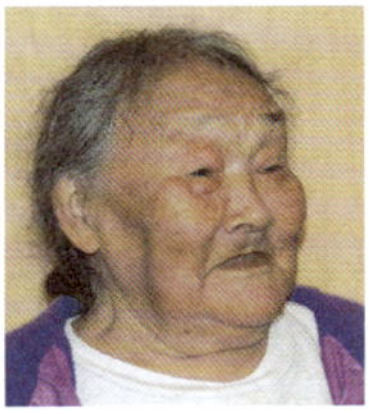

ᓖᑎᐊ ᖄᔫᖅ
Letia Kyak

ᒑᒪᐃᓕ
ᕿᓗᖅᑭᓵᖅ
Gamailie
Kilukishak

ᔪᐊᓇᓯ ᒪᒃᐸ
Joanasie
Muckpa

ᒥᐊᓕ
ᕿᓗᖅᑭᓵᖅ
Mary
Kilukishak

ᐃᓕᓴᐱ
ᐆᑦᑐᕙ
Elisapie
Ootoova

ᑐᓃᓴ ᑯᐹᖅ
ᒪᒃᑖᖅ
Theresa
Koopa Maktar

ᐃᓚᐃᔾᔭ
ᐸᓂᒃᐸᑯᑦᑐ
Elijah
Panipakoocho

ᐋᓂ ᐲᑦᑐᓘᓯ
Annie
Peterloosie

ᔦᐃᑯ ᐲᑦᑐᓘᓯ
Jayko
Peterloosie

ᐸᓂᓗᒃ ᓴᖑᔭ
Paniloo
Sangoya

ᖃᒪᓂᖅ ᓴᖑᔭ
Qamaniq
Sangoya

ᐅᓘᑎ ᓴᖑᔭ
Ruth
Sangoya

ᐸᐅᕐᖓᐃᑦ

“ᒪᑯᐊᓕ ᐸᐅᕐᖓᕈᔪᐃᑦ ᑭᓯᐊᓂ ᓯᓚᒧᑦ ᓯᕐᑭᓂᕐᒧᓪᓗ ᐋᖅᑭᒃᑕᐅᓯᒪᕐᑲᑦᑕᕐᒪᑕ ᒫᓐᓇᓕ ᑕᒪᒃᑯᐊ ᐸᐅᕐᖓᐅᓂᕈᔪᖏᓪᓕ ᐱᕈᑦᑎᐊᖏᓐᓂᖅᓴᐅᕐᑲᑦᑕᓕᖅᑐᑦ ᑕᒪᓐᓇ ᓄᓇᑐᐃᓐᓇᐅᓂᖓ ᐱᕈᖅᐹᓪᓕᖅᓯᒪᓕᕐᑐᐊᖅᑑᓂ ᐸᐅᕐᖓᕈᔪᐃᑦᓕ ᐱᕈᑦᑎᐊᙱᓐᓂᖅᓴᐅᕐᑲᑦᑕᓕᖅᑐᑦ ᑕᐃᔅᓯᒪᓂᐅᓚᐅᖅᑐᒥ ᓯᕐᑭᓐᓂᕐᑲᑦᑕᖏᓐᓂᖅᓴᐅᓕᕐᒪᑦ ᐊᐅᔭᒃᑯᑦ.”

—ᐃᓚᐃᔾᔭ ᐸᓂᒃᐸᑯᑦᑐ

“ᒫᓐᓇ ᑕᐃᒪᓐᓇᐅᒍᓗᐊᖅᑑᓂ ᓲᕐᖁᖅᑰᔨᓗᐊᙱᑦᑐᑦ ᑭᓯᐊᓂ ᓲᕐᓗ ᓯᓚᒧᑦ ᒪᓕᖕᒥᖕᒪᑕ ᑭᑐᑕᖕᑎᕐᓇᐃᑦ ᐱᕈᓇᐊᓖᑦ ᐅᓄᙱᓐᓂᖅᓴᐅᕚᓕᖅᑰᔨᓕᖅᑐ ᐊᐅᔭᒃᑯᑦ ᐅᖅᑰᓂᖓ, ᐅᖅᑰᓂᑯᓗᐊ ᓇᐃᑦᑐᑯᓘᖅᑰᔨᓪᓗᓂ.”

—ᐃᓚᐃᔾᔭ ᐸᓂᒃᐸᑯᑦᑐ

“ᑭᓯᐊᓂ ᑕᐃᒪ ᑕᐃᒃᑯᐊ ᐸᐅᕐᖓᐃᑦ ᐃᓂᖏᑦ ᐊᒻᒪᓗ ᑭᑐᑕᖕᑎᕐᓇᐃᑦ ᐃᓂᖏᑦ ᑐᒃᑐᓂᑦ ᓴᕐᐹᓴᖅᑕᐅᓚᐅᖅᓯᒪᓪᓗᑎᒃ ᑭᖑᓂᖓᒎᓕᖅᑐᖅ ᒥᑭᑦᑑᑎᐅᔭᓂᒃ ᐱᕈᖅᑐᕐᑲᕐᑲᑦᑕᓕᓚᐅᖅᓯᒪᖕᒪᑦ. ᑖᒃᑯᐊ ᐃᓂᖏᑦ ᐸᐅᕐᖓᐃᓪᓗ ᑭᑐᑕᖕᑎᕐᓇᐃᓪᓗ ᐃᓂᖏᑦ ᐱᑐᕐᑲᐅᓂᖅᓴᐅᓕᕐᒥᒃ ᐊᖏᔫᑕᐅᓂᖅᓴᐅᓕᕐᒥᒃ ᐱᕈᑦᑎᐊᖅᐸᓪᓕᐊᒐᒥᒃ ᑕᒪᓐᓇ ᓄᓇᒧᑦ ᐊᒃᑑᑎᑦᑎᐊᕐᒪᑦ ᑖᒃᑯᐊ ᐸᐅᕐᖓᐃᑦ ᒪᓕᒃᓴᕐᑲᓯᐅᑎᓯᒪᔪᑦ.”

—Hᐊᒻ ᑲᓪᓗᒃ

Berries

“The *paurngait* [blackberries/crowberries] don’t grow as much now because they depend on the sunlight, and it’s never really sunny here now. […] They don’t grow big. The fruits are smaller.”

—Elijah Panipakoochoo

“There are more blackberries and blueberries, and they grow earlier now. Blueberries used to grow in the middle of August, but now they grow earlier and there are more, but they don’t bloom or ripen as much as before.”

—Elijah Panipakoochoo

“When there were a lot of caribou around, when they were digging the soil and the plants, they kind of killed the plants. It was years afterwards that they started to grow again. The berries were smaller, and now that they have rooted well to the ground the blueberries and blackberries are getting bigger again, now that they haven’t been disturbed for a while.”

—Ham Kadloo

“ᐅᕙᖕᒐᓕ ᑕᒪᒃᑯᐊ ᓱᕐᕌᖅᑰᔨᒥᖖᒋᑕᒃᑲ ᑭᓯᐊᓂ ᒪᓐᓇᓗ ᐅᓇ ᒥᑦᑎᒪᑕᓕᒃ ᒪᓐᓇ ᑐᓄᓐᓂᖅ ᑕᒪᓐᓇ ᑐᒃᑐᖃᓚᐅᖅᑎᓪᓗᒍ ᑕᐃᔅᓱᒪᓂ ᑐᑦᑐᓂᑦ ᑲᒡᒎᓈᕋᓘᓪᓗᑎᒃ ᖃᕝᕙᒃᑖᓪᓗ ᑕᒪᒃᑯᐊ ᐸᐅᖕᒐᐃᑦ ᐸᐅᖕᒐᖅᑯᑏᓪᓗ ᐊᒻᒪ ᑭᒍᑕᖕᒋᖅᓇᖅᑯᑎᐳᕙᐃᑦ ᑕᒪᒃᑯᐊ ᓂᕆᕙᒃᑕᕗᑦ ᓱᕋᒃᑎᔭᐅᕙᐳᕙᐋᓚᐅᖅᓗᑎᒃ ᑕᐃᔅᓱᒪᓂ ᐱᕈᑦᑎᐊᓚᐅᖅᓯᒪᖖᒋᑦᑐᑦ ᑕᕝᕙᓕ ᓂᕐᕚᑎᓂᑦ ᑐᒃᑐᓂᑦ ᐱᔭᐅᓚᐅᖅᓯᒪᕚᓐᓃᕋᒥᒃ ᐱᕈᑦᑎᐊᓕᕐᒥᕚᑦ. ᑕᒪᔾᔭ ᖃᐅᔨᒪᔭᒃᑲ.”

—ᐋᓂ ᐲᑦᑐᓘᓯ

“ᐅᓄᕐᓂᖅᓴᐅᓪᓚᑦᑖᓕᕐᒪᑕ, ᐃᒪᓐᓇᓗᑭᐊᖅ ᐃᒻᒪᒃᑲᓐᓂᕐᓕ ᐊᓂᐅᕕᓃᑦ ᐊᐳᑎᕕᓃᑦ ᐊᐅᖖᒋᓴᐃᓐᓇᕐᓂᑯᑦ ᐊᐅᔭᒃᑯ ᐊᓂᐅᕙᐅᓂᕋᖅᐸᓚᐅᕋᑦᑎᒍ ᒫᓐᓇᓕ ᑖᑦᑕᓂᓇᑦᓗᓂ ᒫᓐᓇ ᐱᑕᖅᑲᕈᓐᓃᖅᑐᑦ, ᑕᐃᒪᓐᓇᓗᑭᐊᖅ ᓂᖠᓕᓇᕈᑎᖅᑲᖖᒋᓐᓂᓴᐅᓕᕐᓂᖕᒐᓄᑦ ᓄᓇ ᑐᐊᐸᖅᑲᑦᑎᐊᕐᓂᖅᓴᐅᓕᕐᒪᑕ.”

—ᔪᐊᓇᓯ ᒪᒃᐸ

“ᑕᒫᓃᓛ ᐱᕈᓲᖕᒎᖕᒪᑕ ᐋᒑᔅᓯᐅᔭᕌᖕᒐᑦ ᑕᐃᒪ ᐱᕈᕐᒪᑕ ᕿᕐᓂᖅᓯᓇᓱᖕᓂᕐᒥᖕᓄᑦ ᑕᐃᒪ ᒦᓐᓇᐅᖕᒪ ᑭᒍᑕᖕᒋᖅᓇᐃᑦ ᑖᖅᓯᖅᑲᑦᑕᓕᕌᖕᒐ ᑐᖕᒍᕙᖅᓯᖅᑲᑦᑕᓚᐅᕐᒪᑕ ᑐᖕᒍᕙᖕᒋᔅᓱᔾᔭᖅᓗᑎᒃ ᖃᑯᔅᓱᔾᔭᖅᓗᑎᒃ ᑖᖅᓯᖅᑲᑦᑕᓕᕌᖕᒐᑦ ᑭᓯᐊᓂ ᑐᖕᒍᕙᖅᓯᖅᑲᑦᑕᓚᐅᕐᒪᑕ ᒦᓐᓇᓕ ᐊᕗᖅᑕᐅᕙᓂᒃ ᑕᑯᓪᓗᖕᒐ ᑐᖕᒍᕙᖅᑕᓪᓗ ᒪᑯᐊ ᖃᐅᓪᓗᖅᑕᓪᓗ ᐱᖃᑦᑕᓚᐅᕐᒪᑕ ᐊᕐᕌᒎᓚᐅᖅᑐᖅ.”

—ᒥᐊᓕ ᕿᓗᖅᑭᓵᖅ

“When there were many caribou near Pond Inlet, they would step on the blackberries, blueberries, and mountain heather. That’s why there weren’t any blueberries or blackberries, but now that there are no caribou, there are more blueberries and blackberries [again].”

—Annie Peterloosie

“There are absolutely more [blackberries growing], perhaps there used to be hardened snow that would remain in the summer. Where there was hardened snow is now in shade; the land is warmer and there is more gravel.”

—Joanasie Muckpa

“This year the blueberries have been blooming earlier and are soft already. During the month of August the blackberries usually grow, and when the dark season starts to arrive the blueberries grow. Mostly at the end of August, but this year it’s earlier.”

—Mary Kilukishak

ᐱᕈᖅᑐᑦ ᐊᓯᖏᑦ

“ᐊᒻᒪᓗᑦᑕᐅᖅ ᐅᔾᔨᕆᔭᒃᑲ ᓯᒡᔭᖅ ᒪᓐᓇ ᐃᒪᐅ ᖃᓂᒋᔮ ᑐᐊᐸᔮᕈᔫᓪᓗᓂ ᐱᕈᖅᑐᖅᑕᖃᓲᖑᖕᒥᖕᒪ ᐊᖏᔪᐊᓗᖕᓂᒃ ᐅᖃᐅᔭᖃᖅᐳᑎᒃ ᖁᖅᓱᖅᑕᐅᓪᓗᑎᒃ ᑐᖑᔪᖅᑖᓘᓪᓗᑎᒃ ᑖᒃᑯᐊ ᐊᒻᒪ ᐊᔾᔨᒋᑦᑎᐊᙱᑕᖏᑦ ᓯᒡᔭᒦᒡᒍᓪᓗᑎᒃ ᓇᓂᑐᐃᓐᓇᖅ ᓄᓇᑭᕈᔪᒃᑐᒥ ᑭᓯᐊᓂ ᒫᓂ ᓄᓇᑐᐃᓐᓇᒥᓘᓐᓃᑦ ᐱᑕᖃᖃᑦᑕᓕᖅᒥᔪᑦ. ᑭᓱᑐᐃᓐᓇᐃᑦ ᐊᔾᔨᖂᔨᔪᓐᓃᖅᒪᑎᑦ. 2000-ᖄᖅᑎᓐᓇᒍ ᑕᐃᒪᐃᓕᓚᐅᖅᑐᖅ.

—ᐃᓕᓴᐱ ᐆᑦᑐᕙ

“ᑕᐃᒃᑯᐊ ᑕᑯᑲᑕᖕᒥᒐᒃᑭᑦ ᖁᖅᓱᖅᐳᑎᒃ ᒪᐅᓇ ᖃᐅᓪᓗᖅᑕᖃᖅᐳᑎᒃ ᓱᕈᓯᐅᓪᓗᖓ ᑕᑯᖃᑦᑕᓚᐅᖅᑕᒃᑲ ᑕᐅᕙᓂ ᑐᒃᓯᐊᖅᕕᐅ ᓴᓂᐊᓂ ᑕᑯᑕᐃᓐᓇᓕᖅᒥᔭᒐᒃᑲ ᑕᐃᒃᑯᐊ ᑕᑯᖃᑦᑕᓚᐅᖅᓯᒪᔭᒃᑲ ᑯᖅᓱᖅᑕᐃᑦ ᖃᐅᓪᓗᖅᑕᒥ ᐊᑎᖃᖅᐳᑎᒃ.”

—ᐅᓘᑕ ᐊᖅᓇᑲᓪᓚᒃ

“ᑕᒪᒃᑯᐊ ᐱᕈᖅᑐᕈᔪᐃ ᐃᓵᓪᓗ, ᐃᓵᑦ ᓇᐃᑦᑑᑎᑯᓘᖃᑦᑕᓚᐅᖅᓯᒪᒐᓗᐊᖅᒪᑕ, ᐃᒪᓐᓇᑯᓗ ᓇᐃᑦᑑᑎᑯᓘᓪᓗᑎᒃ ᐱᕈᖅᐸᓚᐅᖁᓗᐊᖅᒪᑕ ᒫᓐᓇ ᑕᑭᔪᐊᓘᖃᑦᑕᓕᖅᒪᑕ ᑕᓰᑦ ᓯᓂᖏᓐᓂ.”

—ᔪᐊᓇᓯ ᒪᒃᐸ

“ᐊᒻᒪᓗ ᐅᔾᔨᕈᓱᒃᐳᖓ ᒫᓐᓇ ᐃᒡᓗᒪ ᓴᓂᐊᓂ ᐱᕈᖃᑦᑕᙱᑦᑐᓂᒃ ᑕᑯᒋᐅᖅᑕᓐᓂᒃ ᐱᕈᖅᓯᒪᔪᖃᐅᖅᐳᓂ.”

—ᐅᓘᑕ ᐊᖅᓇᑲᓪᓚᒃ

Other Plants

“I also notice around the beach where there are pebbles, [there are] some plants that grow with lots of big yellow leaves. There are also other plants not native to Pond Inlet. They started growing only around areas with not much soil, but they can be found everywhere now. They started growing before the year 2000.”

—Elisapie Ootoova

“I have noticed yellow flowers with white middles that I would see by the beach, by the old mission building. I haven’t seen them for years, but I see them this year.”

—Rhoda Arnakallak

“I have noticed that all plants bloom earlier. Even the grass is taller now.”

—Joanasie Muckpa

“I have been aware of some plants that don’t normally grow around my house that are now growing around my house.”

—Rhoda Arnakallak

“ᑲᖒᔭᐃᑦᓕ ᐄᖑᖅᑰᔨᒋᔭᒃᑲ ᐱᕈᕐᓂᖅᓵᓗᖅᑰᔨᖃᑦᑕᖅᑐᑦ, ᑲᖒᔭᖃᓗᐊᖅᑰᔨᓚᐅᖅᓯᒪᙱᑦᑐᑦ ᐃᓚᖏᑦ ᑲᖒᔭᐃᑦ. ᑲᖒᔭᑦ ᑕᒪᔾᔭ ᐅᔾᔨᓇᔭᑐᐊᒃᑲ. ᖁᖑᓖᑦᓕ ᐱᕈᑦᑎᐊᓗᐊᙱᑦᑐᑦ ᖁᖑᓕᖃᑦᑎᐊᓗᐊᕈᓐᓇᖅᑰᔨᔪᓐᓃᖅᖢᓂ, ᑕᒪᒃᑯᐊ ᐱᕈᕈᓐᓇᑦᑎᐊᖅᑰᔨᒋᔪᓐᓃᖅᖢᒋᑦ. ᐹᓂ ᖁᖑᓕᖅᑕᖅᐸᓚᐅᕐᑦᑕ ᐃᒥᖅᑕᕐᕕᐅ ᑐᖏᑦᑎᐊᖓᓂ, ᖁᖑᓕᓕᐊᓘᖃᑦᑕᓚᐅᕈᓗᐊᕐᒪᑦᓕ ᐱᕈᒡᒐᖅᑐᐊᓗᐃ.”

—ᐅᓘᑎ ᓴᖑᔭ

“ᐅᔾᔨᓇᙱᔅᓲᔭᓚᐅᖅᑕᒃᑲ ᖃᖓᓕᑭᐊᖅ ᑭᓯᐊᓂ 2000-ᕈᔪᖕᒥᖃᐃ ᑕᒫᓂ ᑕᑯᓕᓵᖃᑦᑕᖅᐸᓪᓕᐊᖅᑰᔨᔪᖓᓕ ᐅᕙᖓ. ᐅᔾᔨᓇᙱᔅᓲᔭᖅᑕᕕᓂᒃᑲᓗᑭᐊᖅ ᐱᑕᖃᐃᓐᓇᖅᑑᒐᓗᐊᓪᓗᑭᐊᖅ ᑕᑯᓕᓵᖅᑰᔨᑦᑎᐊᖃᑦᑕᖅᑭᑦ ᐃᓚᖏᑦ. ᒪᕐᕈᐃᓐᓇᑐᐊᑯᓘᖏᒃᑯᐊᒃ ᑕᐃᒃᑯᐊᒡᓗ ᓄᓇᑦᓯᐊᒦᑦᑐᑦ ᓯᓚᑎᑦᑎᓐᓂᓗ, ᑕᐃᒃᑯᐊ ᖁᖅᓱᖅᑕᑯᓗᐃᑦ. ᐱᑕᖃᐃᓐᓇᖅᑑᒐᓗᐊᓪᓗᑭᐊᖅ ᐅᕝᕙᓗᑭᐊᖅ ᐅᔾᔨᓇᓕᓵᖅᑕᒃᑲ. ᐃᕐᖑᑕ ᐊᐱᓇᒐᒃᑯ ᒪᑯᐊᖃᐃ ᑕᒫᓃᖏᓐᓇᖅᑑᒐᓗᐊᓇᐃ, ᒪᑯᐊᑦᑕᐅᕉᖅ ᑕᑯᓕᓵᖅᑰᔨᑦᑎᐊᖅᐸᐃᑦᓕᒎᖅ ᑕᕝᕙᓚᓚᐅᖅᓯᒪᔪᖅ. ᐊᐱᖅᑯᑎᒋᕙᒃᑲᓗᐊᖅᑭᑦ ᓇᓗᔭᐅᕙᒃᖢᑎᒃ, ᐅᔾᔨᓇᔭᐅᙱᑐᐃᓐᓇᖅᑐᕕᓂᖃᐃ ᐃᓐᓇᐅᖃᑎᕈᔪᓐᓂ.”

—ᑐᓈᓴ ᑯᐹᖅ ᒪᒃᑖᖅ

“I notice the cottongrass is more [abundant] than ever. Back then we didn’t have much, but today we have more of that cottongrass. I notice that those mountain sorrels are less than before. We used to go pick sorrel where the lake is, but there is not much anymore today.”

—Ruth Sangoya

“Around 2000 I started noticing plants that I had never seen before. […] The plant that grows with yellow flowers. […] It was the first time I saw it outside my house. I even went to ask one of the Elders about two years ago. He also said it was the first time he had seen them. He is much older than me. I started asking other people. They didn’t know the name [either]. They never saw one like that before.”

—Theresa Koopa Maktar

“ᐃᓚᖏ ᐅᖅᑲᐅᓯᐃᑦ ᐱᑕᖅᑲᓚᐅᖕᖏᑦᑐᒥ ᐊᒻᒪᓗ ᖁᖑᓕᐅᓂᕋᖅᑕᕗ ᑕᐃᒃᑯᐊ ᖁᖑᓕᖅᐸᓚᐅᖕᖏᑦᑐᖅ ᖁᖑᓕᓕᐊᓘᖅᑲᑦᑕᓕᕐᒪᑦ ᐊᒻᒪᓗ ᐊᓯᐊ ᖃᓄᕋᓕᓂᐅᖕᖏᑦᑑᕕᓗᐊᖅ ᖁᖑᓕᖅᑲᖅᑲᑦᑕᓚᐅᖕᖏᑦᑐᖅ ᖁᖑᓕᖅᑲᖅᑲᑦᑕᕐᓂᖅᓴᐅᓕᖅᑐᖅ ᒫᓐᓇ.”

—ᔪᐊᓇᓯ ᒪᒃᐸ

“There is more mountain sorrel now where they usually grow, at the edge of the roads and where the land is not disturbed.”

—Joanasie Muckpa

ᐆᒪᔪᑦ

“ᓇᑦᑎᖅᑲᑦᑎᐊᕈᓐᓃᖅᑐᐊᓗᒃ. [...] ᓇᑦᑏᑦ ᓴᖅᑭᔾᔭᖅᐸᓚᐅᖅᓯᒪᔪᑦ ᒥᑦᑎᒪᑕᓕᐅᑉ ᓵᖓᓂ. ᐱᑕᖅᑲᕈᓐᓃᖅᑐᖅ. ᒫᖓᑦ ᑕᑯᒃᓴᐅᔪᓐᓃᖅᑐᑦ. ᐊᕐᕌᒍᑕᒫᑦ ᓇᑦᑎᑭᑦᓕᕆᑦᓕᐊᑐᐃᓐᓇᖅᑑᔭᖅᑐᖅ. ᓇᒧᖔᔪᐊᓗᐃᑦ ᐃᒃᑯᐊ?”

—ᐊᐃᑉᐳᓚᕼᐊ ᑯᓄᒃ

“ᑐᒃᑐᖅᑲᕈᓐᓃᖅᑐᖅ. ᓴᐅᓂᕕᓂᖏᑦ ᑕᑯᓇᖅᐸᒃᑐᑦ ᑐᐱᖅᕕᕕᓂᖅᓂ, ᑐᒃᑐᖅᑲᖏᑦᑎᐊᓕᖅᑐᖅ.”

—ᐋᓂ ᐲᑦᑐᓘᓯ

“ᐊᓇᖏᖅᑲᖏᔾᔫᒥᔪᖅ ᑐᒃᑐᖅᑲᕈᓃᖅᒪᑦᑕ ᑕᐃᒪᖓ ᐊᓇᖏᖅᔪᐊᕌᓗᖅᑲᕈᓃᖅᑐᖅ ᑐᒃᑐᖅᑲᖅᑎᓪᓗᒍ ᐊᓇᖏᖅᔪᐊᓕᐊᓘᓚᐅᖅᓱᓂ ᐊᒻᒪᓗ ᖅᑭᒃᑐᓇᐊ ᖅᑭᓚᒥᑯᓘᖅᑰᔭᖅᑲᑦᑕᓕᖅᑐᑦ ᖅᑭᒃᑐᓇᐊᖅᑲᓗᐊᖅᑲᑦᑕᖅᑰᕈᓐᓃᖅᑐᖅ.”

—ᐅᓘᑎ ᓴᖑᔭ

Animals

“There are hardly any seals now. [...] There used to be seals swimming by Pond [Inlet]. There's nothing now. You can't see them from here anymore. There are getting to be fewer seals every year now. Where are they going?”

—Abraham Kunnuk

“When there were more caribou, there were more black flies—the big ones. But now, since there are hardly any caribou, there are less black flies. It seems like there are not a lot of mosquitoes now, and they are here for a shorter period of time. There used to be more mosquitoes.”

—Annie Peterloosie

“ᑕᐃᒃᑯᐊ ᑲᖑᐃᑦ ᐊᐃᑉᐸᖏᑦ ᓂᕐᓖᑦ
ᓂᕐᓕᖅᑕᖃᐃᓐᓇᐅᔭᖅᑐᖅ ᑭᓯᐊᓂ
ᒫᓐᓇ ᐅᓄᕐᓂᖅᓴᐅᖃᑦᑕᓕᖅᕗᑎᒃ
ᐊᒻᒪ ᑕᐃᒃᑯᐊ ᓂᐅᖅᑯᖅᑐᔪᐊᓗᐃᑦ
ᑕᑎᓪᒐᕐᔪᐊᒥᒃ ᐊᑎᖃᖅᑎᑕᐅᕙᒃᑐᑦ
ᒪᕐᕉᒃ ᐱᖓᓱᓪᓘᓐᓃᑦ
ᐃᒪᐃᓪᒍᓚᐅᖅᑐᖅ ᓱᕈᓯᐅᓂᓐᓂᑦ
ᐃᓐᓇᕈᕐᓂᓐᓄᑦ ᐱᔪᖃᓚᐅᖅᐸᒃᕗᓂ
ᐊᑕᐅᓯᑯᓗᖕᒥᒃ ᓲᕐᓗ ᒫᓐᓇ
ᐅᓄᖅᑐᐊᓘᓪᓗᑎᒃ ᑕᒫᓂ ᓲᕐᓗ
ᑕᑎᓪᒐᕐᔪᐊ ᓂᕐᓖᓪᓗ.”

—ᐃᓕᓴᐱ ᐆᑦᑐᕙ

“ᓇᑦᑎᕐᓂᓪᓗ ᕿᓚᓗᒐᕐᓂᓪᓗ,
ᐅᖃᐅᓯᕆᖅᑲᐅᔭᓐᓂᒃ.”

—ᒐᒪᐃᓕ ᕿᓗᖅᑭᓵᖅ

“ᑕᐃᒃᑯᐊ ᖃᖓᓂᑲᑦᓚᑭᐊᖅ
ᐊᕐᕌᒍᓂ ᐱᖓᓱᓗᐊᓂ
ᐱᓕᕐᒪᑕ ᐊᖏᔫᑕᐅᓂᖅᓵᓗᐃᑦ
ᐃᖃᓗᒃᑕᐅᖃᑦᑕᓕᖅᑐᑦ.
ᐃᓛ ᐃᖃᓗᐃᑦᑕᐅᖅ
ᐱᕈᖅᐸᓪᓕᐊᖃᑦᑕᕐᒥᖕᒪᑕ
ᑕᓯᕐᒦᔅᕗᑎᒃ ᒥᑭᓐᓂᖅᓴᐅᓚᐅᖅᕗᑎᒃ
ᑕᕆᐅᕐᒧᐊᕌᖓᒥᒃ
ᐊᖏᓪᓕᒃᑲᓐᓂᖅᕗᑎᒃ ᐃᓛᓐᓂ ᑕᓯᕐᒧᑦ
ᐅᑎᓕᓲᖑᖕᒪᑕ ᑕᐃᒪ ᐊᕐᕌᒍᑦᑕᕌᖓᑦ
ᐊᖏᓪᓕᕙᓪᓕᐊᖏᓐᓇᕐᒥᖕᒪᑕ ᑕᐃᒃᑯᐊ
ᐊᖏᔪᐊᓘᓕᖅᑕᖅᐳᑦ.”

—ᔭᐃᑯ ᐲᑦᑐᓘᓯ

“1947-ᒥ ᒪᐅᖓᕋᑦᑕ ᐊᕐᕖᑦ
ᐅᓄᙱᓐᓂᖅᓴᐅᓚᐅᓯᒪᔪᑦ
ᖃᐅᔨᒪᔭᖃᓗᐊᒃᑲ ᒫᓐᓇᒃᑯᑦ
ᐅᓄᕐᓂᖅᓴᐅᓕᖅᕋᒥᒃ
ᑕᑯᒃᓴᐅᓂᖅᓴᐅᓕᖅᑐᑦ ᓇᓃᔅᕗᓂ,
ᒥᑦᑎᒪᑕᓕᐅ ᐊᓯᐅᓃᔅᕗᓂ

“There are no more caribou left. You can see the bones where they used to pitch the tents, but no caribou at all.”

—Ruth Sangoya

“There are more snow geese and Canada geese. There used to be two or three sandhill cranes when I was a child, and one would occassionally be caught. There are more sandhill cranes and geese now.

—Elisapie Ootoova

“I notice that I hardly see seals and narwhals now.”

—Gamailie Kilukishak

“For the past three years the fish have been bigger. […] Every year they seem to be growing.”

—Jayko Peterloosie

ᑕᑯᒐᔪᖕᓇᕐᓂᖅᓴᐅᓕᖅᑐᑦ. ᑭᓯᐊᓂ ᓇᑦᑏᑦ ᐅᓅᙱᓐᓂᖅᓴᑯᓘᓕᖅᑐᐃ."

—ᔪᐊᓇᓯ ᒪᒃᐸ

"ᐅᒃᐱᒡᔪᐊᓕᐊᓘᕙᓚᐅᖅᓯᒪᔪᖅ, ᑭᓯᐊᓂ ᑕᒫᓃᑉᐸᒍᓐᓃᖅᑐᑦ. ᑎᓇᒐᓂᐊᑦ ᑲᔪᐃᑦ ᑕᒫᓃᑦᑐᑦ. ᐱᑕᖅᑲᖅᐸᓚᐅᖅᓯᒪᖏᑦᑐᑦ. [...] ᐃᖅᑲᓗᐃᑦ ᖃᑖᑐᐊᓘᕙᓚᐅᖅᓯᒪᔪᑦ, ᐊᐅᑉᐸᓗᑦᑎᐊᖅᑲᑦᑕᓕᖅᑐᑦ. ᐃᖅᑲᓗᓕᐊᓘᕙᓚᐅᖅᓯᒪᔪᖅ ᒫᓂ. ᐃᒫᓂ ᐊᓂᓪᓛᔪᑦ ᑕᑯᒃᓴᐅᕙᓚᐅᖅᓯᒪᔪᑦ, ᑕᑯᒃᓴᐅᔪᓐᓇᐃᓪᓕᔪᑦ. ᐊᔾᔨᒋᔪᓐᓃᖅᑕᖓ ᒫᓐᓇ."

—ᓛᑎᐊ ᖃᔮᖅ

"ᐅᔾᔨᕈᓱᖅᑲᑦᑕᖅᓯᒪᙱᓚᑎᓪᓛ? ᐊᖅᖂᒍᓂ ᖃᒪᒃᑕᑦᑎᓐᓂ ᐊᓇᙱᖅᔪᐊᖂᓗᐃᑦ ᑕᐃᒃᑯᐊᖑᕙᓚᐅᖅᓯᒪᔪᑦ ᐊᐅᔭᐅᔪᖅ ᑕᐃᒃᑯᐊ ᐱᑕᖅᑲᓗᐊᖅᑰᔨᖏᓪᒪᑕ ᒥᑭᑦᑑᑎᑯᓗᖕᓂᒃ ᐊᓇᙱᔮᑯᓗᐃ ᑕᐃᒪᑐᐊᖅ."

—ᒥᐊᓕ ᕿᓗᖅᑭᔮᖅ

"When we moved here in 1947 I noticed that there weren't many bowhead whales, and I see them more now, even near camping places. And there is less ringed seal now."

—Joanasie Muckpa

"There used to be a lot of snowy owls, but they are not here anymore. And the brown foxes are here, too; we didn't have them before. [...] Fish used to have very pale flesh, but now they are bright red. There used to be lots of fish here. You could see them popping out of the water, but you don't see that anymore. It's different today."

—Letia Kyak

"I notice that the black flies are smaller than before; before there used to be more big black flies, but this year they are smaller."

—Mary Kilukishak

“ᐅᔾᔨᕈᓯᓚᐅᕐᒥᔪᖓ ᖁᐸᓄᐊᖁᓗᐃᑦ
ᓂᒡᓕᐊᙵᑎᔅᓰᔭᕐᒃᑐᑦ
ᓂᒡᓕᐊᑎᐊᓚᐊᕋᓗᐊᕐᒃᖢᑎᒃ
ᓂᒡᓕᐊᔫᓐᓃᓚᐅᕐᒃᑐᑦ
ᓂᒃᓚᓱᓗᐊᕐᒃᑰᕐᒪᑦ, ᑭᖑᕙᕐᒃᖢᑎᒃ
ᔪᓚᐃᒥ ᓂᒡᓕᐊᔫᓐᓇᕐᒃᑐᑦ
ᖁᙳᐊᒡᓚᒃᖢᑎᒃ ᐅᕐᑲᓗᕋᐅᔭᕐᒃᖢᑎᒃ.
ᔪᓚᐃᒥ ᑕᓴᓚᐅᕐᒃᑖᒃᑲ ᓂᒡᓕᐊᔪᑦ.”

—ᐅᓘᑕ ᐊᕐᓇᑲᒡᓚᒃ

“ᑕᒪᒃᑯᐊ ᓱᕐᕋᕐᒃᑰᔨᙵᑎᕐᕙᕐᒃᐳᑦ
ᓇᐅᔭᐃᒡᓗ ᕿᕐᑯᒡᓗᐃᒡᓗ
ᑭᓯᐊᓂ ᐅᒃᐱᒃᔪᐊᑦ ᑕᐃᒪ
ᐱᑕᕐᑲᕈᓐᓇᐃᒡᓕᔪᐊᓘᕐᒃᑰᔨᖕᒥᔪᑦ
ᑕᑯᓚᐅᙱᓐᓇᒪ.
ᐃᕙᕙᓚᐅᕐᒃᓯᒪᒪᒐᓗᐊᕐᒪᑕ ᑲᖑᒎ
ᖃᓂᒋᔭᖁᓗᐊᓂ ᐃᕙᔪᐊᓘᒡᓗᑎᒃ
ᐊᑭᐊᓂ, ᑕᑯᓚᐅᕐᒃᑲᒻᒥᐊᓗᒍᓐᓃᕐᒃᑐᖓ
ᑲᖑᕐᒥᒃ ᐃᕙᔪᒥᒃ, ᖃᓄᑲᓪᓚᕈᓗᑭᐊᕐᒃ
ᑕᓪᓕᒪᒡᓗᑭᐊᕐᒃ ᐊᕐᕌᒍᑦ
ᐅᖓᑖᓂᐅᕐᒃᑰᕐᒃᑐᖅ. ᑕᓪᓕᒪᑦ ᐊᕌᒍᑦ
ᐅᖓᑖᓂᐅᕐᒃᑰᕐᒃᑐᖅ.”

—ᐅᓘᑎ ᓴᖑᔭ

“ᑕᐃᒪᑐᐊ ᓇᓄᐃᑦ,
ᓇᓄᕐᒃᓗᐊᓚᐅᕐᒃᓯᒪᙵᑦᑑᒐᓗᐊᕐᒃ
ᑕᐅᕙᓃᑉᐸᒃᖢᑕ ᓄᓇᓯᐊᕐᒥ
ᒫᓐᓇ ᓇᓄᓕᐊᓘᕐᒃᑦᑕᓕᕐᒃᑐᕐᒃ.
ᐹᓂᓗ ᐊᕝᕗᓇᐊᕐᕕᒥᕙᒃᑕᑦᑎᓐᓂ
ᑲᖏᕐᒃᖢᖕᒥ ᑐᓂᕈᔪᖕᓂ
ᓇᓐᓄᒃᑐᕐᑲᒻᒪᓇᓚᐅᓇᑦᑐᓂ ᑕᐹᓂ
ᐊᒻᒪ ᑎᑭᕚᕐᒃᑕᐅᒃᑲᓐᓂᓇᑦᑐᑕ
ᑕᐃᒪᓐᓇᐅᓚᐅᕆᕙᕐᒃᐳᕐᒃ.
ᑐᒃᑐᕐᑲᕈᓐᓃᖕᓛᕐᒃᖢᓂ
ᓇᓄᕐᑲᕐᓂᕐᒃᓴᐅᓕᕐᒃᖢᓂ. ᓇᓄᐃᒃᑰᕐᒃ
ᓄᖑᓕᕐᒪᑕ ᓄᖑᓕᕐᒃᑰᙵᑎᑦᑐᑦ.”

—ᐅᓘᑎ ᓴᖑᔭ

“I notice about the birds. They were singing early, but it got too cold and they were really quiet for a while, and then they started to sing in July. I never used to hear them singing in July. […] They are usually singing a lot in June and are quiet in July. This year I heard them in July. It's the first time I was aware of that.”

—Rhoda Arnakallak

“There seems to be no difference with seagulls and fulmars. But it seems to be a long time since I've seen snowy owls. Snowy owls usually have their nests near snow goose nests. Even though I've been going goose hunting, I haven't seen any snowy owls close by […] for the past five years, or more than that.”

—Ruth Sangoya

“There seem to be more polar bears around this area. When we went blueberry picking to Tunirujuit, we even had to kill a polar bear because the polar bear was following us, and then a few years later another polar bear showed up. […] There seem to be no more caribou, but there are more polar bears. And we keep hearing about receding polar numbers.”

—Ruth Sangoya

ᐅᑭᐅᒥ ᐊᓯᔾᔨᖅᑕᕐᓂᖅ

“ᓰᕐᓗ ᐃᒪᓐᓇᐃᒻᒪᓈᖕ ᐊᐅᔭᒃᑯᑦ ᓯᓚᕈᔩᖕᒋᓐᓇᐅᔭᕐᓂᖅᓴᐅᓕᕐᓴᕐᒥ ᓯᕐᑭᓐᓂᖅᑲᑦᑕᓇᐊᕐᑲᕐᓗᐊᖅᖢᓂ ᐅᖅᑰᕐᑲᑦᑕᓇᐊᕐᑲᕐᓗᐊᖅᖢᓂ ᐃᒃᑮᖑᑯᑖᖕᒋᓂᖅᓴᐅᑦᓗᓂ ᐊᐅᔭᒃᑯᑦ ᐊᒻᒪ ᐅᑭᐅᕚᖕᖓᒥ ᐊᐳᑎᕐᑲᑦᑎᐊᓇᐊᕐᑲᓕᕐᓗᐊᖅᖢᓂ ᐊᐳᑎᖅᖖᒋᓰᔭᑯᑖᒃᖢᓂ ᐅᑭᐅᒃᑯᑦ ᓯᓚᑦᑎᐊᕙᐅᕈᔪᐃᓐᓈᓗᒃᖢᓂ ᖃᓂᕈᓐᓇᑎᐊᖖᒋᓐᓂᖅᓴᐅᓕᖅᖢᓂ ᐃᓪᓗᖕᒑᖕᒐᓅᖕᒐᕈᔪᖅᑰᔨᖅᑲᑦᑕᓕᖅᑐᖅ ᑖᓐᓇ ᓯᓚᕈᔩᖅᑲᑦᑕᕐᓂᖕᒐ ᓰᕐᓗ ᐊᐅᔭᖖᒍᖅᑎᖕᒑᖅᖢᓂ ᐅᑭᐅᖖᒍᕐᒥ ᓯᓚᖅᑮᓐᓇᐅᔭᓕᖅᖢᓂ ᐅᑭᐅᓕᒫᖅ ᑕᕝᕙ ᓯᓚᖅᑮᓐᓇᐅᔭᕐᓂᐊᖅᖢᓂ ᖃᓂᕈᓐᓇᖖᒋᑦᓗᓂ ᐊᐳᑎᑭᑦᑐᑐᓘᓗᓂ ᐅᑭᐅᓕᒫᖅ. ᐊᐅᔭᖖᒍᕐᒥ ᒪᕐᑯᕈᔪᐃᓐᓇᐅᔭᓕᖅᖢᓂ ᐊᒻᒪ ᐅᖅᑰᓯᕚᒃᑲᓗᐊᖅᖢᓂ ᐃᒃᑮᖑᑳᑦᓚᒃᐸᒃᖢᓂ ᐅᖅᑰᓯᑳᑦᓚᒃᑐᐊᓘᕚᒃᖢᓂ ᐃᒪᐃᓕᐅᖕᒑᖅᑲᑦᑕᓕᖅᑐᖅ. ᑖᓐᓇ ᓯᓚᐅ ᐅᖅᑰᓂᓇᔭᓇᐊᕐᑲᖅᑕᖕᒐ ᓰᕐᓗ ᓂᒡᓚᓰᖕᓂᓇᔭᓇᐊᕐᑲᖅᑕᖕᒐᓗ ᖃᓄᖅ ᑕᐃᒪᓐᓇᑦᑕᐃᓐᓇᕈᔪᖅᑯᔨᓕᖅᖢᓂ.”

—ᐃᓚᐃᔾᔭ ᐸᓂᒃᐸᑯᑦᑐ

“ᖃᐅᔨᕐᑲᑦᑕᓚᐅᖅᓯᒪᒐᓗᐊᕐᒪᓕ ᐃᒪᓐᓇᐃᒻᒪᓈᒃ. ᒪᐃᒥᓘᓐᓃᑦ ᐃᑖᓐᓂ ᒪᐅᔮᓘᕐᑲᑦᑕᖅᓯᒪᖕᒪᑦ ᖃᓐᓃᓐᓇᕈᔫᒃᑐᐊᓘᑦᓗᓂ ᐊᐳᑎᐊᓗᐊ ᐃᔾᔪᓯᔪᐊᓘᖕᒪᑦ ᐊᓄᕚᖅᑐᐊᓘᖕᒪᑦ ᓄᓇᒧᑦ ᑕᒪᐅᖕᒐ ᐃᓱᓕᑦᑐᐊᓘᑦᓗᓂ ᐊᐳᑎᖅᑯᖅᑐᓂᖅᓵᓘᕐᑲᑦᑕᓚᐅᖅᓯᒪᖕᒪᑦ ᐃᓪᓗᐃᑦᓘᓐᓃᑦ ᒪᑯᐊ ᕐᑯᓕᖕᒋᑦ ᐊᐳᑎᐊᓘᕐᑲᑦᑕᓚᐅᖅᑐᑦ

Seasons

“During the summer, the weather is different than before. Back then, usually summertime was sunny—cloudy at times. But nowadays it's cloudy or raining. During the winter back then, the land used to get more snow, but now we hardly get snow in the wintertime. [...] The weather has switched, and summer is winter.”

—Elijah Panipakoochoo

“But in the winter we have hardly any snowfall, which results in less snow. In June when the sun gets hot, the snow melts it right away. We used to have thick snow, but these past few years the snow comes later. I think this is why plants grow earlier and dry faster. I remember it used to be in April or May that we would have lots of snow. As we would have more snowfall, the wind would blow all the snow on the ground. Even the roof would have lots of snow. We would have so much snow, and now we have hardly any these past few years. The snow used to be up to our legs.”

—Elisapie Ootoova

ᐊᐳᑎᖅᑲᑦᑎᐊᖏᓐᓇᔅᓲᔭᓗᒃᕗᓂ
ᒫᓐᓇᓕ ᐊᐳᑎᖅᑲᑦᑎᐊᖏᑦᑐᐊᓘᑦᑐᓂ
ᐊᐳᑎᖅᑲᑦᑎᐊᖐᓐᓇᖅᑐᐊᓘᑦᑐᓂ
ᐊᖅᖀᒍᓇ ᖃᔅᓯᓄᑭᐊ
ᐊᐱᑐᐃᓐᓇᖅᕗᓂ ᐊᐳᑎᐊᓗ
ᓂᐅᒍᑦ ᓈᒻᒪᒃᑐᐊᓘᖅᑲᑦᑕᓚᐅᖅᒪ
ᐃᖏᖅᕋᐅᓯᖅᑕᐅᕙᒃᕗᓂᓘᓐᓃ ᐊᓄᓇᒍ,
ᑕᐃᒪᐃᓚᐅᖅᓯᒪᖖᒋᑦᑐᑐᖅᑳᓘᓕᖅᑐᖅ."

—ᐃᓕᔅᐱ ᐆᑦᑐᕋ

"The winter has been arriving later and the ice melts earlier, in the beginning of July. Ice melts in July. It varies; melting [times] and the temperature cooling varies."

—Gamailie Kilukishak

"ᐅᑭᐊᖅᓇᓵᖅᓂᖅᓴᐅᖅᑲᑦᑕᖅᑐᖅ
ᐊᔾᔨᐆᖅᑲᑦᑕᖏᑦᑐᖅ ᓯᑯᓗ
ᐊᐅᒃᓵᓕᔪᐊᓘᕙᒃᕗᓂ ᑕᒫᓂ
ᔪᓚᐃ ᐱᖏᐊᖅᓂᖔᓂ ᐃᓛᓐᓂᑦᑐ
ᓄᖑᓕᖅᕗᓂ ᔪᓚᐃ ᐊᐅᓲᖑᑦᑐᓂ
ᐊᔾᔨᐆᖅᑲᑦᑕᖏᑦᑐᑦ, ᐊᐅᒃᐸᑦᑎᐊᓂᖔ
ᐊᒻᒪᓗ ᓂᓪᓚᖅᐸᑦᑎᐊᓂᖔ
ᐊᔾᔨᐆᖅᑲᑦᑕᖏᑦᑐᖅ."

—ᒐᒪᐃᓕ ᖅᑭᓗᖅᑭᓵᖅ

ᓯᓚ/ᐊᓄᓇ

"ᓯᑯ ᐊᓕᒍᐊᓘᔮᖅᐸᓕᖅᑐᖅ;
ᐃᖏᖅᕋᕓᒃᓴᐅᖅᑲᑦᑕᖏᑦᑐᖅ,
ᐊᐳᑎᖅᑲᖅᐸᒍᓐᓃᖅᒪᑦ
ᑎᑦᑕᐅᑐᐃᓐᓇᖅᑲᑦᑕᓕᖅᒪᑕ
ᐊᓄᓇᒍᑦ. [...] [ᐊᓄᓇᒍᖅ]
ᑎᑦᑕᐅᑎᑕᐅᑐᐃᓐᓇᖅᐸᓕᖅᒪᖅ. [...] [ᓯᑯ]
ᓵᓐᓂᖅᓴᐅᕙᓕᖅᑐᖅ ᖁᐊᓵᓘᑦᑐᓂᓗ.
ᑕᓲᑦ ᐊᐳᑎᓕᐊᓘᔭᖀᖓᑕ
ᐃᔾᔪᓂᖅᓵᓘᕙᒃᑐᖅ, ᓲᖅᑲᐃᒻᒪ
ᐅᒃᓯᒃᑎᕈᑎᖅᑲᖅᒥ."

—ᐊᐃᑉᕋᕼᐊ ᑯᓄᒃ

Climate/Weather

"The ice is like glass; you can hardly drive on it, because there is no snow from blowing wind. [...] [The wind] just blows the snow. [...] [The ice] is freezing thinner and it's more slippery. When the lake is covered in snow the ice is thicker, because it's got insulation."

—Abraham Kunnuk

“ᑕᐃᒪᓐᓇ ᓱᓕ ᓄᕕᐱᕆᒥ ᓯᑯᓲᖑᔅᖤᖅᐳᖅ, ᐊᒃᑑᕙᒥ ᓯᑯᕙᓪᓕᐊᓪᓗᓂ ᓄᕕᐱᕆᒥ ᓯᑯᓲᖑᑦᑕᖅᐳᖅ. ᓯᑯ ᐃᔾᔪᔪᐊᓘᖃᑦᑕᓚᐅᖅᓯᒪᒐᓗᐊᕐᒪᑦ ᓵᓐᓂᖅᓴᐅᖃᖅᑕᔅᖤᓕᖅᐳᖅ. ᐊᒻᒪᓗ ᐅᐱᕐᖔᒃᑯᑦ ᐊᐅᓪᓚᖅᖢᑕ ᐃᒻᒪᒃᑎᓐᓂᐊᓗᐃᒻ ᒪᑯᐊ ᐊᐳᑎᖃᕈᓐᓃᖅᖢᓂ ᓂᓚᐃᓐᓇᐅᓕᖅᖢᓂ ᓯᑯ ᓂᓛᓘᖃᑦᑕᓚᐅᖅᓯᒪᖕᒪᑦᓕ ᐃᒻᒪᒃᑎᓐᓃᓪᓗ ᐃᑎᔪᐊᓘᓕᖅᖢᑎᒃ ᓱᓕ ᐊᖕᒪᖕᒋᔅᖢᓂ. ᒫᓐᓇ ᐃᒻᒪᒃᑎᓐᓃᓪᓗ ᐃᓗᑭᔅᖢᑎᒃ ᐊᖕᒪᓕᕇᖅᐸᒃᖢᑎᒡᓗ. ᓯᑯᓵᑦ ᐊᕐᑭᑦᑐᐊᓘᓂᖅᓴᐅᖅᑰᔨᖃᑦᑕᓕᖅᑐᑦ ᐊᐳᑎᕈᔫᖅᑰᔨᓂᖅᓴᐅᓪᓗᓂ ᐊᕐᑭᑦᑑᓪᓗᓂ. ᑕᐃᔅᓱᒪᓂᓕ ᓱᕈᓯᐅᑎᓪᓗᖑ ᐃᓐᓇᐅᓕᕐᓂᓐᓄᑦ ᓂᓛᓘᖃᑦᑕᓚᐅᕐᓗᐊᕐᒪ ᒫᓐᓇ ᐊᕐᑭᕈᔪᖕᓂᖅᓴᐅᖃᑦᑕᓕᖅᑐᖅ. ᓇᓂᑐᐃᓐᓇᖅ ᒫᓂ ᓯᑯᒥ ᐊᐅᒃᑲᓐᓂᖃᖃᑦᑕᕐᒪᑦᓕ ᒫᓂ ᓂᕈᑭᓐᓂᕐᓂ ᑭᓯᐊᓂᓕ ᒪᓐᓇᑐᐃᓐᓇᑐᓕᒫᖅ ᐊᖕᒪᖅᓴᕋᐃᑦᑐᑯᓘᖃᑦᑕᖅᑐᖅ ᑭᓪᓚᕐᓄᑦ. ᓂᕈᑭᓐᓃᑦ ᑐᐊᓐᓃᑦ ᐊᐅᒃᑲᕐᓂᐅᖕᒪᑕ ᒪᓐᓇᓗ ᐊᐅᒃᑲᒡᒍᔪᐊᓘᓪᓗᓂ ᐃᖕᒋᕐᕋᓂᕐᒧᑦ ᓄᖑᑕᐅᔪᖅ ᒪᐅᓇᑐᐃᓐᓇᕐᓕ ᓯᑯᑐᐃᓐᓇᒃᑯᑦ ᐊᖕᒪᖅᑐᕐᓂᖅᓴᐅᖃᑦᑕᖅᑐᑦ ᑭᓪᓚᖃᐅᕐᓂᖅᓴᐅᖃᑦᑕᖅᑐᑦ.”

—ᐃᓕᓴᐱ ᐆᑦᑐᕙ

“ᒫᓐᓇ ᑕᐃᒪᓐᓇᐃᖅᑰᔨᔪᓐᓃᕐᒪ ᓇᐅᒃᑯᕈᓘᔭᑐᐃᓐᓇᖅ ᓯᓚᓗᖅᑰᔨᖃᑦᑕᓕᕐᒪ ᓯᓚᑦᑎᐊᕙᐅᓪᓗᓂᓗ ᐅᓐᓄᒑᖓᑦ

“The ice freezes in November, and in October it starts freezing. Ice used to be so thick, and now it is thinner. And in the spring we would travel on the sea ice and there would be deep meltwater [on] top of the ice. It would be slush free deep, and without melt holes yet. Now, meltwaters are shallower and holes form in them right away. Nowadays the nearly formed ice is not as hard and solid as it used to be. Nearly formed ice [used to be] hard and solid. From when I was a child to the time I became an adult, ice was solid. Now it seems less hard. In narrow places, there are melt holes formed by currents.”

—Elisapie Ootoova

“The weather has been really unpredictable. Some days it’s nice out, but then in the evening it changes to bad weather.”

—Annie Peterloosie

ᓯᓚᕐᑎᐊᕙᐅᑦᑖᓗᒃᔪᓂ
ᓯᓚᕐᑎᐊᕙᐅᓕᑳᑦᓚᒃᔪᓂ, ᖃᓄᖅ
ᓄᕗᔭᖕᒐ ᖃᓄᐃᓕᐅᕈᓘᔭᖅᑕᕐᒪ,
ᑕᒪᔾᔭ ᐊᓯᔾᔨᖅᐸᑦᑎᐊᖅᑰᔨᒥᔭᕋᓕ."

—ᐋᓂ ᐲᑦᑐᓘᓯ

“ᐊᒻᒪ ᓱᕈᓯᐅᑎᑦᑐᖕᒐ ᒪᒃᑯᒃᓯᓂᓐᓄᑦ
ᓱᕐᑭᓂᖅ ᐅᖅᑰ... ᐃᑖ ᐊᐳᑦ
ᐊᐅᒃᐸᑦᓕᐊᑎᑦᑐᒍ ᐃᓯᑦᑎᒍᑦ
ᐋᓂᐊᖅᑐᐊᓘᖅᑕᑦᑕᓚᐅᖅᓯᒪᖕᒥᒐᑦᑕ
ᐃᑦᑐᖕᓂᕋᖅᑕᐅᑦᑐᑕ
ᐅᑦᑐᓄᑦ ᐱᖕᒐᓱᓂᒡᓘᓐᓃᑦ
ᐃᓯᕗᑦ ᐋᓂᐊᖅᑐᐊᓘᑦᑐᑎᒃ
ᐊᖑᑎᑦ ᐱᑐᐊᖅᔪᑎᒃ,
ᑕᐃᒪᐃᖅᑲᑦᑕᖅᑐᐊᓘᓚᐅᖅᓯᒪᖕᒪᑕ.
ᑕᑉᕙ ᐱᔨᓐᓃᕋᒥ ᑕᐃᒪᐃᓕᒃᑲᓐᓂᔾᔫᕋᓂ
ᐅᐱᕐᖕᒑᓇᔭᖕᒐᓂ. ᒫᓇ
ᐃᑦᑐᒃᑐᖅᑲᕈᓇᐃᑦᑎᔪᖅ
ᐃᒡᒐᖅᖖᒋᒃᑲᑐᐊᕐᑐᓂ
ᐃᑦᑐᖅᑲᑦᑕᕈᓐᓃᖅᑐᑦ. ᐃᒻᒪᖄ
ᐊᐅᒫᔮᑐᐊᖅᑲᑦᑕᕈᓐᓃᕋᒥᒃ
ᐃᑦᑐᖅᑲᑦᑕᕈᓐᓃᖅᑐᑦ ᒫᓇ ᑭᓯᒧᒃᑭᐊ
ᐊᐳᑦᑐᑭᐊᖅ ᐊᔾᔨᕙᓐᓃᕐᓂᖕᒐᓄᑦ
ᐊᐳᑎᖅᑕᑎᐊᖕᒋᓐᓂᖕᒐᓄᑦᑐᑭᐊᖅ
ᖃᓄᑭᐊᖅ ᓇᑐᒐᑐᐊᖅᔪᑎᒍᑦ
ᑭᓯᐊᓂ ᐃᑦᑐᒃᑐᖅᑲᖅᑕᑦᑕᕈᓐᓃᖅᑐᖅ
ᑕᐃᒪᓐᓇ ᐃᓯᖕᒥ ᐋᓂᐊᖅᑐᐊᓘᑦᑐᑎᒃ
ᐋᓂᐊᖅᑐᐊᓘᖅᑕᑦᑕᓚᐅᕐᑕ
ᐃᓯᑦᑎᒍᑦ ᖃᑭᖅᓯᑲᑕᓕᒌᖕᒐᑕ
ᑕᑉᕙ ᐋᓂᐊᖅᑐᐊᓘᑦᑐᓂ,
ᐅᐃᑦᑐᓐᓇᖅᐸᒃᔪᑕ
ᐊᓯᐃᑖᒃ ᐅᐃᑦᑐᓐᓇᖅᓯᒐᑦᑕ
ᖃᓄᐃᖕᒋᑦᑎᐊᓕᖅᐸᒃᔪᑕ
ᐃᓅᓕᓴᓚᐅᖖᒋᔅᔪᑕ
ᖃᓄᐃᑦᑐᓐᓂᖅᐸᒃᔪᑕ, ᒫᓇ
ᐃᑦᑐᒃᑐᖅᑲᕈᓇᐃᑦᑎᔪᖅ ᓯᓚᐅ
ᐊᔾᔨᒥᔫᓐᓃᕐᓂᖕᒐᓅᔫᒃᓴᐅᔫᖅ."

—ᐃᓕᓴᐱ ᐆᑦᑐᕙ

“When I was a child, the sun was hot [...] and when the snow was melting we would get snow blindness for about three days, especially the men. When they were all better, they would not get snow blindness again all spring. Now, nobody seems to get snow blindness, even without sunglasses. Perhaps it's not as warm now, or the snow is different, or there is less snow. I'm not sure. I know that there is less snow blindness now. We would have snow blindness, and when we were finally able to open our eyes again, we were all fine without treatment. Now, you don't see snow blindness.”

—Elisapie Ootoova

“The current has changed a lot. Before, when it was a full moon it was windy. It would be windy and the current would be moving more, but now the current has changed. It's slow now. [...] Now, even when there is a full moon, the current doesn't change much. I think that's why there is not much thicker snow and ice [...] [is] thinner and melts faster, because it's not that salty anymore. All the salt is on the bottom.”

—Jayko Peterloosie

“ᐃᖏᕐᕋᓂᖅ ᐊᔾᔨᒋᕙᒍᓐᓃᖅᑕᖓ. ᑕᐃᔅᓱᒪᓂᓕ, ᑕᖅᑭᖅ ᐃᓗᐃᑦᑑᓕᕌᖓᑦ ᐊᓄᕌᖓᔪᒃᐸᓚᐅᖅᓯᒪᖕᒪᑦ. ᐊᓄᓇᖓᔪᒃᑐᓂ ᐃᖏᕐᕋᓂᖅᑐᓪᓗᓂᓗ, ᐃᖏᕐᕋᓂᖅ ᐊᓯᔾᔮᖅᓯᒪᓕᖅᑐᖅ. ᓱᒃᑲᐃᓕᖅᑐᖅ. [...] ᒫᓐᓇ, ᑕᖅᑭᖅ ᐃᓗᐃᑦᑑᓕᕋᓗᐊᕌᖓᑦ, ᐃᖏᕐᕋᓂᖅ ᓱᕐᕋᑦᑎᐊᖃᑦᑕᕈᓐᓃᖅᑐᖅ. ᑕᐃᒪᐃᓐᓂᖓᓄᑦ ᐊᐳᑎᖃᑦᑎᐊᖅᐸᒍᓐᓃᖅᑐᖅ ᓯᑯᖃᑦᑎᐊᖅᐸᓐᓇᓂᓗ [...] ᓵᓐᓂᖅᓴᖅ ᐊᐅᒃᓴᕋᐃᓐᓂᖅᓴᖅ, ᑕᓇᐅᖑᑦᑎᐊᕈᓐᓃᖅᑐᓂᓗ. ᑕᓇᐅᖅ ᐃᖅᑲᖓᓃᓕᕋᒥ.

—ᔭᐃᑯ ᐲᑦᑐᓘᓯ

“ᐅᐱᕐᖔᒃᑯᓪᓕ ᓲᖃᐃ ᓄᓇ ᐊᐳᑎᐊᓘᖕᒪᑦ ᐊᐅᒃᐸᓪᓕᐊᔪᐊᓘᓕᕌᖓᑦ ᑰᒃᑐᐊᓘᖃᑦᑕᓚᐅᖅᓯᒪᖕᒪᑦ ᑰᒃ ᑖᓐᓇ ᑰᒃᑐᐊᓘᓗᐊᕐᓂᑯᒧᑦ ᒪᑯᐊ ᓄᓇᐅ ᐃᓚᖓ ᓱᐱᖅᓯᐅᑎᑦᑐᓂ, ᐃᒪᓐᓇᐅᖃᑦᑕᕐᒪᑦ ᒫᓐᓇ ᑰᖑᔮᕐᔪᑐᐃᓐᓇᖅᖢᓂ ᐊᐳᑦ ᓄᖑᑦᑕᓈᖃᑦᑕᕐᒪᑦ. ᒪᓐᓇ ᐅᐱᕐᖔᒃᑯᑦ ᓄᓇᒋᒡᒍᒐᑦᑎᒍᑦ ᑕᕝᕙᓕ ᐃᖏᕐᕋᓪᓗᑕ ᓂᒍᓇᐊᖃᖅᐸᒃᖢᑕ ᑰᕌᓘᑉ ᓄᖑᑖᓗᐊ ᓯᑯᖃᕈᓐᓃᖃᑦᑕᓚᐅᖅᓯᒪᔪᖅ. ᒫᓐᓇ ᓱᖁᑕᐅᖖᒋᓐᓂᖅᓴᐅᖃᑦᑕᖅᑐᖅ ᑰᕋᔮᑐᐃᓐᓇᖃᑦᑕᓕᕐᓂᖓᓄᑦ ᓲᖃᐃ ᐊᐳᑎᖃᑦᑎᐊᖖᒋᒧᑦ. ᑰᑐᐃᓐᓈᓘᒐᓗᐊᑦ ᓯᓕ ᑰᖏᑉᐹᓪᓕᖅᑐᑦ ᒪᖁᒃᑐᐊᓘᔭᕌᖓᑦ ᑰᒃᑐᐊᓘᕙᓚᐅᖅᑐᖅ ᑭᓯᐊᓂᓕ ᐊᐅᓕᓵᖅᑎᓪᓗᒍ ᑰᑭᑉᐸᓪᓕᖅᑐᑦ. ᑕᒪᒃᑯᐊ ᓯᕐᒥᖕᒥᖖᒑᖅᑐᑦ ᐊᐅᒃᐸᓪᓕᐊᓂᖏᓐᓄᑦ ᑰᒃᑐᐊᓘᒐᓗᐊᖅ ᓯᓕ.”

—ᐃᓚᓴᐱ ᐆᑦᑐᕗ

“In the spring, of course, when the snow was melting, rivers would flow stronger. The rivers would burst and wash away parts of the land. Now the rivers are not as strong and snow melts faster. As this is our land in the spring, we have to pass the rivers, and they would wash the ice away. They are tolerable now; they flow less and there is less snow. The huge rivers flow less until it starts raining, and then they get stronger again, but when [the ice] first melts they are not as strong. The rivers are still flowing from glaciers.”

—Elisapie Ootoova

“I have noticed the land is drier than before and the smaller lakes have less water. This I have noticed; the lakes are shrinking.”

—Joanasie Muckpa

“ᑕᕝᕙᑐᐊᕐᓕ ᐅᔾᔨᕆᔭᕋ ᓄᓇ ᑕᒪᓐᓇᓕ ᐸᓂᖅᐸᖕᒪᑦ ᑕᐃᒪᖖᒐᓂ ᑕᓰᑦ ᒥᑭᑦᑐᑦ ᐃᒪᑭᑦᑐᑯᓘᓕᕐᒪᑕ ᒫᓐᓇᒃᑯᑦ ᑕᕝᕙ ᐊᓯᔾᔨᕈᑎᖕᒐ ᖃᐅᔨᔭᑐᐊᕋ, ᑕᓰᑦ ᒥᑭᑦᑐᑦ ᐃᒪᐃᓕᖅᑐᑦ.”

—ᔪᐊᓇᓯ ᒪᒃᐸ

“The glaciers that are across from us are smaller now. They don’t touch the ocean now, and they seem to be lower.”

—Joanasie Muckpa

“ᐊᒻᒪ ᑕᕝᕙ ᐅᔾᔨᕆᓂᖅᐹᓚᔭᕋ ᐊᑭᐊ ᓯᕐᒥᐊᓗᖕᒋ ᐊᐅᒃᑐᐊᓗᐃ ᐅᓇᓃᑦᑐᑯᓘᓕᖅᕗᑎᒃ ᐊᒻᒪᓗ ᐃᒫᓄᑦ ᓯᕐᒥᖕᓄᑦ ᐃᒫᓄᐊᖕᒐᔪᖃᕐᓈᖅᑐᖅ.”

—ᔪᐊᓇᓯ ᒪᒃᐸ

“I have noticed the glaciers that are across from us are melting away slowly. They were touching the ocean, but they are way up now.”

—Ruth Sangoya

“ᐄ ᐅᔾᔨᕆᓯᒪᔭᒃᑲ ᐊᐅᒃᐹᓪᓕᖅᑐᐊᓗᐃᑦ ᐊᒃᑯᐊᓗ ᓯᕐᒥᐊᓗᐃᑦ ᑕᐃᒃᑯᐊ ᐅᓄᖕᒐ ᖃᓂᖅᑰᔨᓚᐅᕈᓗᐊᕐᒪᑕ ᐃᒫᓄᑦ, ᐹᓂᑲᓪᓚᒃ... ᐊᐅᒃᐹᓪᓕᖅᑐᓪᓛᓗᒃ.”

—ᐅᓘᑎ ᓴᖑᔭ

“[There was a landslide] at Salmon Creek. I think the permafrost [has] thawed, and all the grass and rocks went down probably six or eight feet.”

—Abraham Kunnuk

“[ᑎᓲᓚᐅᖅᓯᒪᖕᒪᑦ] ᕿᓚᓗᒃᑲᓂ. ᓯᑯᐊᓗᒃ ᓄᓇᐅᑉ ᐃᑭᐊᖕᒐᓃᑦᑐᖅ ᐊᐅᓕᕐᒪᑦ, ᓄᓇ ᐅᔭᕌᓗᐃᓪᓗ ᑎᓲᒃᑐᐊᓘᓚᐅᖅᓯᒪᖕᒪᑕ ᐃᓯᒐᐃᑦ ᐊᕐᕕᓂᓖᑦ ᐅᕝᕙᓘᓐᓃᑦ ᐊᕐᕕᓂᓖᑦ ᐱᖕᒐᓱᓂᒃ.”

—ᐊᐃᕝᕙᓚᕼᐊ ᑯᓄᒃ

“The sun is more to the west than before. I think the sun has been getting lower to the west. It’s been going like this slowly. […] The weather is almost like [it is] in the south now. It changes a lot now. The weather is more unpredictable now. […] When I was a child the Elders could predict the weather, but it’s not like that anymore. The Elders didn’t have to depend on the weather station; they could predict the weather. Now it’s not like that anymore.”

—Paniloo Sangoya

“ᓯᕿᓂᖅ ᐅᓐᓄᐊᕗ ᐊᓯᔾᔨᖅᓯᒪᓕᖅᑐᖅ ᑖᓐᓇ ᐅᓪᓗᕐᑲᓚᐅᖅᓯᒪᒐᓗᐊᕋᑦᑕ ᓯᕿᓂᖅ ᒫᓂ ᒫᓐᓇ ᒫᓃᓕᖅᕗᓂ ᖁᒻᒧ ᖁᒻᒧᒃᓯᕆᐊᖃᑦᑕᓕᕐᒪᑦ ᓴᖅᑮᔭᖅᑎᓪᓗᒍ ᑕᐃᒪᐃᓕᕙᓪᓕᐊᖒᕐᔪᐃᓐᓇᖅᑰᖅᑐᖅ ᐊᕐᖒᒍ ᖃᔅᓯᐊᓗᐃᑦ ᖄᖕᒋᖒᖕᒐᑕ ᑭᓯᐊᓂ ᐊᕐᖒᒍᑕᒫᑐᐊᖑᙱᑦᑐᒥ. ᓯᓚᕗᑦᓕ ᖃᑦᑐᓈᑦ ᓯᓚᖕᒐᑐᕐᓚᐅᔭᖅ ᐋᖅᑭᒃᓯᒪᓕᕐᒪᑦ,

ᐱᐅᔪᐊᓘᓂᐊᙳᐊᖅᖁᓂ
ᐱᐅᔪᓐᓃᑳᓪᓚᒃᖁᓂ ᑕᐃᔅᓱᒪᓂ
ᑕᐃᒪᐃᓚᐅᙱᒃᑲᓗᐊᖅᖁᓂ
ᑕᐃᒪᐃᑉᐸᓪᓕᐊᑐᐃᓐᓇᖅᑐᖅ
ᑕᐅᙵᓯᓕᒫᖅ ᒫᓐᓇᓕᓵᖕᒍᙱᑦᑐᖅ.
ᓄᑕᕋᐅᑎᓪᓗᖓᓕ
ᖃᐅᔨᒪᔭᐅᓂᖅᓴᐅᓚᐅᖅᑐᖅ
ᓯᓚᓯᐅᑎᖃᙱᑦᑐᒃᑯᓐᓂ ᓯᓚᐅ
ᖃᓄᐃᓐᓂᐊᕐᓂᖓ ᐱᐅᓯᓂᐊᕐᓂᖓᓗ
ᓄᑕᕋᐅᑎᓪᓗᖓ ᑕᐃᔅᓱᒪᓂ
ᐅᓪᓗᒥᓕ ᓇᓚᐅᑦᑖᒐᒃᓴᐅᔪᓐᓃᖅᑐᖅ,
ᐊᔾᔨᐅᙱᓕᖅᑐᖅ."

—ᐸᓂᓗᒃ ᓴᖑᔭ

"ᐅᑭᐅᒃᑯᓪᓗ ᑕᐃᔅᓱᒪᓂᒥᑦ
ᖃᐅᒪᖃᑦᑕᕐᓂᖅᓴᐅᓕᖅᑐᖅ
ᑖᕋᓗᐊᖅᖁᓂ
ᑕᑯᓐᓇᕐᓂᖅᓴᐅᖃᑦᑕᓕᕐᒪᑦ
ᐅᖓᓯᐊᕐᔪᒃᑐᑯᓗᖑᑦ ᑕᐃᔅᓱᒪᓂᓕ
ᓯᓚᖅᑭᕋᓗᐊᖅᑎᓪᓗᒍ ᒪᓃᓚᖅ
ᑕᑯᒃᓴᐅᔪᓐᓇᓚᐅᙱᑦᑐ ᑕᐅᖓ
ᑕᑯᓐᓇᕐᓂᖅᓴᐅᓕᖅᑐᖅ ᒫᓐᓇ
ᑖᖅᓯᓗᐊᕌᓗᖅᑲᑦᑕᕆᓐᓃᕐᓂᖅᓴᐅᒐᑦᑕ
ᑕᐃᔅᓱᒪᓂᐅᓚᐅᖅᓯᒪᔪᒥᒃ, ᐊᕌᒍᐃ
50 ᐅᖓᑖᓂᖃᐃ ᐱᓪᓗᓂ.
ᒫᓐᓇᑦᑎᐊᕈᓗ ᑕᐃᒪᐃᓕᓵᖅᑰᖏᑦᑐᖅ
ᑕᐃᒪᐃᑉᐸᓪᓕᐊᓲᔭᓗᖅᑰᖅᑐᖅ
ᐃᓅᓚᐅᖅᑎᓐᓇᑕ ᑕᐅᙵᑦ
ᖃᐅᒻᒪᒃᐸᓪᓕᐊᕌᕐᔪᑯᓗᑐᐃᓐᓇᖅᑰᖅᑐᖅ
ᐊᕌᒍᐃᑦ ᖃᔅᓯᐊᓗᐃᑦ ᐊᓂᒍᕋᖓᑦ
ᐅᔾᔨᕐᓇᖅᓯᕙᒃᖁᑎᒃ."

—ᐸᓂᓗᒃ ᓴᖑᔭ

"ᐃᒻᒪᑲᓪᓚᒃ ᐊᒃᓱᐊᓗᒃ
ᓂᒡᓚᓱᖃᑦᑕᓚᐅᖅᓯᒪᒐᓗᐊᕐᒪ
ᐊᓂᓪᓗᓂ ᐅᑭᐅᒃᑯᑦ
ᐃᒻᒪᒃᑲᓐᓂᖅ ᒪᓐᓇᓗ ᖁᐊᕈᓯ...

"I have noticed that there is more light during the winter months. [...] We can see the [sea] ice now during the winter. Before it was too dark. Fifty years ago we couldn't see the ice during the dark season. It's been going on like that for many years, before I was born, too."

—Paniloo Sangoya

"It used to be bitterly cold in the winter, but nowadays it doesn't seem to be that cold anymore."

—Rhoda Arnakallak

"Everybody says that it is global warming [...] but nowadays it's been cold, really cold. [...] It was warmer before. They say it's getting warm, but it's the other way. [...] Even though it seems colder than before, the ice is thinner."

—Ruth Sangoya

"To me it seems like the rain and the wind come together. Usually [...] the storms come in the beginning of September or late September, but today it seems earlier."

—Theresa Koopa Maktar

ᐃᒪᓐᓇ, ᒫᓐᓇᓕ ᑕᐃᒪᓐᓈᓗᒃ ᓂᒡᓚᓰᓯᕐᒥᑑᖅᑲᑦᑕᕈᓐᓅᖅᑐᖅ ᐅᑭᐅᒃᑯᑦ."

—ᐅᓘᑕ ᐊᕐᓇᑲᑦᓚᒃ

"ᐅᖅᑰᓯᕚᑦᓕᐊᓂᓴᖅᑕᐅᒐᓗᐊᖅᓱᓂᓗ ᐅᖅᑰᔭᓐᓇᖏᑦ. ᐅᖅᑰᓂᖅᓵᐅᒃᑲᓐᓂᖅᑰᔮᖅᑲᑦᑕᓚᐅᕈᓗᐊᕐᒪᑦ ᐅᐱᕐᖔᒃᑯᑦ, ᐊᐅᒪᔮᖅᑐᐊᓘᕙᒃᓱᑕᓗ ᐅᖅᑰᓗᐊᓗᑦ. ᐅᐱᕐᖔᖑᔅᓯᓴᓗᐊᕐᒪᑦ ᕿᐅᔾᔨᕈᔭᐃᓐᓇᖅᓱᑕ ᐅᖕᖓᓗᑭᐊᖅ ᐃᒡᓗᕐᔭᐊᒦᓕᕐᓂᑯᓪᓗᑦ, ᖃᓄᑭᐊᖅ. ᓄᓇ ᐃᓛᒃ ᓯᓚ ᐅᖅᑰᓯᕚᑦᓕᐊᓗᐊᖅᑰᔨᒋᖕᒋᑕᓴ ᑭᓯᐊᓂ ᓯᑯ ᓵᓐᓂᖅᓵᓲᖅᑲᑦᑕᓕᕐᒪᑦ."

—ᐅᓘᑎ ᓴᖑᔭ

"ᐊᓄᓂᓕ ᓯᓚᐅ ᐊᔅᔨᖓ, ᐊᓄᓂᓗ ᐱᖅᓯᐅᑎᑦᑐᓂ ᐊᑕᐅᓯᕐᒥ, ᐃᖑᖅᑰᔨᒋᓕᖅᑕᓴᓕ. ᓯᑎᐱᓂᒥ ᐊᓄᖄᕐᔭᐊᖄᓗᓚᐅᖅᓱᓂ ᒫᓐᓇ ᓯᑯᕙᖅᑐᐊᓘᖕᒪᑦ ᐅᖅᑦᑎᐊᓂᐊᖓ ᓇᓗᓇᖅᓯᔭᐊᓗᒃ."

—ᑐᓃᓴ ᑯᐹᖅ ᒪᒃᑖᖅ

ᐃᓄᐃᑦ ᐱᖅᑯᓯᑐᖃᖓᑦᑕ ᐊᒃᑐᖅᑕᐅᓂᖓ

"ᑕᐃᔅᓱᒪᓂ ᓱᕈᓯᐅᓚᐅᖃᒪᐃᓛᒃ ᐃᑲᓂ ᓇᑦᑐᐊᓅᕐᓱᑕ ᐊᑖᑕᒪ ᓇᓂᐊᖅᑲᑦᑕᓚᐅᖅᓯᒪᔪᒃ ᑕᐅᕗᖓ, ᓇᓄᖅᑲᑦᑕᓚᐅᖅᑐᖅ ᐊᑕᐅᓯᖅ ᒪᕐᕈᖕᓅᓐᓂᑦ ᐊᖁᔪ ᐃᓗᐊᓂ. ᒫᓂᓗ ᓇᓄᒃᑐᖅᑲᖅᑲᑦᑕᓚᐅᖅᒥᔭᖅ

Impacts on Traditional Ways of Life

"Before, in a year we would catch a polar bear or two, and in another camp they would catch one or not see one at all, [...] but now there are more polar bears, not only in Pond Inlet but at other campsites. When they would catch a whale they would cache it [...] during the summer and go pick it up in wintertime, [...] but now whatever they catch is gone because the polar bears always get the cached meat. So people hardly cache the meat anymore."

—Elisapie Ootoova

"There's an old prediction that the North will get warmer and the south colder. Maybe it's coming true now. I am also aware of the ice melting faster, and it's not getting as thick as it used to. It stays thinner. Hunters cannot go to some areas now because the ice is not thick enough. [...] Even the sealskins, they are not supposed to be moulting like that. They normally shed in June, but [now] even in the wintertime the fur comes off easily. And I have heard that some of the seals have no fur, maybe from the heat."

—Rhoda Arnakallak

ᐃᑖᓐᓂ ᐃᒐᕐᔪᐊᕐᒥᐅᑕᐅᓕᖅᖢᑕ
ᓇᓐᓄᒃᑐᕐᑲᕐᑲᑦᑕᓚᐅᙱᑦᑐᖅ
ᑭᓯᐊᓂ ᓇᓐᓄᒃᑐᕐᑲᓚᐅᖅᐸᒃᖢᓂ,
ᓇᓄᕐᑲᓚᐅᙱᑦᑐᖅ,
ᑐᒥᓯᐊᖅᑐᕐᑲᖅᑐᕉᖅ ᓇᐅᒃᑯ
ᑐᓴᑕᐃᓐᓇᖅᐸᒃᖢᑕ, ᒫᓐᓇ
ᓇᓄᓕᐊᓘᓕᖅᑐᖅ ᑲᑉᐱᐊᓇᖅᓯᔪᐊᓗᒃ
ᑕᒪᓐᓇᓕᒫᖓ. ᒫᓂᓘᓐᓃᑦ ᓄᓇᒥ
ᐅᕙᓃᑦᑐᑕ ᑎᑭᑕᐅᕐᑲᑦᑕᓚᐅᕐᒥᔪᒍ
ᒪᕐᕈᐃᖅᓯᖅᖢᑕ ᕿᓚᒥᒍᓗᒃ
ᐱᓇᓱᐊᕈᓰᒡᓗᓐᓃᑦ ᒪᕐᕉᒃ ᑕᐃᑲᓃᒃᑲᑦᑕ
ᓇᓄᐊᓗᖕᓄᑦ ᐅᐸᒃᑕᐅᕐᑲᑦᑕᓚᐅᖅᑐᒍᑦ,
ᓇᓄᓕᐊᓘᓕᖅᑐᒥᒃ ᐅᕐᑲᕈᓐᓇᖅᑐᖓ
ᓇᓄᓕᐊᓘᓕᖅᑐᖅ. ᐅᒥᐊᖅᑐᖅᑐᓪᓗ
ᓇᓄᕐᒥᒎᖅ ᑕᑯᖅᑲᐅᔪᑦ
ᓲᕐᑲᐃ ᐱᒋᐊᔾᔩᙱᖦᖢᑎᒃ
ᓇᓄᓕᐊᓘᓕᕐᓂᖓᓂᒃ ᕐᑲᐅᔨᒪᕗᒍᑦ
ᑕᒪᓐᓇᓕᒫᖅ ᓲᕐᓗ ᐃᒃᐱᐊᕐᔪᖕᒥᐅᓂ
ᑲᖏᖅᑑᒐᐱᒡᓗᓐᓃᑦ ᑕᒪᓐᓇ
ᐊᕗᖓᓕᒫᕌᓗᒃ ᐃᒡᓗᓕᒡᓗᓐᓃᑦ
ᑕᒫᓂ ᓄᓇᕗᑦ ᐃᓗᐊᓂ
ᓇᓄᐃᑦ ᐅᓄᖅᑐᐊᓘᓕᖅᑐᑦ.
ᑖᓐᓇ ᐅᔾᔨᕆᔭᕗᑦ. ᑕᐃᔅᓱᒪᓂ
ᒥᑦᑎᒪᑕᓕᖕᒥᐅᑦ ᓴᓐᓂᕈᒻᒥᒎᓐᓃᑦ
ᑕᒫᓂ ᕿᓚᓗᒐᕈᑎᒃ ᕿᖕᓂᓗᒍ
ᒪᐅᖓᓘᓐᓃᕐᒪ, ᒪᐅᖓᓅᓐᓃᑦ
ᕿᖕᓂᖅᑲᑦᑕᓚᐅᖅᑐᑦ ᐅᖕᒐᓯᒃᑐᒥ
ᐅᑭᐅᕐᒪᑦ ᑕᕝᕙ ᐱᔭᐅᒋᕐᒥ ᐱᓪᓗᓂᔅᔪᒃ.
ᒪᐅᖓᓅᓐᓃᑦ ᐊᐅᓪᓚᖅᓯᒪᓪᓗᑕ
ᕿᖕᓂᖅᑕᕐᒥᔫᒐᓗᐊᑦ ᓇᑦᑎᕐᓂᒃ
ᓂᕿᒃᓴᕆᓪᓗᒍ ᐃᒍᓇᒃᓴᑦᑎᓐᓂᒃ
ᐅᑭᐅᕐᒪᑦ ᐱᔭᐅᒋᐊᕐᒪᑦ ᐱᑕᕐᑲᙱᑦᑐᖅ
ᓇᓄᕐᒥᑦ ᓂᕆᔭᐅᔪᕕᓃᑦ,
ᕿᓐᓂᔪᓐᓇᐃᑦᓕᔪᑦ ᓄᓇᓕᖕᓂ ᑭᓯᐊᓂ
ᕿᓐᓂᕐᑲᕈᓐᓇᖅᑑᒐᓗᐊᑦ. ᓇᓄᕐᓂᑦ
ᓂᕆᔭᐅᕐᑲᑦᑕᖅᑐᑦ.”

—ᐃᓕᓴᐱ ᐆᑦᑐᕙ

“We bleach our skins in February. They don't bleach very well now [compared to how] they used to. They dry too fast.

—Theresa Koopa Maktar

“We used to go to Iqaluit to go fishing, but now the ice is breaking up so early; it's dangerous to travel.”

—Rhoda Arnakallak

“ᑐᓴᐅᒪᓐᓇᕐᒪᑦ ᒪᓐᓇᒨᖅ
ᐅᖅᑰᓯᖕᒑᓛᖅᑐᐊᓘᖕᒪᑦ ᑕᐅᖕᓇᒨᖅ
ᓂᓪᓚᖅᑎᑦᑐᒍ ᐊᓯᐃᓛᖅᑲᐃ
ᐅᖅᑰᓯᓕᖅᐳᒍ. ᐅᔾᔨᕈᓯᓐᓇᔪᖕᒥᔪᖕᒐ
ᐊᐅᒃᓴᕋᐃᓕᖅᑐᖅ ᓯᑯ ᐊᒻᒪ
ᐃᔾᔪᖅᑎᓗᐊᕈᓇᐃᑦᓕᔪᖅ
ᑐᖁᑮᓐᓇᐅᔭᖅᑐᖅ ᑐᖁᑭᑦᑐᖅ ᑕᐃᒪᐃᓇ
ᐅᑭᐅᖑᓕᓗᐊᖅᑎᑦᑐᒍ ᐊᒻᒪ ᕿᓰᑦ
ᒪᒫᓪᓗᐊᙱᑦᑐᑦ ᐅᐱᖖᒑᒃᑯᑦ ᔩᓂᒥ
ᔪᓚᐃᒥ ᒪᒫᓰᖑᖕᒪᑕ ᐊᐅᔭᙳᕌᖕᒐᑦ
ᒪᒫᕈᓐᓃᓰᑦ ᐅᑭᐅᖑᓕᓗᐊᒃᑯᑦ
ᐅᑭᐊᒃᓵᖑᓕᓗᐊᒃᑯᑦ ᓯᓕ ᕿᓯᖕᒋᑦ
ᐱᐅᑦᑎᐊᖅᑰᔨᖅᑲᑦᑕᕈᓐᓃᖅᑐᑦ, ᐊᓯᐊᓂᒃ
ᒥᖅᑯᖅᑖᕋᓯᓰᖑᖕᒪᑕ. ᐊᒻᒪ ᑐᓴᓚᐅᖅᑐᖕᒐ
ᐃᓚᖕᒋᑦ ᒥᖅᑯᖅᑲᖅᑲᑦᑕᕈᓐᓃᖅᑐᑦ
ᒫᓐᓇ ᐃᓚᖕᒋᑦ ᒥᖅᑯᐃᔭᖅᓯᒪᔪᐃᑦ
ᐆᓇᓗᐊᖅᑐᓘᓇᓯᕆᑦᑐᑎᒍᑦ.”

—ᐅᓘᑕ ᐊᕐᓇᑲᓪᓚᒃ

“ᐅᔾᔨᕆᖕᒥᔭᕋ ᕖᕝᕗᐊᕆᒥ ᕿᓯᖕᒥᒃ
ᓴᓇᓪᓗᓂ ᖃᐅᓪᓗᖅᑖᓄᖕᒥᒃ
ᕿᓯᐅᑎᖃᕈᒪᓪᓗᓂ ᕖᕝᕗᐊᕆᒥ
ᐊᓂᒃᑲᓗᐊᖅᖢᒍ ᖁᐊᖅᓯᐊᖅ ᒪᓐᓇ
ᖃᑯᖅᓯᓇᓗᐊᕈᓃᖅᑐᐊᓗᐃᑦ
ᐸᓂᔭᑦᑐᐊᓗᒃ ᐅᑭᐅᒃᑯᑦ
ᐸᓂᑲᐅᖅᑐᔭᑦᑐᐊᓘᓕᖅᑐᑦ.”

—ᑐᓅᓴ ᑯᐹᖅ ᒪᒃᑖᖅ

“ᑕᐃᒪ ᐅᐱᖖᒑᒃᑯᑦ
ᐃᖃᓗᓕᐊᖃᑦᑕᓚᐅᕋᓗᐊᕋᑦᑕ
ᓇᖏᐊᕐᓇᕈᔪᐃᓐᓇᐅᔭᖅᑐᑦ ᑕᑉᐸᐅᖕᒐ
ᐃᖃᓪᓕᐊᓚᐅᙱᕐᕋᖅᐳᒍ ᐊᒻᒪ
ᖃᓄᖅ ᓯᑯ ᐅᑭᐅᖑᓕᓗᐊᒃᑯᑦ
ᑖᓇ ᓇᖏᐊᕐᓇᕈᔪᐃᓐᓇᐅᔭᖅ
ᑐᖁᑭᕈᔪᐃᓐᓇᐅᔭᖅ ᑖᓇ
ᐅᔾᔨᕆᓗᐊᖅᑰᔨᖅᑲᑦᑕᕐᕋᖅᑯ.”

—ᐅᓘᑕ ᐊᕐᓇᑲᓪᓚᒃ

ᐸᖕᓂᖅᑑᖅ
Pangnirtung

ᐸᖕᓂᖅᑑᖅ

ᐸᖕᓂᖅᑑᖅ 50 ᑭᓛᒥᑐᓂᒃ ᐊᑦᑎᖕᓂᖅᓴᐅᔪᖅ ᐅᑭᐅᖅᑕᖅᑑᑉ ᓇᓗᓇᐃᖅᑕᐅᓯᒪᓂᖓᓂᑦ, ᖃᓂᖦᖢᑎᒃ ᐸᖕᓂᖅᑑᑉ ᓯᒡᔭᖃᙵᕐᕕᕕᓂᐊᓂ ᐅᓯᐊᓘᑉ ᖃᓂᒋᔭᖓᓂ (N 66.15°, W 65.70°; ᖁᑦᑎᖕᓂᖓ 23 m). ᐅᑭᐅᖅᑕᖅᑑᑉ ᕿᑎᖅᐸᓯᐊᓃᑦᑐᖅ ᐅᖅᐱᐊᓛᓂᒃ ᐱᕈᖅᑐᖃᖅᖢᓂ, ᓄᓇᐅᓪᓗ ᐃᑭᐊᖓ ᖁᐊᖑᐃᓐᓇᖅᐸᒃᖢᓂ. ᓄᓇᓕᙳᓚᐅᖅᓯᒪᔪᖅ 1962-ᒥ ᐃᓄᒋᐊᖕᓂᖃᖅᖢᓂᓗ 1,425-ᓂᒃ 2011-ᒥ ᑲᓇᑕᒥ ᑭᒡᓕᓯᓂᐊᖅᑕᐅᔪᓂᑦ.

ᐸᖕᓂᖅᑑᕐᒥ, ᐃᓄᑐᖃᐃᑦ ᐅᔾᔨᕈᓱᒃᓯᒪᔪᑦ ᐅᖅᐱᒐᑦ ᐊᕚᓚᕿᐊᓪᓗ ᐱᑕᖃᕐᓂᖅᓴᐅᓕᕐᓂᖏᓐᓂᒃ. ᐊᐱᖅᓱᖅᑕᐅᔪᑦ ᐅᔾᔨᕈᓱᓗᐊᖅᓯᒪᔪ ᐅᖅᐱᒐᑦ ᐱᕈᖅᓵᓕᕙᓕᕐᓂᖏᓐᓂᒃ ᐊᖏᓂᖅᓴᐅᓕᕐᓂᖏᓐᓂᓪᓗ. ᐃᓚᖏᑦ ᐱᕈᖅᑐᑦ ᐱᕈᕐᓂᖅᓴᐅᕙᓕᖅᖢᑎᒃ ᓲᕐᓗ ᑲᖒᔭᑦ ᐃᕖᓪᓗ. ᓄᑖᑦ ᐱᕈᖅᑐᑦ ᑕᑯᔭᐅᕙᓕᕐᒥᔪᑦ, ᐱᖃᓯᐅᖦᖢᒋᑦ ᖁᖅᓱᖅᑐᑦ ᓄᓇᕋᖏᑦ ᐊᒥᓱᓂᒃ ᐊᒥᑦᑐᓂᒃ ᐅᖃᐅᔭᓖᑦ ᐊᓯᖏᓪᓗ ᐱᕈᖅᑐᑦ. ᐃᓚᖏᑦ ᐊᐱᖅᓱᖅᑕᐅᔪᑦ ᐅᔾᔨᕈᓱᒃᓯᒪᔪᑦ ᓄᓂᕙᒃᑕᐅᕙᒃᑐᓂᒃ, ᐱᓗᐊᖅᑐᒥᒃ ᐸᐅᕐᖓᕐᓂᒃ ᐱᕈᕐᕕᒋᕙᙱᑕᖏᓐᓂ ᐱᕈᖅᐸᓕᕐᓂᖏᓐᓂᒃ. ᐅᖃᓚᐅᕆᓪᓗᑎᒃ ᐸᐅᕐᖓᑦ ᐱᕈᖅᓵᓕᕙᓕᕐᓂᖏᓐᓂᒃ ᐃᑦᑎᖃᙱᓐᓂᖅᓴᐅᕙᓕᕐᓂᖏᓐᓂᓪᓗ.

ᐊᐱᖅᓱᖅᑕᐅᔪᑦ ᐅᖃᖃᑦᑕᓚᐅᖅᑐᑦ ᐅᐱᕐᖓᒃᓵᒃᑯᑦ ᐊᐅᓚᓴᕆᐊᕐᓇᙱᓐᓂᖅᓴᐅᓕᕐᓂᖓᓂᒃ ᓯᑯ ᐊᐅᒃᓵᓕᕙᓕᕐᓂᖓᓄᑦ. ᐅᖃᓚᐅᕆᓪᓗᑎᒃ ᓇᑦᑏᑦ

Pangnirtung

Pangnirtung is located 50 kilometres south of the Arctic Circle, on the broad reaches of an old beach below the mountains of the Pangnirtung Fjord on the northern side of Cumberland Sound (N 66.15°, W 65.70°; elevation 23 m). It is part of the Middle Arctic and characterized by prostrate and dwarf-shrub tundra vegetation, underlain by continuous permafrost. The community was established in 1962 and its population was estimated at 1,425 people in the 2011 Statistics Canada Census.

In Pangnirtung, Elders have observed greater abundance of shrubs, such as willows and dwarf birch. Participants particularly noticed that willows were blooming earlier and growing taller. Some species have also tended to become more abundant, such as Arctic cottongrass and American dunegrass. New plant species have been observed, including dandelions and other species of flowering plants and grasses. Some interviewees observed changes in berry plants, in particular blackberries, which appear to be growing and spreading in new areas. They

ᕐᑭᓯᑦᑎᐊᕆᒃᐸᒍᓐᓃᓕᕐᓂᖕᒋᓐᓂᒃ ᓯᑯᕐᑲᑯᑖᖕᒋᓐᓂᖅᓴᐅᕙᓕᕐᓂᖕᒐᓄᑦ, ᓇᑦᑏᑦ ᐆᑦᑑᑯᑖᖕᒋᓐᓂᖅᓴᐅᕙᓕᕐᓂᖕᒋᓐᓄᑦ. ᓂᕐᑖᑦ ᑲᖑᐃᓪᓗ ᐅᓄᕐᓂᖅᓴᐅᓕᕐᓂᕋᖅᑕᐅᔪᑦ, ᓄᑖᓂᓪᓗ ᐅᕙᓗᓕᖕᓂᒃ ᒥᑭᑦᑐᓂᓪᓗ ᑎᖕᒥᐊᓂᒃ ᑎᑭᑦᑐᕐᑲᖅᐸᓕᕐᓂᖕᒐᓂᒃ.

ᐸᖕᓂᖅᑑᕐᒥᐅᑦ, ᐅᔾᔨᕆᓱᒃᓯᒪᓕᕐᒥᔪᑦ ᐊᓯᖕᒋᓐᓂᒃ ᐊᕙᑎᒧᑦ ᐊᓯᔾᔨᖅᓯᒪᔪᓂᒃ, ᓯᑯᐃᖅᓵᓕᓂᖅᓴᐅᕙᓕᖅᓱᓂ, ᖁᐊᖑᐃᓐᓇᖅᐸᒃᑐᓪᓗ ᐊᐅᒃᐸᓕᕐᓂᖕᒋᓐᓂᒃ, ᓄᓇᓗ ᑰᖕᓃᑦᑐᑦ ᑎᓱᐊᖅᐸᓕᕐᓂᖕᒋᓐᓂᒃ ᑰᖕᓂ ᐊᐅᔪᐃᑦᑐᕐᒥᓪᓗ ᒥᕐᖑᐃᖅᓯᕐᕕᖕᒥ. ᐊᐱᖅᓱᖅᑕᐅᔪᑦ ᐅᔾᔨᕆᓱᒃᓯᒪᔪᑦ ᒪᕐᑯᕐᑲᑦᑕᖖᒋᓐᓂᖅᓴᐅᓕᕐᓂᖕᒐᓂᒃ. ᐊᓄᓚ ᓴᖖᒋᓂᖅᓴᐅᓵᓕᖅᑐᖅ, ᓇᑭᑐᐃᓐᓇᖅ ᐊᓄᓕᓯᑳᓪᓚᒍᓐᓇᖅᓱᓂ. ᐊᐱᖅᓱᖅᑕᐅᔪᑦ ᐅᖃᓚᐅᕐᒥᔪᑦ ᓯᕐᒦᑦ ᓱᒃᑲᓂᖅᓴᒥᒃ ᐊᐅᒃᐸᓪᓚᐊᕙᓕᕐᓂᖕᒋᓐᓂᒃ ᐊᐅᓪᓛᓇᐊᒃᓴᕐᓗ ᐊᑦᑕᕐᓇᕐᓂᖅᓴᐅᓕᕐᓂᖕᒐᓂᒃ.

also noted that blackberries are ripening earlier and tend to be less juicy.

Participants described that it is more difficult to go ice fishing in the spring because the ice is melting faster. They also reported that seal fur quality is declining because the sea ice period is shortening, thus leaving the animals less time to bask in the sun. More Canada geese and snow geese are said to come to the area, and new species of shellfish and small birds are now being observed.

Pangnirtungmiut have observed other environmental changes, including earlier sea ice breakup, the thawing of permafrost, and increased erosion along riverbanks and in Auyuittuq National Park. Participants have generally noticed less rain in the area. The wind seems stronger, coming all of a sudden from any direction. Interviewees also noted that glaciers are melting more rapidly and that travelling has become more hazardous.

ᐃᓐᓇᐃᑦ ᐊᐱᖅᓱᖅᑕᐅᔪᑦ
Elders Interviewed

ᓕᐊ ᐊᒃᐸᓕᐊᓗᒃ
Leah Akpalialuk

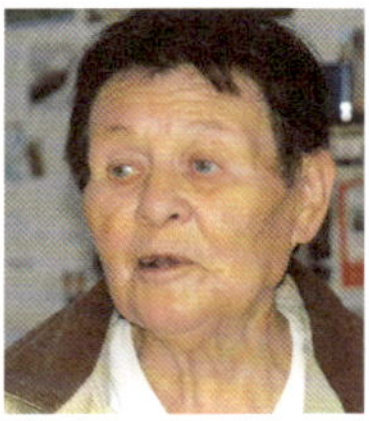

ᒥᐊᓇ ᐹᑎ
Mary Battie

ᓕᐅᐸ ᐊᒃᐸᓕᐊᓗᒃ
Leopa Akpalialuk

ᑕᐃᓯ ᑎᐊᓚ
Daisy Dialla

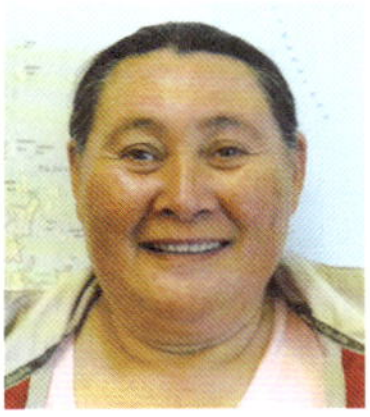

ᑮᓚ ᐋᑯᓗᒃᔪᒃ
Geela Akulukjuk

ᓕᐊ ᐃᕕᒃ
Leah Evik

ᐄᕕ ᐊᓂᖅᓂᓕᐊᖅ
Evee Anilniliak

ᐊᐃᒐ ᐃᓱᓪᓗᑕᖅ
Aiga Ishulutaq

ᑕᐃᔭᐅᔪᒪᙱᑦᑐᖅ
Anonymous

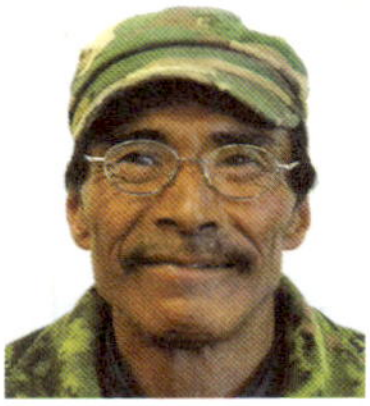

ᔭᐃᑯ ᐃᓱᓪᓗᑕᖅ
Jaco Ishulutaq

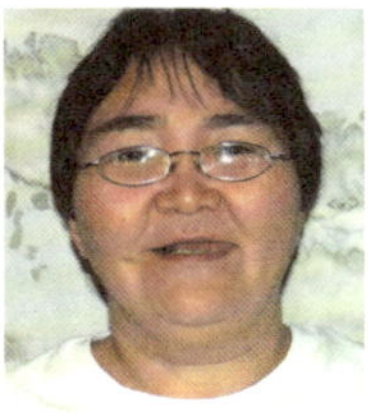

ᐅᓚᐃᑲ
ᐄᓯᒪᐃᓕ
Ooraika
Iseemailee

ᐃᓅᓯᕐᒃ
ᓇᓴᓕᒃ
Inusirq
Nashalik

ᓕᓯ ᒥᐊᓕ
ᑲᒃᑭᒃ
Leesee
Mary Kakee

ᐲᑦᑐᓘᓯ
ᖄᐱᒃ
Peterloosie
Qappik

ᓴᐃᓚ ᑭᓴ
Saila
Kisa

ᑕᐅᑭ ᖄᐱᒃ
Taukie
Qappik

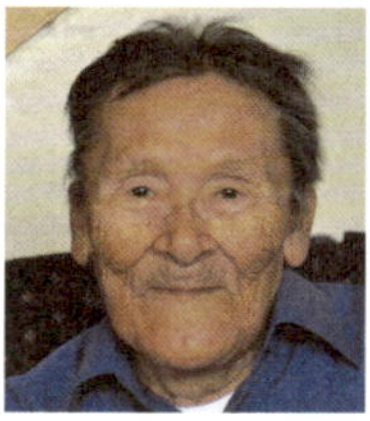

ᔮᐃᒥᓯ ᒪᐃᒃ
Jamasie
Mike

ᐸᐅᓗᓯ ᕖᕕ
Pauloosie
Veevee

ᓴᐅᑦᓗ ᓇᑲᓱᒃ
Sowdloo
Nakashuk

ᐸᐅᕐᖓᐃᑦ

“ᑭᓯᐊᓂᓕ ᑕᕝᕙ ᐃᒫ ᖃᐅᔨᒪᒐᒃᑯᑦ ᓯᓚᒧᑦ ᒪᓐᓇ ᑕᕝᕙ ᓯᓚᓗᖃᑦᑕᖏᑦᑐᐊᓘᖕᒪᑦ ᐃᒥᕈᑦᑐᑯᓘᕙᓪᓚᐃᒻᒪᑕ. ᒫᓐᓇ ᖃᐅᔨᒪᓇᖅᑑᔫᔮᒻᒪᕆᒃᑐᖅ ᐸᐅᕐᖓᒃᓴᑯᓗᐃᑦ ᐃᓚᖏᑦ ᕿᕐᓂᖅᓯᓕᕋᓗᐊᖅᑐᑎᒃ ᑕᒫᓂᓕ ᒫᓂ ᐱᕈᖅᑐᐊᓗ, ᖃᓄᖅ ᐳᕙᓪᓚᙱᒋᐊᖏᑦ ᐃᓱᒪᒋᔭᒃᑲ ᖃᓅᒐᓗᐊᑭᐊᖅ. ᒫᓐᓇ ᐊᐅᓪᓚᖅᓯᒪᓚᐅᕐᒥᒐᑦᑕ ᑐᖕᒍᔪᖅᓯᔪᑯᓘᓕᕐᒥᔫᒐᓗᐊᑦ ᐃᓚᖓ ᐊᒻᒪ ᐃᓚᖏᑦ ᕿᕐᓂᖅᓯᔪᑯᓘᒐᓗᐊᖅᑐᑎᒃ ᐃᓚᖏᑦ ᓱᓕ ᐸᐅᕐᖓᑦᓴᑐᐃᓐᓇᓪᓚᕆᑯᓘᒻᒪᑕ. ᐄ ᓯᓚᓗ ᐊᑦᓱᐊᓗᒃ ᐅᖅᑰᖅᔪᐹᓘᕙᑦᑐᓂ ᓯᓚᓗᖃᑦᑕᙱᑖᓗᒋᒻᒪᒋᑦ ᓯᓚᓘᑎᕙᖕᒋᓪᓚᕆᒻᒪᒋᑦ.”

—ᓕᐊ ᐊᒃᐸᓕᐊᓗᒃ

“ᐊᕐᕌᒍᑕᒫᑦ ᐊᕐᕌᒍ, ᐊᕐᕌᒍᑕᒫᖅᐸᒍᓐᓃᕐᒥᔪᖅ. [...] ᐋᕐᕌᒍᑦ ᐱᖓᓱᑦ ᖄᖏᖅᐸᓪᓚᐃᔪᖅ ᐱᑕᖃᓚᐅᖅᓯᒪᖏᑦᑎᐊᒻᒪᕆᒃᑐᖅ. [...] ᐊᕐᕌᒍᒥ ᑕᒫᓂ ᓯᕗᕙᖅᑐᑦ. ᑐᖁᓕᔫᔮᓕᕇᖅᑐᑦ, ᑲᔪᕈᓕᕇᕐᖓᑕ.”

—ᐅᓚᐃᑲ ᐄᓯᒪᐃᓕ

“ᑭᓯᐊᓂᓕ ᐸᕝᕕᓴᖅᑕᖅᐸᓘᓕᕐᒪᑕ ᓂᕐᓖᑦ ᐊᒥᓱᕈᖅᑐᖅᐸᓘᒻᒪᑕ, ᓂᕐᓕᕐᓄᑦ ᐸᐅᕐᖓᖃᙱᑎᑕᐅᔪᐊᓘᓕᕐᒪᑦ. [...] ᐃᓚᖏᓐᓄ ᐊᕐᓇᓄ ᐱᐅᒋᔭᐅᙱᑐᕐᔪᐊᕌᓗᐃ ᓂᕐᓖᑦ. ᐊᕐᕌᓃᓛᒃ ᖃᐅᔨᒋᐅᕋᒪ ᑕᐃᒫ ᖁᑭᐊᑐᐃᓐᓇᖃᑦᑕᒻᒪᕆᑦᑐᑦ ᓂᕐᓕᖃᖁᔨᙱᑦᑐᑯᓗᐊᐲᑦ.

Berries

“The young *paurngait* [blackberries/crowberries] are not going to bloom enough with lack of water. I’ve been camping recently, and I’m noticing that they start to turn blue or black, but just at the *paurngaq* stage because they don’t have enough moisture. [...] There is way too much sunlight, and they are not getting enough irrigation.”

—Leah Akpalialuk

“Every year we used to have berries, but not every year now. [...] Maybe three years ago there was nothing. [...] This year [2009] they are early. It seems like they are already dying, already turning brown.”

—Ooraika Iseemailee

ᐱᔪᓐᓇᖏᓐᓂ ᑲᐅᔾᒪᔪᐊᓘᒋᑦᓗᑎ
ᐸᐅᕐᖓ ᑖᓐᓇᑕᐃᓐᓇᐅᖏᒻᒥᒻᒪ
ᐸᐅᕐᖓᕐᑯᑎᖏᑦ ᑲᔫᑉᐸᑕ
ᕐᑭᕐᓂᑕᐅᑎᐅᑉᐸᑕ ᐸᐅᕐᖓᕐᑯᑎᑯᓗᖏᑦ
ᑲᔪᕈᔫᑉᐸᑕ ᕐᑲᑯᕈᔫᑦᓗᑎᒃ
ᑕᐃᒪ ᐸᐅᕐᖓᕐᑲᑦᑕᖅᑯ
ᐳᕙᓚᔫᑎᐊᓗᖕᓂᒃ
ᐸᐅᕐᖓᕐᑲᑦᑕᖅᐳ ᑕᒪᒃᑯᓂᖓᖕᒑᖅ
ᕐᑲᐅᔨᒪᔪᐸᑦᓰᖀᔫᒻᒥᒻᒪᑕ. ᐊᒻᒪᓗ
ᐱᐅᓯᔭᐅᒐᓗᐊᕆᑦᓗᑎ ᓂᕐᑖᑦᓗ
ᐱᐅᓯᔭᐅᖏᑭᑦᓗᑎᒃ."

—ᐸᐅᓗᓯ ᕚᕕ

"ᒪᑯᐊᓕ ᒫᓐᓇᐅᓕᖅᑐᕐᓖ ᕐᑲᓄ
ᑕᕝᕙ ᐱᕈᖅᑐᓕᖅᐹᓘ, ᒪᑯᐊᓕ
ᐸᐅᕐᖓᐃᑦᓗ ᐱᕈᓗᐊᖀᓗᑦᓯᒪᙱᑦᑐᑦᓕ
ᐃᒫ ᐱᕈᖅᑐᖅᐹᓘᓗᑏ
ᒫᓐᓇᓕ ᐅᖅᑰᓯᓂᖅᓴᐅᒻᒪ ᓯᓚ
ᐱᕈᕐᓂᖅᓵᓘᓕᖅᑐᐃᑦᓖ ᓄᓇᕿᐃᑦ
ᑕᑯᓯᒪᔭᒃᑲ ᑕᕝᕘᓇ."

—ᓴᐃᓚ ᑭᓴ

"ᐃᒫᑦᓕ ᐸᐅᕐᖕᒑᓗᐃ
ᕐᑭᑭᕐᕿᕐᑲᕐᓇᒻᒪᕆᑉᐸᔫᒻᒪᑕ
ᐃᒫ ᑐᑦᓕᕿᖅᑐᒋ
ᐸᐅᕐᖓᓕᕙᔾᔪᐊᖀᓘᕙᔫᒻᒪᓂᓖᒃ,
ᒫᓐᓇᓕᒎ ᑕᐃᒪᐃᙱᑦᑎᐊᓕᖅᑐᕉᖅ
ᐃᒫ ᑐᓴᖅᑎᑕᐅᕙᑦᑐᖓ
ᕐᑲᐅᔨᓚᐅᖅᓯᒪᔪᓐᓃᖅᑐᖓ
ᑐᓴᓚᐅᖅᓯᒪᔪᓐᓃᖅᑕᕿ."

—ᓴᐅᑦᓗ ᓇᑲᓱᒃ

"The Canada geese seem to be increasing now, and they eat a lot of *paurngait*. There are lots more Canada geese. [...] Some women hate Canada geese. They seem to know what's better, too. [...] The women want to take those [better-tasting] berries, but the Canada geese want to eat those berries, too. There is a competition between geese and women to get those berries first."

—Pauloosie Veevee

"There is an abundance of growth these past couple of years. Even the *paurngait*, they don't usually grow that large or big, but because of the warmth we are starting to see a lot more growth on the ground."

—Saila Kisa

"There used to be a lot of berries before, but it's not like that anymore."

—Sowdloo Nakashuk

ᐱᕈᖅᑐᑦ ᐊᓯᖏᑦ

"ᐋ ᑕᐃᒃᑯᐊ ᑐᖖᒍᖅᑐᑦᑎᐊᕙᒃᐹᓗᐃᑦ ᐱᑕᖃᖅᐸᔪᖏᑦᑎᐊᕐᒥᔫᒐᓗᐊᑦ ᓇᐹᖅᑐᖖᒍᐊᐸᓗᑯᓘᔭᖅᑐᑦ ᐱᑕᖃᖅᐸᔪᖏᒃᑲᓗᐊᕐᒥᖕᒪᑕ ᑖᒪᓂ ᑐᖖᒍᑦᑐᑐᐊᓘᕙᓕᕐᒥᖕᒪᑕ ᐱᑕᖃᐸᔪᖏᑦᑎᐊᖅᑑᒐᓗᐊᑦ ᑕᐃᒃᑯᐊ."

—ᑕᐃᓯ ᑎᐊᓚ

"ᑕᒪᓐᓇ ᐊᑕᖏᖅᑑᔮᖅᑐᑦ ᓯᓚᕐᔫᑉ ᐃᓛ ᐱᕈᖅᑐᖏᑦ ᑕᐃᒫᑦᑕᐅᖅ ᒪᓐᓇ ᐃᒫ ᑭᓯᐊᓂ ᐱᕈᕐᕕᖕᒐ ᓈᓪᒪᑦᓯᑉᐸᑦ ᓂᓪᓚᖅᐸᓪᓕᐊᓕᖅᑲᑦ ᐃᒫ ᐱᕈᕐᓂᖅᓴᐅᓂᐊᕐᒥᔪᑦ."

—ᐃᓅᓯᖅ ᓇᓴᓕᒃ

"ᒫᓐᓈ ᖃᐅᔨᓯᒪᓪᒥᔪᖓ ᑕᐃᒃᑯᐊ ᖁᖅᓱᖅᑐᑯᓗᐃᑦ ᓄᓇᖁᑯᓗᐃᑦ ᐃᒍᑦᑕᒧ ᓂᖀᐅᖏᑦᑐᑯᓗᐃᑦ ᐱᕈᖅᐸᔪᓐᖏᑎᐊᖅᑐᑦ ᒫᓂ ᐱᕈᖅᑲᑦᑕᓕᕐᒥᔪᑦ, ᐊᓪᓛ ᒫᓂ ᓴᓂᑦᑎᓐᓂ ᐱᕈᖅᑲᑦᑕᓚᐅᖅᑐᑦ ᐅᐱᖖᒑᖅ."

—ᐄᕕ ᐊᓂᕐᓂᓕᐊᖅ

"ᐅᖅᑯᓯᕙᓪᓕᐊᔪᒥᒃ ᑐᓵᔪᖅᐹᓘᕙᒃᑲᒪ ᐃᖅᑲᐃᓂᐅᓴᑐᐃᓐᓇᖅᐸᑦᑕᖃᓛ ᑕᒪᓐᓇ ᐅᖅᑰᓯᕙᓪᓕᐊᔪᖅ. ᐃᒫᒃ ᑕᕝᕙ ᐸᐅᖕᒐᐃᑦ ᐆᑦᑑᑎᒋᓗᒋᑦ ᑕᕝᕙ ᐳᕝᕙᓚᓄᐊᖅᑐᑯᓘᔭᖅᑐᑦ ᓱᒃᑰᓂᖅᓴᐅᑦᑕᓕᖅᑯᑦ ᓯᓚᓗᖅᑲᑦᑕᖕᒋᓪᒪᑦ. ᐃᒪᐃᓪᒪᑕ, ᑕᕝᕙ ᓯᖀᓐᓂᖅᑲᑦᑕᖕᒋᑦᑐᒥ ᑕᖃᓂᕐᒥᒃ ᑕᐃᓰᓇᒐᑦᑎᒍᑦ ᓯᖀᓐᓂᖅᑕᐅᖕᒋᓐᓇᖕᒋᑦᑐᖅ. ᐱᕈᖅᓯᒪᔪᖅᐸᑯᓘᕙᑦᑐᖅ ᑕᐃᒪᐃᑦᑐᐃ ᐃᒫ ᖃᓄᓗᐊ ᖃᓄ

Other Plants

"There are green plants that never used to grow here. [...] Not flowers. It's like green grass growing straight. We never had those, but now they are around."

—Daisy Dialla

"There seems to be more abundance this year of any plants that grow up here."

—Inusirq Nashalik

"I'm noticing some yellow flowers that bumblebees tend to feed on. There weren't any of those [before], and now I notice that they have started to grow outside my house."

—Evee Anilniliak

ᓯᕐᑭᓂᕐᒧ ᐆᓯᒪᓐᒋᑦᑎᐊᖅᑐᑎᒃ
ᐊᒻᒪ ᓄᓇᒥ ᑕᑯᓯᒪᒻᒥᔪᖓ ᒫᓐᓂ
ᐱᕈᖅᑐᑯᓗᒻᒥ ᑕᑯᓲᓇᓐᒋᑦᑎᐊᖅᑕᓐᓂ
ᐊᔾᔨᒌᓐᒋᑦᑐᑯᓘᓐᓂᒃ ᐃᒫ
ᓄᓇᕐᑯᓗᓐᓂᒃ ᑕᑯᓯᒪᒐᒪ ᑕᐃᒃᑯᐊ
ᐲᖅᓯᕕᒋᓚᐅᖅᓯᒪᔭᕐᓗᐊᒃᑲ
ᑐᕐᑯᓕᔪᑯᓘᓚᐅᕐᒪᑕᓕ ᐅᐱᓐᓇᕐᓂ.
ᐃᒫ ᓄᓇᕐᑯᓗᐃᑦ ᑕᐅᓇᖓᑦ ᑕᐃᒪ
ᐃᒫᓐᒋᓛ ᐅᕐᑲᐅᔭᓕᑯᓘᑦᓗᑎᒃ
ᓄᕗᓕᑯᓗᐃᑦ ᓄᓇᕐᑯᓗᒻᒥ
ᐃᒪᐃᑦᑐᐸᓗᑲᓴᑯᓘᔮᖅᑐᖅ ᓄᕗᐃᓐᓇᖓ
ᐃᒪᐃᑦᑐᓕᑯᓗᐃ ᓄᕗᐃᓐᓇᖓᑦ ᐃᒫ
ᐊᑯᓪᓕᕐᓃᑯᓗᐃᑦ ᑕᑉᐱᑲᓂ ᓄᕗᐊᓂ,
ᑕᕝᕙ ᑖᒃᑯᐊ ᓲᔨᓂᓯᒪᓗᐊᖑᐊᖅᑕᒃᑲ,
ᐊᒻᒪ ᑕᐃᒃᑯᐊᑦᑕᐅ ᕐᑭᔾᔭᒃᑖᓃᑦᑐᑦ ᒪᓐᓇ
ᐅᐱᕐᖕᒑᖅ ᐸᐅᖓ ᕐᑭᐱᓴᐅᓂᕐᖅᑕᒧ
ᐃᖅᑲᓪᓕᐊᖅᓯᒪᑲᐃᓐᓇᔪᒐᑦᑕ
ᑕᑉᐱᓐᓇ ᐅᐸᓐᓇᓵᖅᑕᖅᐹᓗᒋᒐᑦᑎᒍ
ᐱᕈᖅᑖᖅᑐᕐᔪᐊᒥᓂᐊᓘᔪᒻᒪ.
ᑕᐃᒃᑯᐊ ᕐᑭᔪᑦᑖᒦᑦᑐᑦᑕᐅ
ᑕᑯᓪᓚᓇᒋᐅᔫᒥᔭᒃᑲ ᐃᓛᒃᑯ ᐱᕈᖅᑐ
ᕐᑭᔪᑦᑖᓂ ᑕᕝᕙᓂ ᕐᑭᔪᑦᑖᖅᑕᖓ
ᑕᒫᖖᒑ ᐅᐃᔾᔮᓗᕐᑭᓯᒪᖖᖑᐊᖅᑐᑯᓗᐃᑦ
ᐃᒪᐃᑦᑑᓲᖑᒐᓗᐊᕐᒪᑕᓕ ᕐᑲᑯᖅᑕᐃ
ᕐᑭᔪᑦᑖ ᓄᕗᐊᓂ, ᑕᐃᒃᑯᐊ
ᐅᐃᔾᔮᓗᕐᑭᓯᒪᖖᖑᐊᖅᑐᑯᓘᑦᓗᑎ
ᕐᑲᑯᖅᑕᐅᒐᓗᐊᖅᑐᑎ
ᐊᔾᔨᖏᖖᒋᑕᑯᓗᓐᒋᓐᓂ ᑕᐃᒪ
ᐋᓪᓚᒋᔪᔭᒃᑲᑦᑕᐅ ᑕᐃᒃᑯᐊ
ᒫᓐᓇ ᐅᐱᕐᖕᒑ, ᑕᕝᕙ
ᑖᒃᑯᐊ ᐋᓚᒋᓯᒪᓗᐊᖅᑕᒃᑲ
ᐅᕐᑰᓯᕙᓕᐊᓂᖓᓄᑦ ᐊᓱᐃᓪᓛᒃ."

—ᔩᓚ ᐋᑯᓗᒃᔪᒃ

"I notice that a lot of the plants are getting bigger on the land. But there are one or two particular species that I have never seen before. [...] [One has] leaves like a tree, and at the top a flower that blooms. I found that very unique. [...] The second one is like heather, but I saw something different: instead of having growth at the tip, it grew in a spiral, [...] a white spiral."

—Geela Akulukjuk

"The heat from the sun is actually killing off the growth, because it's so hot and immediate that it's not giving the land enough time to adjust to the heat. So it's actually creating some problems for growth, and also overgrowing some species."

—Jaco Ishulutaq

“ᓲᕐᓗ ᒪᑯᐊ ᓱᐳᖅᑐᑯᓘᕙᑦᑐ
ᐃᓛᓐᓂᒃᑯᑦ ᐅᐸᓗᐃᔭᐅᖃᑦᑕᓕᖅᑐᑦ.
ᐃᒫ ᐆᓇᑲᓪᓚᓗᐊᕐᓂᑦᑎᓄ ᓱᕿᓂᕐᒧ.”

—ᔭᐃᑯ ᐃᓯᓪᓗᑕᖅ

“ᐄ ᑕᒪᒃᑯᐊᓕ ᑕᒫᓂ
ᐱᕈᖃᑦᑕᔪᖏᑦᑐᑦ ᐱᕈᖅᓯᒪᔪᑦ
ᒪᑯᐊᓗ ᐸᐅᓐᓇᐃᑦ. ᑕᐃᒃᑯᐊᓗ
ᐳᔪᐊᓗᔅᓴᐅᓂᕋᖅᑕᕗ ᑕᒪᓐᓇ
ᑕᐃᒪᐃᑦᑐᖃᖅᐸᔪᖏᒃᑲᓗᐊᕐᒥᒻᒪ
ᑕᐃᒪᐃᑦᑐᖃᖅᑐᖅᐹᓘᒻᒪᑦ ᐅᐱᕐᖔᖅ.
ᐊᖅᑯᑎᒻᒪᕆᐊᓘᒻᒪᑕ ᑕᒪᒃᑯᐊ ᐃᓪᓗᓪᓗ
ᐊᑯᓐᓂᖏᓐᓂ ᐊᔾᔨᒌᖏᑦᑕᖅᑐᑦ
ᐊᕐᕌᒍᒥ ᐱᕈᖃᑦᑕᓕᖅᑐᐃ. ᑕᒫᓂ
ᐱᕈᖃᑦᑕᔪᖏᑦᑐᐃᑦ. ᑕᒫᓂ
ᐃᓪᓗᖃᐅᓕᖅᑎᓪᓗᑕ, ᑕᒪᓐᓇᓕ
ᑕᐃᒪᐃᑦᑐᓂᒃ ᐱᕈᖅᑐᖃᖅᐸᔪᖏᒻᒪᑦ.”

—ᓕᐊ ᐃᕕᒃ

“ᒫᓐᓇ
ᒥᖅᑯᔅᓴᕋᐃᓐᓂᖅᓵᓘᔭᖅᐸᓕᕐᒥᔪᑦ.”

—ᓕᓯ ᒥᐊᓕ ᑲᒃᑭᒃ

“ᒫᓐᓇ ᐃᔾᔪᖅ ᑕᒪᓐᓇ ᓄᓇ
ᓄᓇᕈᖅᐸᑦᓕᐊᑐᐃᓐᓇᖅᑐᑦ
ᑕᐃᒪᓐᓇᐅᕙᔪᒻᒪᑦ
ᐱᕈᖅᓯᒪᓂᖅᓴᐅᓕᖅᑐᑦ
ᓇᒥᑐᐃᓐᓇᖅ ᖃᑯᖅᑕᐅᕙᔪᔫᒐᓗᐊᑦ
ᐊᐅᔪᐃᑦᑐᖃᕐᕕᑦ ᐅᔭᕋᕈᔪᐃᑦ ᐃᓚᖏᑦ
ᐃᔾᔪᖃᖏᑦᑎᐊᓚᐅᖅᓯᒪᔪᑦ
ᒪᓐᓇ ᐃᔾᔪᖃᓕᖅᑐᑦ.”

—ᓕᐅᐸ ᐊᒃᐸᓕᐊᓗᒃ

“I’m starting to notice plants that didn’t grow [before]. There weren’t many mushrooms growing here in Pang, but this season I’ve noticed a lot here; even between houses I have noticed mushrooms growing.”

—Leah Evik

“*Suputiit* [Arctic willow catkins] are blooming much quicker.”

—Leesee Mary Kakee

“I’ve seen a lot of tundra growth. As a child I saw areas that were rocky and sandy, and most of it is covered now.”

—Leopa Akpalialuk

“It seems like [the grasses] are getting taller. The growth is the same, but this year we are having weird weather. This time of the summer they would [usually] turn a really nice green and they would grow, but everything is dying already. I think it’s too dry.”

—Ooraika Iseemailee

“ᐃᓛᑦ ᑕᑭᑦᓕᓂᖅᓴᖅᔫᔭᖅᑐᑦ.
ᑕᑭᓂᖕᒋᑦ ᑕᐃᒪᐃᑦᓗᐊᖅᑑᔮᖅᑐᑦ,
ᐊᖅᖀᒍᒥ ᑕᒪᑐᒪᓂ ᓯᓚᕗᑦ
ᑲᒪᓇᖅᑐᐊᓘᒻᒪᖅ. ᑕᐃᒪᐃᓕᖅᑎᑦᓗᒍ
ᐊᐅᔭᒃᑯᑦ ᐆᔭᐅᔭᑦᑎᐊᕚᓘᕙᒻᒪᑕ
ᐱᕈᑦᑎᐊᖅᑐᑎᒡᓗ, ᑭᓯᐊᓂ
ᑐᖅᑯᓕᔪᐊᓘᔭᕇᖅᑐᑦ.
ᐸᓂᓗᐊᖅᐸᑦᓚᐃᔪᖅᓕ.”

—ᐅᓚᐃᑲ ᐄᓯᒪᐃᓕ

“ᖃᓄ ᐱᑕᖅᑲᖏᓂᖅᓴᐅᔫᔭᕈᔪᓕᖅᑐᑦᓕ
ᒪᑯᐊ ᐸᐅᓐᓇᐃᑦᓘ
ᐱᕈᖅᑐᖅᐹᓘᖅᑲᑦᑕᓕᖅᒪᑕ
ᐅᖅᑲᐅᔭᖅᑲᓗᐊᖅᑰᔨᖕᖏᓐᒪᑦᓕ ᐅᓇ
ᐸᓐᓂᖅᑑ. ᐅᖅᑲᐅᔭᑭᕈᔪᑦᑐᐊᓘᔮᖅᑐᖅᓕ
ᑕᑯᕙᑦᑕᖅ ᒪᑯᐊᓗ ᐸᐅᓐᓇᐃ ᒪᑯᐊᓗ
ᐳᐊᓗᖖᒍᐊᑦ ᑕᒪᒃᑯᐊ ᐃᒍᑦᑕᒧᑦᓗ
ᓂᖅᑮᑦ, ᑕᒪᒃᑯᐊ ᓴᖅᑭᑦᑐᖅᐹᓘᔭᖅᑐᑦ
ᐃᓛ ᐱᕈᓗᐊᖅᑑᔭᖅᐸᑦᑐᑦ.”

—ᓴᐃᓚ ᑭᓴ

“ᒫᓐᓇᐅᔪᖅ ᐱᕈᖅᓂᖅᓵᓗᐃ
ᐅᖅᑰᒐᒥᖅᑲᐃ ᑕᐃᒫ ᐱᕈᖅᑐᖅᐹᓗᐃᑦ.
ᑕᑉᐹᓂᓕ ᐹᓂᖕᒑᖅ ᐃᖅᑲᓗᒑᖅᔪᖕᒥ
ᐱᑕᓕᐊᓗᐃ. ᐱᕈᑦᑎᐊᖅᑲᑦᑕᕈᓐᓃᔪᒻᒪᑕ
ᐱᕈᑦᑎᐊᖅᑲᑦᑕᕈᓐᓃᑲᐃᓐᓇᔪᒻᒪᑕ
ᒫᓐᓇ ᐱᕈᖅᓂᖅᓵᓗᐃ.
ᐅᖅᑰᓂᖅᓴᐅᖅᑲᑦᑕᖅᓯᒪᒻᒪ
ᐊᖅᖀᒍᓂ ᐃᓛ ᐊᖅᖀᒎᓚᐅᖅᑐᓂ
ᐊᖅᖀᓂᓗ ᐅᖕᒐᑖᓂᓗ ᓱᒻᒪᒃᑭᐊ
ᐅᖅᑰᖕᒋᓐᓂᖅᓴᐅᖅᑲᑦᑕᖅᓚᐅᖅᒪᑦ
ᐱᕈᖕᒋᓐᓂᖅᓴᐅᖅᑲᑦᑕᖅᓯᒪᓚᐅᖅᑐᑦ.”

—ᑕᐅᑭ ᖄᐱᒃ

“Right now there is less *suputiit* because there is more growth of *paurngait* [blackberries/crowberries]. [...] The weeds are growing more. The *ivitsugak* [American dunegrass], and also the cottongrass. They are more in abundance now. Because of these weeds growing more, there are less of other species, especially *uqaujait* [willows] and *suputiit*.”

—Saila Kisa

“There is more growth in the summer, and I think it's a result of the heat and sun. [*Avaalaqiat*] [birches] were declining in growth over the years, but now they are really starting to grow larger. It just has to do with the weather.”

—Taukie Qappik

Nunavut / Pangnirtung

ᐆᒪᔪᑦ

“ᐊᒻᒪ ᒪᑯᐊ ᐃᒫ ᐅᓕᓐᓂᕐᒧᑦ ᓴᙱᑦᑎᑉᐸᓪᓕᐊᓂᖓᓄᑦ ᒪᑯᐊ ᓴᖅᑭᑉᐸᓪᓚᐅᖏᑦᑐᑦ ᒪᐅᖓ ᐆᒪᔪᖃᓛᑦ ᓴᖅᑭᖅᑲᑦᑕᐸᓪᓕᐊᑐᐃᓐᓇᖅᑐᑦ ᑎᓂᔅᓴᐅᖅᑲᑦᑕᓕᖅᑐᑦ ᐊᓪᓚ ᐃᒫ ᓴᙱᓕᖅᐸᓪᓕᐊᓂᖓᓄᑦ ᑖᔅᓯᒪ ᐃᒪᖅᔪᐊᖄᓘᑉ ᐅᐊᑦᑎᐊᕈ ᓴᖅᑭᑉᐸᓪᓚᐅᖏᑐᐃᑦ ᓴᖅᑭᖅᑲᑦᑕᖅᐸᓪᓕᐊᔪᑦ ᐆᒪᔪᐃᑦ ᐃᓛ ᒪᑯᐊ ᐃᖅᑲᓗᖃᓛᕈᓗᐃᑦ.”

—ᐃᓅᓯᖅ ᓇᓴᓕᒃ

“ᑭᓯᐊᓂ ᒪᓗᒋᓚᐅᖅᑕᒃᑲ ᐅᐱᕐᖔᖅ ᐊᖏᔫᑎᐊᓘᒻᒪᑕ ᐸᓐᓂᖅᑰ ᐃᖅᑲᓗᖏᑦ ᓄᕙᐊᓂᓕ ᐃᓛ ᐱᖅᑲᑦᑕᖅᑕᕗᑦ ᓂᖅᑭᖏᑦ ᖃᑯᖅᑐᐊᓘᖅᑲᑦᑕᖅᐸᓪᓚᐅᕐᒪᑕ ᖃᑯᓗᐊᓰᖑᖏᒃᑲᓗᐊᕐᒪᑕ ᐸᓐᓂᖅᑰᑉ ᐃᖅᑲᓗᖏᑦ. ᑲᖐᒋᕙᓪᓚᐅᖅᑕᒃᑲ ᑕᐃᒃᑯᓂᖓ ᖁᑖᓕᒫᓅᓂᖁᖅᑕᓂᒃ ᑕᐃᒪᐃᑦᑐᓂᒃ ᓂᕆᔭᐊᓅᖅᑲᑦᑕᕐᒪᖔᑕ, ᑕᐃᒪᐃᑦᑐᓕᖅᐹᓅᖅᑲᑦᑕᕐᖓᑦ ᑕᒪᓐᓇ ᐅᐱᕐᖔᖅ. ᐃᖅᑲᓗᐃᑦ ᐊᖏᔫᑎᖅᐹᓅᓚᐅᕐᖓᑕ ᐅᐱᕐᖔᖅ ᐸᓐᓂᖅᑰᑉ ᐃᖅᑲᓗᖏᑦ. [...] ᐃᖅᑲᓗᒋᓰᓇᙱᑖᓗᒻᒥᓂᒃ ᐃᖅᑲᓗᖅᑲᖅᐸᓚᐅᕐᖓᑦ. ᐅᑯᐊ ᑭᓯᐊᓂ ᖁᑖᓕᒪᐅᔾᔭᑎᖏᑦ ᖄᖓᓃᑦᑐᕈᓗᐃᑦ ᐅᑯᐊᓗ ᓴᓂᖃᖓᓃᑦᑐᕈᓗᐃᑦ ᐃᒪᐃᑦᑐᕈᓗᐃᑦ ᖃᑦᑎᒃᑐᒪᓖᖄᓗᐃᑦ ᐃᒫᒃ ᐃᒪᐃᓕᐅᙳᐊᖅᓯᒪᔪᕈᓗᐃᑦ ᖄᖓᓃᑦᑐᑦ ᑖᒃᑯᐊ ᖁᑖᓕᒪᐅᔾᔭᑎᒋᔅᓴᖏᑦ. [...] ᒫᓂ ᖄᖓᓃᑦᑐᒪᓵᓗᐃᑦ.”

—ᐄᕕ ᐊᓂᕐᓂᓕᐊᖅ

Animals

“There seem to be more different species [...] because the high tides are getting higher and stronger, so there are more species of different seashells that we have not seen before. [...] I’ve been observing it for a long time.”

— Inusirq Nashalik

“This year I noticed that the fish are a lot bigger and the meat was whitish in colour instead of reddish. For the past couple of years we’ve been having little salmon fish. Thousands of them. [...] They are called *quliiligaq*. They have kind of rough skin, and usually a fin on the top. [...] [There have been] more in the past couple of years.”

—Evee Anilniliak

“ᑕᐃᒃᑯᐊ ᖃᐅᒪᔪᕋᓛᕌᓗᐃ ᐃᒃᑯᐊᓕ ᐃᖅᑲᐅᒪᔾᕗ ᐊᖏᔫᑎᐅᑦᑐᑎᒃ ᖃᒡᓕᔭᒃᓴᐅᕙᔪᖏᒻᒪᑕ ᒪᑯᐊᓕ ᓯᓚᒥ ᓂᕆᓪᓗᓂ ᑕᒪᒃᑯᐊ ᖃᐅᒪᔪᕋᓛᕌᓗᐃ ᑕᒪᐅᖓ ᓂᕆᔪᒧ ᓄᖅᑳᑦᑐᒻᒪᕆᐊᓗᐃ ᑕᒪᔾᔭᐅᕙᓕᕐᒪᑕ. ᑕᐃᒃᑯᐊ ᐲᔪᕆᓯᒪᕙᒃ. ᒫᖖᒎᕐᓂᑰᕙᓪᓚᐃᔪᑦ ᒪᑯᐊ ᐅᒥᐊᕐᔪᐊᒧᑦ ᓯᑯᓯᐅᑎᐅᕙᒃᑐᒧᑦ ᓂᐅᕋᖅᑕᐅᕙᒃᑐᐃ ᑕᒪᐅᖓᐅᕙᓪᓚᐃᓂᑯᑦ ᑕᐃᑰᓇ.”

—ᔭᐃᒥᓯ ᒪᐃᒃ

“ᑕᐃᒃᑯᐊ ᓄᖑᑉᐸᑦᓕᐊᓂᕋᑦᑕᐅᓕᑦᑐ ᓇᓄᐃᑦ ᓄᖑᑉᐸᑦᓕᐊᖏᑦᑐᐊᓘᒻᒪᑕᓕ ᒫᓂᓖ, ᐅᐊᑦᑎᐊᕈᓕ ᓇᓄᕐᓂᒃ ᑕᑯᖃᑦᑕᖅᐸᔪᖏᒃᑲᓗᐊᕋᑦᑕ ᑕᕝᕙᓂ ᐸᓐᓂᖅᑑᒥᐅᑕᐅᓪᓗᑕ. ᒫᓇᓕ ᐊᕐᕌᒎᓕᒫᖅ ᑕᑯᖃᑦᑕᓕᕋᑦᑕ ᓇᓄᕐᓂᒃ ᐱᑕᖃᓲᖑᓕᕐᒪᓂᓖᒃ ᐊᕐᕌᒎᓕᒫ, ᑕᐃᒪ ᓄᖑᓐᓂᐅᓴᖏᑦᑐᐊᓗᐃᑦ ᐃᓱᒪᒋᕙᒃ.”

—ᔭᐃᒥᓯ ᒪᐃᒃ

“ᐃᓛ ᑕᐃᒃᑯᐊ ᐃᒍᑦᑕᐃᑦ ᐃᒃᑯᐊ ᑕᐃᒃᑯᐊᖑᕙᓪᓕᐃᖏᑦᑐᐊᓗᐃ ᑕᐃᒃᑯᐊ ᑐᑦᑐᖃᓪᓛᖃᑦᑕᖅᑎᓪᓗᒍ ᓂᐱᖃᖏᑦᑐᐊᓗᐃ ᑐᑦᑐᒦᒐᓯᑦᑐᐊᓘᖃᑦᑕᔪᔪᑦ. ᑕᐃᒫᒃ ᐊᑦᑐᕆᐊᑦᓴ ᑲᑉᐱᐊᓇᖅᑐᒻᒪᕆᐊᓘᖃᑦᑕᑲᐃᓐᓇᓚᐅᖅᓯᒪᒻᒥᔪᑦ ᓂᐱᖃᖖᒋᑦᑐᐊᓗᐃᑦ ᑕᕝᕙᓂ ᑐᑦᑐᒥ ᐋᑦᑐᕐᓂᐊᖅᑐᓂ ᐋᑦᑐᐃᖃᑕᐅᓂᐊᕋᓗᐊᖅᑐᓂᓗ ᑕᕝᕙᓃᓐᓂᐅᓴᔪᐊᓗᑐᐃᓐᓇᐃᓗ ᐱᖃᑦᑕᓚᐅᖅᓯᒪᒻᒪᑕ. ᒫᓇᓕ ᐃᒃᑯᐊ ᓂᐱᓕᖅᐹᓗᕆᑦᑐᑦ ᐃᒍᑦᑕᖅᐹᓗᐃᑦ.”

—ᓕᐊ ᐊᒃᐸᓕᐊᓗᒃ

“I notice little black flies, smaller than we usually have. [...] I think the small black flies are coming from the south, from the ships that come in.”

—Jamasie Mike

“There are polar bears appearing every year now. [...] They come into the community now and appear out of nowhere. I don’t really want to hear people say they are going extinct, because they are not. When I was a child I never really saw polar bears—maybe once every five years. I’ve noticed a lot more polar bears. [...] There are too many.”

—Jamasie Mike

“There is a species of bee that is quiet—no buzzing. They are usually around caribou. A lot of times it gets difficult to skin a caribou because there are so many of those bees around. There are a lot of noisy bees right now. [...] There are more *kitturiat* [mosquitoes] and *milugiat* [black flies].”

—Leah Akpalialuk

“ᑕᒪᒃᑯᐊᕈᓗᐃᓪᓚ ᑕᒪᔾᔭ
ᐸᐅᕐᖕᒐᖅᑑᓴᑦᑐᑐᐊᕆᔫᑎᒃᑲ
ᓂᕐᓚᕈᓗᐃᑦ,
ᒫᓂᕐᒥᐅᑕᐅᖕᒋᑦᑐᐊᓘᒐᓗᐊᕐᖕᒐᑕᓚ
ᓂᕐᓚᕈᓗᐃᑦ ᑕᒫᓃᑦᑐᐊᓘᓚᕐᖕᒐᑕ
ᐅᐊᑦᑎᐊᕈᓚ ᓂᕐᓚᖃᖅᐸᔪᖖᒋᒻᒪ
ᑕᐅᕝᕙᓂ ᐱᓇᓱᒃᑕᐅᖃᑦᑕᖅᑐᖃᐃ
ᑯᐸᐃᒥᐅᔪᑦᓴᐅᔪᖅ ᒪᐅᖕᒐ
ᓄᓇᑖᕐᓂᑯᐊᓗᒻᒪᑕ.”

—ᒥᐊᕆ ᐹᑎ

“ᑐᓴᐅᑎᒃᑯᓪᓗ ᐄ
ᑕᒫᓃᖃᑦᑕᕐᐸᔪᖕᒋᑦᑐᓂ
ᐱᑕᖃᖃᑦᑕᖅᓯᒪᓚᖅᑐᑦ ᑕᐃᒃᑯᐊ
ᐱᓗᐊᖖᒍᐊᖅᑐ ᓇᑦᑐᕋᓚᐊᓗᐃ,
ᓇᑦᑐᕋᓚᐊᓗᐃᑦ ᑕᒫᓃᖃᑦᑕᖅᓯᒪᓚᕐᒪᑕ
ᑕᒫᓃᑉᐸᓚᖅᑐᑦ. ᓇᑦᑐᕋᓛᑦ ᑕᐃᒃᑯᐊ
ᑎᒻᒥᐊᖅᐹᓗᐃ ᑐᓴᔪᒻᒥᔪᖕᒐᓚ
ᑲᖕᒋᖅᑐᒑᐱᒻᒥ ᑕᐃᒪᐃᑦᑐᖅ
ᓇᑦᑎᖅᓯᒪᔪᖅ, ᓇᑦᑎᓰᒻᒪᓈᑦ.
ᑭᓰᒐᓗᐊᕐᓂᑭᐊᖅ ᐃᓛ
ᑕᐃᒃᑯᐊ ᐅᕙᖕᒐᓛ ᖃᐅᔨᒪᔭᒃᑲ
ᑕᐃᒃᑯᐊᓪᓗᐊᖕᒍᐊᑦ ᑕᕝᕙᑐᐊᖅ.
ᑕᑯᔭᐅᓚᐅᖅᓯᒪᖕᒋᑦᑐᒥᒃ ᑎᒻᒥᐊᒥᒃ
ᑕᑯᔪᖃᕐᓂᕋᐃᕙᑦᑐᒐᓗᐊᑦ ᐅᐱᕐᖕᒎᒃᑯ.”

—ᐲᑦᑐᓘᓯ ᖄᐱᒃ

“ᑭᑦᑐᕆᐊᖃᖖᒋᓐᓂᖅᓴᐅᔫᑎᒋᒐᒃᑯᓛ
ᐊᐅᔭᐅᓰᖕᒍᒻᒪ ᐅᖅᑰᓗᐊᒎᕙᓪᓚᐃᔪᖅ
ᐸᓂᖅᐸᓪᓚᐃᓂᖅᑐ ᑭᑦᑐᕆᐊᑦ
ᐱᑕᖃᖖᒋᓂᖅᓴᐃᑦ ᐊᐅᔭᐅᓰᖕᒍᒻᒪᑦ
ᑭᓯᐊᓂ ᒪᑯᐊ ᒥᓗᒋᐊᖄᓗᐃ
ᐃᒍᑦᑖᓗᐃ ᑕᒪᒃᑯᐊ ᖄᒥᐊᓗᐃ
ᑕᒪᒃᑯᐊᖕᒍᓂᖅᓵᓗᐃ.”

—ᑕᐅᑭ ᖄᐱᒃ

“There used to be no Canada geese here. They don’t belong here. [...] Now they got settled up here and they are eating our berries. I don’t want to see any Canada geese here because they eat all the berries.”

—Mary Battie

“According to what I have heard and seen, there is some wildlife that never used to be here before. Bald eagles [...] have been seen around here now. I heard in Clyde River an eagle killed a seal. [...] Sometimes I hear from hunters that they have seen a bird they have never seen before.”

—Peterloosie Qappik

“There are fewer mosquitoes this year. I believe it’s from the heat. The creeks and marsh areas are too dry. But *iguttait* [bumblebees], *milugiat*, and *kaumajait* [house flies] are in abundance.”

—Taukie Qappik

ᐅᑭᐅᒥ ᐊᓯᔾᔨᖅᑕᕐᓂᖅ

“ᑕᐃᒫ ᓯᑯᐃᖅᓴᕋᐃᓐᓂᖅᓴᐅᑦᑕᓕᖅᑯᖅ ᐊᒻᒪ ᓯᑯᓇᓵᕐᓂᖅᓵᓘᑦᑕᓕᖅᑯᖅ. ᑕᕝᕗᑐᐊᖑᔮᖅᑐᖅ.”

—ᐊᐃᒐ ᐃᓱᓪᓗᑕᖅ

“ᐅᐱᕐᖔᕈᖅᓴᕋᐃᓐᓂᓴᑦᓚᕆᐊᓘᓕᖅᑐ ᐅᐊᑦᑎᐊᕈᓂᑦ ᐊᔾᔨᙱᑦᑎᐊᓕᖅᑕᖓᑦᑕᐅᖅ ᓯᓚᕗ ᖃᐅᔨᒪᑦᑎᐊᖅᑕᕋᑦᑕᐅᖅ. [...] ᐃᓛᓐᓂᒃᑰ, ᐃᓛᓐᓂᑯᑦ ᐅᐱᕐᖔᕈᖅᓴᕋᐃᓐᓂᖅᓴᐅᔮᖅᐸᓕᕐᒥᔪᖅ, ᐅᐊᑦᑎᐊᕈᓕ ᐊᔾᔨᖃᑦᑕᖅᑑᔮᖅᐸᔪᙱᒃᑲᓗᐊᕐᒪᒍ ᓯᓚᕗᑦ ᑕᐃᒪᐃᑉᐸᒍᓐᓂᕐᖓ. ᐅᐊᑦᑎᐊᕈᓕ ᕿᒧᑦᓯᖃᑦᑕᑎᓪᓗᒋᑦ ᔫᓂᒥ ᕿᒧᑦᓯᖃᑦᑕᑐᐊᓘᕙᔪᒐᓗᐊᕐᖓᑕ ᐸᓐᓂᖅᑑᒥ, ᐊᓪᓛᑦ ᑎᑭᑉᐸᔪᔪᑦ ᔫᓂ ᓄᖑᓕᖅᑎᓪᓗᒍ ᕿᒧᑦᓰᑦ. ᓯᑯᓐᓇᓵᕐᓂᖅᓴ ᒪᓐᓇ ᓯᑯᐃᔭᖅᓴᕋᐃᑦᑐᖅ. ᔫᓂᒥᓕ ᓯᑯᐃᔭᖅᐸᓕᕋᑦᑕ. [...] ᐅᑭᐅᕐᓇᓵᕐᓂᖅᓴᐅᖃᑦᑕᓕᖅᑐᖅ ᐅᑭᐊᒃᓵᒃᑯᑦ ᓯᑯᓇᓵᖅᑐᐊᓘᖃᑦᑕᓕᕐᖓᑦᓕ ᒪᓐᓇ ᓯᑯᓴᕋᔮᑉᐸᔪᓗᐊᕐᖓ ᐅᑭᐅᕐᓇᓵᓂᖅᓴᐅᑎᒋᖃᑦᑕᓕᖅᑯᓕ ᒪᓐᓇ ᖃᐅᔨᒫᔭᒃᑯᑦᓕ. ᑕᒫᓂ ᑭᓯᐊᓂ ᓯᑯᓰᖑᓕᕐᖓᑦ ᑎᓯᐱᕆᒥ, ᓄᕕᐱᕆᒥᓕ ᓯᑯᕙᔪᓗᐊᕐᖓ. ᐃᓛᓐᓂᑦ ᓯᑯᓴᕋᐃᑎᓪᓗᒍ ᐊᑦᑑᕙᒥ ᓯᑯᔭᐊᓘᕙᔪᔪᑦ, ᑕᒫᓂᓕ ᓯᑯᓂᐊᓴᓪᓗᐊᖅᐸᒍᓐᓃᕋᑦᑕ.”

—ᐄᕕ ᐊᓂᕐᓂᓕᐊᖅ

Seasons

“The ice tends to go out a lot quicker, a lot earlier; then it takes a long time for the ice to form up.”

—Aiga Ishulutaq

“[We have an] earlier spring, usually around March or April now. Before it was like June or late May. [...] Sometimes I notice that the fall comes in earlier, but only sometimes. In June there used to be dog teams coming up on the sea ice, but now, these days, the sea ice melts by the end of May. [...] [We have] later winters. The winter doesn't usually come in until December. Back then it was November.”

—Evee Anilniliak

“In the past you didn’t see a lot of ice melting or breaking during the spring, and that is dangerous.”

—Jaco Ishulutaq

“This spring [2008] we had a major melt in the community. There was a lot of rainfall, and most of the rivers were raging.”

—Jamasie Mike

“In the springtime, when the snow usually starts to melt, during the day it tends to melt and by evening it tends to form up into ice again. But [today] it doesn’t freeze at night. It just melts all at once. It’s getting to be warmer in springtime than in the ’60s to ’80s.”

—Jamasie Mike

“These days we don’t usually get snow until December, like good-quality snow that we can ride our snowmobiles on. [...] Back then snow would start coming in late October or early November.”

—Leah Evik

“ᑖᒃᑯᐊ ᓯᑯᖓ
ᐊᐅᔅᓴᕋᐃᓐᓂᕐᒃᓴᐅᑦᑕᓕᕐᒃᑯ ᑕᒫᓂ
ᑕᐃᒪᐃᑦᑑᕐᒃᑲᑦᑕᕐᒃᓯᒪᖏᑦᑑᒐᓗᐊᕐᒃ
ᐅᐊᑦᑎᐊᕈ. ᑕᕝᕙ ᑖᒃᑯᐊ
ᐃᓱᒫᓗᓐᓇᓪᓚᓇᑦᑐᑦ ᑕᒪᒃᑯᓄᖓ
ᓄᑲᕐᒃᑎᐅᔪᓄᑦ.”

—ᔮᐃᑯ ᐃᓱᓪᓗᑕᕐᒃ

“ᓯᓚ ᒪᓐᓇ ᑕᕝᕙ ᐅᐱᕐᖔᕐᒃ
ᐊᒃᓱᕐᔪᐊᕌᓗᒃ ᐅᕐᒃᑰᔦᕐᒃ.
ᐅᕐᒃᑰᕐᒃᑲᑦᑕᓚᐅᕐᒫ ᐊᒻᒪᓗ ᐃᓚᖓ
ᓱᒃᑲᓕᔪᐊᓗᒻᒥ ᐊᐳᑎᓕᐊᓘᓪᓗᓂ
ᐅᐱᕐᖓᒃᓵᕐᒃ ᐊᐅᑉᐸᓪᓕᐊᓕᐊᓚᐅᕐᒪᑦ.
ᓱᒃᑦᑐᐊᓗᒻᒥᒃ ᐊᒻᒪ
ᓯᓚᓗᒃᑐᐊᓘᓕᕐᒃᑐᓂ. ᑕᐃᒫᒃ ᑕᒪᔅᓴ ᑰ
ᒪᑯᐊ ᓱᕋᓚᕐᑭᓚᐅᕐᒃᑐᑦ.”

—ᔮᐃᒥᓯ ᒪᐃᒃ

“ᑕᐃᒪᐃᑐᐃᓐᓇᕐᒃᑑᒐᓗᐊᕐᒃ
ᐅᐱᕐᖓᓇᐊᕐᑲᑦᑕᕐᓂᖓ
ᑭᓯᐊᓂ ᐅᐊᑦᑎᐊᕈᓕ ᐃᒪᕐᒃ
ᐊᐅᓪᓴᕐᒃᐸᔪᒐᓗᐊᕐᖓᑦ ᒪᓐᓇ ᐊᐳᑦ
ᐅᓐᓄᐊᖑᓕᕋᐃᒻᒪᑦ ᕐᑯᐊᕐᒃᐸᒃᑐᓂ
ᑕᐃᒪᐃᑦᑑᕙᔪᒐᓗᐊᕐᒪᑦ ᒫᓇᓕ
ᒪᙳᒐᐃᒐᒥ ᕐᑯᐊᕐᒃᑲᑦᑕᕈᓐᓃᕐᒃᑐᓂᓗ
ᑕᐃᒪᐃᑦᑑᔮᓕᕐᒃᑐᕐᒃ ᒪᓐᓇ
ᐊᐳᑎ. ᐅᐊᑦᑎᐊᕉᕙᔪᒪᓪᓖ
ᐅᕐᒃᑰᓯᑲᓪᓚᓂᕐᒃᓴᐅᕐᒃᑲᑦᑕᕐᒃᑐᕐᒃ.”

—ᔮᐃᒥᓯ ᒪᐃᒃ

“ᐄ ᓰᔪᓯᒫᕋ
ᐊᐱᒍᓐᓇᐃᓪᓕᑲᔅᐸᓪᓕᕐᖓᓪᓕ
ᕐᑯᕝᕙᐊᓯᕝᕙᐅᓪᓗ ᐅᓪᓗᖏᓐᓄᑦ
ᑎᑭᓐᓂᐅᓴᕙᓕᕐᒪᑦ
ᐊᐱᓪᓚᓪᓗᓂ ᐊᐱᓚᐅᕐᒃᓯᒪᓇᓂ,
ᐊᐳᑦᑎᐊᕚᓗᕙᔪᒐᓗᐊᕐᖓᓪᓕ
ᐅᐊᑦᑎᐊᕈ. ᑕᒫᓂᓕ ᕐᑲᓐᓂᕐᑲᑦᑕᓕᕋᐃᒐᒥ
ᐊᑦᑑᐸᐅᑎᓪᓗᒍ ᓄᕙᐃᑉᐸ

ᑕᒪᓐᓇ ᐊᐱᑦᓚᓂᓕᓲᓂᕙᔪᒻᒪᒍ,
ᑕᐃᒪᐃᒍᓐᓇᐃᓕᒻᒪᑦᓕ ᐊᐱᑲᐃᓐᓇᑕᖅᑯ
ᐊᓄᓂᒧᑦ ᓱᐱᔭᐅᑦᑕᖅᑯ. ᐃᒪᐃᓲᖅᒍᓕᖃᑦᑕ
ᐊᐱᑦᓚᓂᒃᑲᓗᐊᖃᐃᓚᑦᑕ
ᐊᐱᑲᐃᓐᓇᖃᐃᓚᑦᑕ ᐊᐱᒍᓐᓇᐃᑦᓕᑕᖅᑯ
ᓯᑭᑰᖄᖅᑐᐊᓘᑕᖅᑯ ᐊᐱᑲᐃᓐᓇᖕᓚ
ᓲᓐᓇᐃᑦᓕᑕᖅᑯᑦ.”

—ᓕᐊ ᐃᕖᒃ

“ᓱᕈᓯᐅᑎᑦᑐᑕᓕ ᑕᐃᒫ
ᐊᓄᓂᐊᓗᖕᒥᒃ ᖃᐅᔨᖃᑦᑕᔪᖕᒋᑦᑐᒍ,
ᓯᓚᑦᑎᐊᕚᓘᕙᔾᒪᑦᓕ. ᐃᒫ
ᓯᓚᖅᑭᑦᑎᐊᖕᒋᓐᓇᖅᑰᔨᔭᐊᓘᑦᑐᓂ
ᓄᕗᔭᓚᐃᒻᒪᑦ ᑖᖅᓯᕈᔾᒪᓂᑉᐸᐅᒃ.
ᓯᓚᑦᑎᐊᕚᓘᖃᑦᑕᓚᐅᖅᓯᒪᔭᖅ
ᐅᐱᕐᓛᒃᑯᑦ ᑕᒫᓂ ᔫᓐ ᑕᒪᒃᑯᐊ
ᔪᓚᐃ ᑖᒪᓂ ᐱᓱᓚᖅᑐᓂ.
ᖃᐅᔾᔨᓯᒪᔪᕗᑦᓕ ᑕᐃᒫᒃ
ᑲᑎᙵᓖᑦᑐᑕ ᐅᖃᖃᑦᑕᐅᑎᓲᖅᒍᑦᑐᑕ,
ᑕᐃᒪᐃᑦᑑᒍᓐᓅᖅᒪᑦᓕ ᓯᓚ
ᐊᔾᔨᖕᒋᑦᑎᐊᓕᖅᑕᖕᓚ
ᑕᐃᔅᓱᒪᓂᐅᖃᑦᑕᔪᖕᒪᑦ
ᓯᑯᐃᖅᓵᑎᐊᖅᑐᐊᓘᕙᑦᑐᓂ,
ᕿᒧᒃᓰᑦ ᔫᓂᐅᑎᑦᑐᒍ ᕿᒧᒃᓯᖅᐸᔪᔪᑦ
ᕿᒧᒃᓯᖃᑦᑕᔪᔪᐃᑦ ᐅᕝᕙᓘᓐᓂᑦ ᐅᕗᖕᓚ
ᐸᖕᓂᖅᑑᒡᖅᐸᔪᔪᐃᑦ, ᐆᒪ ᐊᓯᐊᓂ
ᓄᓇᖃᖅᑎᑦᑐᑕ, ᐱᕙᓪᓕᐊᓚᒥ ᐊᓄᓂ
ᐊᓪᓛᑦᑕᐅᖅ ᐱᕙᓪᓕᐊᖃᓯᐅᑎᓯᒪᐅᒃ
ᐊᓯᔅᓴᖅᓯᒪᓪᓚᓂᒃᑐᖅ
ᑕᐃᔅᓱᒪᓂᐅᖃᑦᑕᓚᐅᖅᓯᒪᔪᒧᑦ,
ᖃᐅᓴᒪᔭᑦᑎᒍᓪᓗ ᐅᖃᕈᓐᓇᖅᑕᕗᑦ
ᑖᓐᓇ. [...] ᐄ ᐊᓄᖀᓕᑲᓪᓚᑦᑐᐊᓘᑦᑕᖅᑯ
ᐊᓯᐊᓂ ᐊᓄᖀᖅᑲᐅᓚᐅᓗᐊᖅᑐᓂ
ᐃᒪᐃᓕᐅᖅᓯᒪᓂᖅᓵᓘᓕᖅᑐᖅ
ᐅᐊᑦᑎᐊᕈᓂ ᐃᑯᐊᓕᐊᕈᓐᓇᖕᒋᓐᓂᖅᓴᖅ.
ᐅᐊᑦᑎᐊᕈᓕ ᐃᑯᓪᓕᐊᖅ ᐃᒫ
ᓯᓚᑦᑎᐊᕚᓗᒃ ᓴᐃᓕᔭᐊᓘᕙᔭᔭᖅ.”

—ᓕᐊ ᐃᕖᒃ

“Back when I was a child we used to have beautiful, sunny days throughout June, July, and August. [...] But these days the wind usually shows up and stays for a week or so, and then just disappears. [...] The wind changes; it just tends to show up out of nowhere.”

—Leah Evik

“In October we would start having the formation of sea ice, and by the end of the month people would be able to travel, but today you are waiting up to December for the formation of the sea ice.”

—Leesee Mary Kakee

ᓄᓇᕗᑦ / ᐸᖕᓂᖅᑑᖅ

“ᐊᒃᑑᕙᐅ ᓄᙳᐊᓂ ᓯᑎᐱᓇᐅ ᑕᒫᓂ ᓄᙳᐊᓂ ᓯᑯᐊᖅᑕᓕᕋᒥ ᐊᒃᑑᕙ ᓯᑯᐊᓪᓚᓇᒃᑲᒥ ᑕᐃᒪ ᓄᖑᐊᓂ ᐊᓪᓛᑦ ᐊᑐᕋᔅᓴᕈᖅᑐᓂ, ᓯᑯᓯᐅᖅᕙᒃᓴᐅᓕᖅᑐᓂ ᑕᐃᒪᐃᑉᐸᔪᔪᖅ ᓄᕕᐱᓇ ᑕᒪᓐᓇ ᓱᔾᔭᖕᒐ ᐊᑲᐅᓗᐊᖕᒋᒃᑲᓗᐊᖅᑎᓪᓗᒍ ᐊᒃᑑᕝᕙ ᐱᓱᕝᕕᑦᓴᐅᓕᖅᐸᔪᔪᖅ. ᒫᓐᓇ ᐸᓐᓂᖅᑑᒦᒃᑲᑦᑕᐅᓐᓃᑦ ᑎᓯᐱᓇᒥᒃ ᑎᑭᑉᐸᓕᖅᑐᖅ ᓯᑯᓚᐅᕋᓂ ᐊᑦᓯᕙᔾᔪᐊᕌᓗᒃ ᐊᓯᔾᔨᖅᑐᖅ.”

—ᑕᓯ ᒥᐊᑕ ᑲᒃᐱᒃ

“In the past the winters used to be very cold, and for the past few years it seemed like it was warming up, but last year the temperatures seemed like they had been in the past. But it seems to be going up and down. [...] The ice was thinner this winter. Even if it was very cold, the ice wasn’t thick enough.”

—Taukie Qappik

“ᐊᓯᔾᔨᖅᓯᒪᓕᔪᔫᒐᓗᐊ ᐅᖅᑰᓂᖅᓴᐅᖅᑲᑦᑕᓕᔪᒐᓗᐊᖕᒐ ᐅᑭᐅᒃᑯᑦ ᑕᒫᓂ ᕚᕝᕙᐊᓇ ᒫᔅᓵ ᑳᒃᑯᐊ ᐅᖅᑰᓂᖅᓴᐅᖅᑲᑦᑕᔫᒪᑕ ᑕᐃᔅᓯᒪᓂ ᐅᐊᑦᑎᐊᕈᒃᑲᓐᓂ ᐃᒃᑮᓇᖅᑐᐊᓘᕙᔪᒐᓗᐊᖅᑐᓂ ᑕᕝᕙ ᐅᑭᐅᖅ ᐃᒃᑮᓇᕐᓂᖅᓴᐅᖅᑲᑦᑕᓚᐅᕐᒥᔪᖅ ᑳᒃᑯᐊ ᐅᔅᓴᓇᓯᒪᔅᓴᒃ ᑕᐃᒫᒃ.”

—ᑕᐅᑭ ᖄᐱᒃ

ᓯᓚ/ᐊᓄᓇ

Climate/Weather

“ᐋ ᖃᓐᓇ ᑰᖅᑲᑦᑕᖕᒋᒻᒪᑦ ᐱᓗᐊᖅᑐᖅ ᑰᖕᒐ ᒪᓇ ᐃᒪᒥᓂᐅᒐᓗᐊᖅ ᐃᒪᖅᑲᕈᓇᐃᑦᑕᖅᑯᖅ ᒪᓇ ᐸᓂᖅᑐᐊᓘᖕᒪᑦ ᓄᓇ, ᐃᔅᔪᖅ ᐃᓚᖕᒐ ᖁᓪᓗᐊᓯᒪᒪᓇᑦᑑᔭᓕᖅᑐᖅ ᐸᓂᓗᐊᓘᑦ.”

—ᐊᐃᒐ ᐃᓱᓪᓗᑕᖅ

“This year it’s very dry. I see some places where there used to be streams. They are no longer there. [...] This year there is no rain at all.”

—Aiga Ishulutaq

“ᐃᓛ ᐊᔅᓯᕆᒍᓐᓃᑦᑎᐊᖅᐸᑦᑕᖕᒐ ᓱᓕᖅᑲᐃ ᐊᔅᓯᕆᒍᓐᓃᖅᑕᖕᒐ ᐅᐱᖕᒐᑳᓪᓚᑦᑐᐊᓘᕚᑦᑐᓂᓗ

“Right now I notice that when there is snow, it seems like it will stay for a long period, but

ᑕᕝᕙ ᐃᒫ ᐊᐳᑎᑯᓘᓚᐅᖅᑐᓂ
ᓂᓪᓚᓱᔾᔭᐊᑦᑕᓚᐅᖅᑐᓂ
ᑲᑦᑐᓇᐊᖏᑦᑕᐅᑎᒋᓪᓗᓂᐅ
ᓱᑭᑑᖅᕕᑦᓴᐅᔭᓐᓃᑲᐅᑎᒋᓪᓗᓂᐅᓲᖑᓕᕐᒪᑦ.
[...] ᐊᐱᔭᖅᐹᓗᖅᑲᑦᑕᔪᒐᓗᐊᖅᑐᒍᓪᓕ
ᐅᐊᑦᑎᐊᕈ ᓱᕈᓯᐅᑎᓪᓗᑕ ᐊᒃᓱᖅᐹᓗᒃ
ᐊᐳᑕᐅᓕᖅᐸᓚᐅᖅᓯᒪᔫᒐᓗᐊᖅ,
ᐃᒫᒃ ᐊᐳᔅᓴᓇᔅᔭᐊᖄᓗᒻᒥ, ᐃᒫᒃ
ᐊᐳᑎᕙᔅᔭᐊᑯᓘᓕᖅᑲᑦᑕᓚᐅᖅᓯᒪᔭᖅ
ᐃᒫ ᐊᓪᓛᑦ ᒪᑯᐊ
ᐊᐱᔭᐊᓘᖃᑖᖅᑐᒥᓂᐅᔭᖃᐃᒻᒪ
ᒪᑯᐊ ᓯᕙᖓᕿᔭᓐᓂᖏᑦ
ᐱᙳᐊᖅᕕᒋᔭᖅᑐᖅᐸᔪᒐᑦᑎᒍᑦ
ᐊᓪᓛᑦ ᑕᐃᒪᐃᑦᑐᒥᒃ ᐊᐱᔭᐊᓗᒻᒥᒃ
ᖃᐅᔨᓇᖅᓯᓚᐅᖅᓯᒪᔭᓐᓃᖅᑐᖅ."

—ᑕᐃᓯ ᑎᐊᓚ

"ᐃᑖ ᑕᒪᒃᑯᐊ ᐊᓐᓂᓇᔫᒥᕗᑦᑕᕗ
ᐸᓐᓂᖅᑑᒥ ᑕᐃᒪᙱᑦ
ᓱᕈᓯᐅᓂᑦᑎᓐᓂ ᒪᓐᓇ ᖅᑳᖓ
ᐃᒫ ᐊᐅᒌᖅᐸᔪᖏᓪᒥᔫᒐᓗᐊᑦ,
ᑕᒪᔅᓴ ᖅᑯᑦᑕᐅᕗᑉᐳᑦ
ᐊᐅᔭᐃᑦᑐᓇᕗᔫᓪᒥᔭᖃᓗᐊᖏᑦ
ᐊᐅᒃᓴᖃᐃᑦᑐᐊᓘᕗᑉᐳᑦ
ᐊᐅᔭᐃᑦᑐᖅᑲᖅᕕᒥᓂᖏᑦ. ᑕᒪᔅᓴ
ᓇᓗᓇᖅᐸᖏᓚᑦ ᖅᑯᖅᑕᐅᓂᓇᕗᒃᐸᐃᑦ
ᑕᒪᔅᓴ ᐊᐅᔭᐃᑦᑐᒥᓂᐅᒐᓗᐊᑦ
ᓱᕈᓯᐅᑎᓪᓗᑕ. ᐊᒻᒫᑦᑕᐅ ᐊᓪᓛ
ᐸᓐᓇ ᐊᐅᒌᖅᐸᔪᖏᓪᒥᔫᒐᓗᐊᖅ,
ᑕᐃᒪ ᖅᑯᖅᑕᐅᓂᓇᑉᐸ
ᐊᐅᔭᐃᑦᑐᓇᔭᒥᓂᖃᓗᐊᖏᑦ ᐃᒫᓕ
ᑕᑯᔅᓴᖅᑕᖅᑲᖏᑦᑎᐊᖅᒪᓇᑉᐸᓕᖅᖓᓪᓕ
ᐃᒫ ᑕᑯᔅᓴᖅᑲᖕᒎᒐᓗᐊᖅᓕ. [...]
ᓅᖑᑦᑐᐊᓗᐃᑦ."

—ᑕᐃᓯ ᑎᐊᓚ

overnight the snow will be gone. [...] When I was a kid I remember we used to get a lot more snow, because I used to play in the fresh snow. It was really nice, but now it seems like we get a lot less snow."

—Daisy Dialla

"On top of the mountain used to be all snow and glaciers, but it's all gone. Melted. You can see where the glaciers used to be. Even on top. [There used to be ice there] all year round."

—Daisy Dialla

"The wind seems to be coming from anywhere the wind might want to come from. One thing I notice is that in the older days we used to have four main winds: north, south, east, and west. Now the wind can come from anywhere, not any particular direction. Seems to be the wind is getting stronger."

—Inusirq Nashalik

Nunavut / Pangnirtung

“ᓇᑭᑐᐃᓐᓇᖅ ᐅᔾᔨᕆᓯᒪᔭᒃᑯ ᖃᐅᔨᒪᔭᒃᑯᑦ ᐃᓛ ᐅᐊᑦᑎᐊᕈ ᑕᐃᒪᐃᑦᑑᕋᔪᖕᒋᑦᑑᒐᓗᐊᖅ ᖃᐅᔨᒪᒐᓚᓕᕐᒥᔪᖕᒐ ᓇᑭᑐᐃᓐᓇᖅ ᐊᓄᕌᓲᖕᒍᓕᖅᑐᖅ ᓇᑭᐊᓗᑐᐃᓐᓇᖅ. ᑕᐃᒪᐃᑉᐸᔪᖕᒋᑦᑑᒐᓗᐊᖅ ᐅᑯᐊᑑᕋᔪᒻᒪᑕᓛ ᐅᐊᖕᓇᕐᓗ ᓂᒋᕐᓗ ᐱᖕᒐᖕᓂᕐᓗ ᐊᕋᖕᒐᓂᕐᓗ ᑕᕝᕙᑑᕋᔪᒐᓗᐊᕐᒪᑕᓛ, ᒫᓐᓇ ᐊᓯᖕᒐᓂ ᖃᐅᔨᖅᑲᑦᑕᖅᓯᒪᒻᒥᔪ ᓇᑭᑐᐃᓐᓇᓕᒫᖅ ᐊᓄᕌᖃᑕᓲᖕᒍᓕᖅᑐᖅ. ᐊᓄᕌᒃᑲᐅᓂᖅᓴᖅ ᓴᙱᑕᖅᐸᑦᑕᐊᑐᐃᓐᓇᖅᑐᖅ ᐊᓄᓇᖕᒐ ᓴᖕᒋᔩᑕᖅᐸᑕᐊᑐᐃᓐᓇᖅᑐᖅ ᓇᒥᑐᐃᓐᓇᖅ.”

—ᐃᓅᓯᖅ ᓇᓴᓕᒃ

“ᐃᒫ ᐃᓱᓘᕕᖅᑲᑦᑕᑦᑐᐊᓘᒥᒻᒪᑦᑕ ᐅᐊᑦᑎᐊᕈ, ᐅᐊᑦᑎᐊᕈ ᐃᓱᓗᐃᔪᐊᓘᖅᑲᑦᑕᖅᐸᔪᒐᓗᐊᕐᒪᑦᑕ ᒫᓐᓇᑕ ᐃᓱᓘᕕᑦᓴᖅᑲᕈᓐᓃᕐᑦᑕᑕ ᐸᓐᓂᖅᑑᒥ ᑕᒫᓂ.”

—ᐃᓅᓯᖅ ᓇᓴᓕᒃ

“ᑕᐃᔅᓯᒪᓂᓕ ᓲᖅᑯᐃᓯᓕᖅᑎᑦᓗᖕᒐᓕ ᒪᐅᔭᕈᖅᑐᐊᓘᕋᔪᒐᓗᐊᕐᒪᑦᓛ, ᒫᓐᓇᓕ ᒪᐅᔭᑕᓗᐊᕈᓐᓇᖅᑑᔭᕈᓐᓃᖅᑐᖅ, ᑕᐃᔅᓯᒪᓂᓕ ᒪᐅᔭᑕᖅᑐᐊᓘᕋᔪᒻᒪ.”

—ᔭᐃᑯ ᐃᓱᑦᓗᑕᖅ

“ᖃᐅᔨᓯᒪᔭᒃᑯᑦᑕ ᐊᕙᓂᑕ ᐅᐸᖅᑲᑦᑕᖅᑕᓐᓂ, ᐅᖃᐅᓯᓇᔪᓐᓇᖅᑕᖃᑕ ᐃᑭᐊᖕᒐ ᐊᐅᑉᐸᑦᑕᐊᖑᔩᕐᔪᓚᐅᖅᑐᖅ ᓄᓇᐅᑉ ᑕᐅᓐᓇ ᓂᓚᐅᒻᒪᑦ ᐃᑭᐊᖕᒐ ᖁᐊᖕᒍᐃᓐᓇᖅᑐᖅ. ᓲᕐᓗ ᑕᐅᓄᖕᒐ ᐊᐅᒃᐸᑦᑕᐊᖕᒍᔭᑕᖅᑐᖅ ᐊᒻᒧᖕᒑᖅ. ᓲᕐᓗ ᒪᓐᓇ ᐃᓚᖕᒐ

“We used to get thunder, but now we don’t get that anymore. It’s all gone! Thunder and very fast rains. There used to be lightning, too, but nowadays we don’t get that in Pang anymore.”

—Inusirq Nashalik

“When I was a youngster we used to have a lot more snow, and it used to get so heavy with snow that it would be hard to travel on. Today we don’t seem to have that type of snowfall anymore.”

—Jaco Ishulutaq

“That’s the type of weather today: when it’s bad it’s bad, and when it’s good weather it’s good for too long. There are no breaks in between.”

—Leah Akpalialuk

“I’m noticing that because the permafrost is thawing, much of the area where you have *ippiugaq* [slope of a plateau] is starting to sink. It’s sucked in most of the plateaus. And also, one of the conditions [that is causing] the permafrost to be thawing much more quickly is that we have less rain. When it rains a lot [...] the rain goes beyond the soil, and once it goes down it freezes. [...] With less

ᐃᑉᐱᒑ ᒪᑯᐊ ᐱᓗᐊᖅᑰ, ᐃᓚᖏᑦ
ᑲᑕᑦᑎᐸᓪᓕᐊᖑᔭᓕᖅᑐᑦ.
ᑕᕝᕙ ᓯᓚᓗᖅᑲᓗᐊᕌᓗᖏᑉᐸᑦ
ᑕᐃᒪᐃᖅᑲᑦᑕᕋᔭᖅᑐᖅ ᓄᓇ
ᖃᐅᓯᓕᓲᖑᒻᒪ, ᓯᓚᓗᖅᑲᑦᑕᖅᑎᓪᓗᒍᓕ
ᑕᐅᓄᖓ ᖃᐅᓯᓕᓲᖑᒻᒪ ᑕᐅᓄᖓ
ᐊᒻᒪ ᖁᐊᓲᖑᒻᒪ ᖃᐅᓯᐅᓪᓗᓃ,
ᓯᓚᓗᖅᑲᑦᑕᓗᐊᕌᓗᖏᑉᐸᑦ ᐃᒻᒪᖄ
ᑕᐅᓄᖓ ᐊᐅᑉᐸᓪᓕᐊᒍᓐᓇᓂᔅᓴᐅᒻᒪ
ᐊᓂᓐᓇᑎᓪᓗᒍ ᑕᒪᓐᓇ ᖄᖓ.”

—ᔭᐃᑯ ᐃᓱᓪᓗᑕᖅ

“ᑕᐃᓖ ᓯᓚᖅᑲᖅᑲᑦᑕᕋᓗᐊᖅᓂᕋᑦᑕᓕ
ᐃᓖ ᐅᓪᓗᒥᓕ ᓯᓚᖅᑲᖅᑲᑦᑕᓕᖅᑭᑕ
ᐃᓖᖏᓐᓈᓗᑐᐃᓐᓇᖅ
ᐊᑕᐅᓲᓐᓇᓕᔪᑲᑦᑳᓗᑐᐃᓐᓇᒥᒃ
ᓯᓚᑦᑎᐊᕙᕈᖅᑕᕋᐃᒐᑦᑕ
ᓯᓚᑦᑎᐊᕚᓗᑐᐃᓐᓇᐅᖅᑲᑦᑕᖅᑐᓂ
ᓯᓚᑲᒻᒪᕈᕋᐃᒐᒥ
ᓯᓚᑲᒻᒫᓗᑐᐃᓐᓇᐅᕙᒃᑐᓂ.”

—ᓕᐊ ᐊᒃᐸᓕᐊᓗᒃ

“ᐊᑦᓯᕕᔾᔪᐊᕌᓗᒃ ᓄᓇᐅᑉ ᐊᐳᑎᖓ
ᑲᑕᒑᓴᕋᐃᑦᑐᑯᓘᑦᓗᓂᐅᒐᓗᐊᖅᒪᑦ
ᐸᓂᖅᑐᑯᓘᔭᖅᑐᓂ, ᑕᕆᐅᑉ
ᐊᐳᑎᖓ ᐊᕕᑲᓪᓚᑦᑖᕈᓐᓇᖅᑐᖅ
ᐊᒻᒪ ᑲᑎᔪᓐᓇᖅᑐᓂ ᖃᐅᓯᐅᔫᖅᓗᓂ
ᑭᓯᐊᓂ ᓄᓇᐅ ᐊᐳᑎᒐᓗᐊᖓ ᑕᕆᐅᑉ
ᐊᐳᑎᒋᔫᔭᖅᑕᖓ.”

—ᓕᓯ ᒥᐊᓕ ᑲᒃᑭᒃ

“ᑕᕝᕙᓂ ᐃᕗᓘᒥᓲᖑᒻᒪᕆᒃᑐᖅ
ᓯᓚᓗᒃᑲᕐᔪᐊᕌᓘᒻᒪᑦ ᓯᓚᓗᓚᕋᐃᒐᒥ,
ᐄᖑᔮᖅᑰ ᐃᕗᓘᒥᓲᖑᓚᖅᑐᖅᓚ.”

—ᓕᓯ ᒥᐊᓕ ᑲᒃᑭᒃ

rain, there will be more thawing of the permafrost, and as a result the ground is sinking.”

—Jaco Ishulutaq

“The snow on the land is supposed to be dry…but the snow on the sea ice is supposed to be mushy and wet and you can shape it. […] Today the snow on the land has the same consistency as you would have on the sea ice.”

—Leesee Mary Kakee

“In our area, we are starting to experience thunderstorms […] [and] a lot more rain than before. Some are even comparing it to the southern way of rain and thunderstorms.”

—Leesee Mary Kakee

“ᖃᖅᑲᐃᑦ
ᐊᐳᑎᓕᖅᐹᓘᕙᓚᐅᖅᓯᒪᒪᒐᓗᐊᖕᒐᑕ
ᑭᓯᐊᓂᓕ ᐱᑕᖃᖏᑦᑐᖅ ᒫᓐᓇ.
ᐊᓂᐅᒐᖃᕐᓂᖅᓴᐅᕙᓚᐅᖅᓯᒪᒻᒥᔫᒐᓗᐊᖅ.
ᐊᐱᕐᑎᐊᖃᑦᑕᕈᓐᓃᖕᒐᑦ ᐊᒻᒪ
ᓯᓚᓗᓗᐊᖅᐸᒐᓂ. ᑕᐃᒫᖃᐃ ᖃᖅᑲᐃᑦ
ᐊᐳᑎᖏᑦ ᐊᐅᓕᖅᑯᑦ. ᑕᐃᔭᕗᓪᓕ
ᐊᐅᔪᐃᑦᑐᕐᓂ.”

—ᐅᓚᐃᑲ ᐄᓯᒪᐃᓕ

“ᑭᓯᐊᓂᓕ ᒫᓐᓇ ᐊᐅᔭᖅ ᓰᕐᓗ
ᖃᓪᓗᓈᑦ ᓄᓇᖓᑦᑕ ᓯᕿᓂᖓᓄᑦ
ᐆᓇᕐᓂᖓᓄᑲᓴᐅᔫᖅᑐᖅ ᑭᓯᐊᓂ
ᐊᐅᔭᖅ ᒫᓐᓇ ᐅᐱᖕᒑ ᐊᐅᔭᖅ
ᐆᓇᖃᑦᑕᕆᐅᕐᓂᖓ ᑕᒡᒐ ᐋᓪᓚᓪᓚᕆᒃ
ᒫᓐᓇ. ᑕᐃᒪᐃᓪᓚᕆᒃᑐᖅ ᒫᓐᓇ 2009.
ᑕᐃᒫ ᓯᕿᓂᖅ ᓴᖅᑮᔭᑐᐃᓐᓇᕈᓂ
ᓇᒧᖕᒐᕕᑦᓴᑕᖃᑦᑎᐊᖅᑑᔮᖕᒋᒻᒪᕆᑦᑐᖅ
ᐸᓐᓂᖅᑑᑉ ᐊᓯᐊ ᐆᓇᖅᑐᑯᓗᐊᐱᒃ.
[...] ᐃᒫ ᐊᓚᒃᑲᐅᒥᑎᓪᓗᒍ ᓯᕿᓂᖅ
ᐊᓄᖄᖅᑎᓐᓇᒍ ᐃᓯᕋᓗᐊᖅᑐᓂᓗ
ᑐᐱᕐᒧᑦ ᐊᓂᒐᓗᐊᖅᑐᓂᓗ. ᐃᓛ
ᖃᓪᓗᓈᓃᑦᑑᔮᓇᕈᔪᑎᐊᑭᑦᑐᐊᓘᔮᕐᒪᓇᐃ,
2009-ᒥ ᓰᓕᑦᑎᐊᖅᑯᖕᒐ ᑕᒡᒐ.”

—ᐸᐅᓗᓯ ᕖᕕ

“ᐅᐱᖕᒑᓕᒫᖅ
ᓯᓚᓗᖃᑦᑕᕋᔭᖕᖑᐊᕋᓗᐊᖅᑐᓂ
ᒫᓐᓇ ᐃᒪᐅᓂᓕᒫᖕᒐᓂ ᐸᓐᓂᖅᑑᕐᓕ
ᓯᓚᓗᖃᖖᒋᑦᑎᐊᖅᑐᖅ, ᓯᓚᓗᖃᖏᑦᑐᖅ.
ᐋ ᐊᔾᔨᒋᖕᒋᑦᑎᐊᖅᑕᖕᒐ ᒫᓐᓇ
ᑕᒡᒐᓂ 2009-ᒥ ᐃᒻᒪᖄ ᐊᓯᕈᕐᓂᖅᐸ
ᓇᓪᓕᐅᑎᓯᒪᒻᒪ. ᓯᓚᓗᖃᖖᒋᑐ
ᓯᓚᓗᒋᐊᖅᑑᔮᑦᑕᖁ ᓯᓚᓗᖃᖖᒋᑦᑐᖅ.
ᐃᓐᓇ ᐸᓐᓂᖅᑑᕐᒥᐅᑕᐅᓪᓗᑕ
ᐃᓐᓇ ᑰᓕᒃ ᖃᐅᕚᒪᔭᓯ ᐃᓐᓇ
ᓄᕗᓕᐊᓘᑉ ᓇᓛᓃᑦᑐᖅ

“There used to be a lot of snow on top of the mountains, but [there is] nothing now. There were more [permanent] snow beds before. We are hardly getting any snow or rain. Maybe that is why the snow on top of the mountains is melting. We call them *aujiuttuq*, the ice that never melts.”

—Ooraika Iseemailee

“The sun seems to be like the down-south sun. It’s so hot. Especially here, outside of Pang, you can’t really go [out] because it’s so hot. [...] There is nothing for shade, even when you go inside the tent. That’s how I have felt this summer.”

—Pauloosie Veevee

“It always seems like it is going to rain, but [there are] only a few drops and then it stops.... When it becomes misty over there [at Kullik] we know there’s going to be rain in Pang. It was like that a few days ago, but after a few drops it stopped. It seems like it is lying.”

—Pauloosie Veevee

ᑕᑯᑦᓴᐅᕙᒃᑐᖅ ᐸᖕᓂᑑᕐᒥᐅᓄᑦ
ᖃᐅᔨᓵᖕᒍᕙᔪᒻᒪᑦ ᓯᓚᓗᓐᓂᐅᓴᑎᑦᑐᒍ
ᓂᑦᑕᔪᓐᓇᐃᑦᑕᖅᑲᑦ ᑕᐃᓐᓇ
ᓂᑦᑕᔪᓐᓃᖅᑲᑦ ᐃᑉᐸᔅᓴᖅ
ᑕᐃᒪᐃᓚᐅᕐᖓ ᐃᑉᐸᔅᓵᓂᓗ
ᑕᐃᒪᐃᓚᐅᕐᖓ ᑕᐃᓐᓇ
ᖃᐅᔨᓴᐅᑎᕗ ᑕᐃᓐᓇ ᑕᐃᑲ
ᓯᓚᓪᓗᓐᓂᐊᖅᑐᐊᓘᓕᖅᑯᒍ
ᑕᐃᓐᓇ ᓂᑦᑕᐃᒍᓐᓃᑦᑕᕐᖓᑦ,
ᓴᓪᓗᖅᑲᑦᑕᒻᒪᕆᓕᖅᑐ ᑕᐃᓐᓇ
ᖃᐅᔨᓴᐅᑎᓪᓚᕆᐅᒻᒪ ᑕᐃᓐᓇ
ᓯᓚᓗᒻᒧᑦ."

—ᐸᐅᓗᓯ ᕖᕕ

"ᒪᑯᐊ ᑕᓲᑦ ᑕᐅᑦᓱᒧᖓ ᑕᕆᐅᖅᒧᑦ
ᐃᓚᒋᔭᐅᒐᔭᙳᐊᙱᑦᑐᑦ
ᓯᑯᓗᐊᕿᓐᓇᐃᑦᓕᓂᖕᒐᓄᑦ
ᐃᓚᒋᔭᐅᒻᒪᑕ ᓱᒻᒪᑭᐊᖅ.
ᓄᓇ ᐃᓗᖕᒑᖕᒐᓄᑦ ᐅᓇᙵᑦ
ᓂᓪᓕᓇᓗᐊᕋᓐᓃᕆᐊᖕᒐ
ᐃᓱᒪᒋᕙᓪᒥᔭᖅ."

—ᐲᑦᑐᓘᓯ ᖄᐱᒃ

"[ᖁᐊᖅ ᓄᓇᐅᑉ ᐃᑭᐊᖕᒐᓃᑦᑐᖅ]
ᐊᐅᒃᑐᐊᓘᓕᖅᑯᖅ. ᑕᑯᒃᓴᑦ;
ᑲᑕᒃᐸᓪᓕᐊᓯᔪᑦ. [...] ᓲᖃᐃᒻᒪ
ᐃᑭᐊᖕᒐᓃᑦᑐᑦ ᐊᐅᓕᖅᑐᑦ. ᑕᑯᒃᓴᐃᑦ."

—ᐲᑦᑐᓘᓯ ᖄᐱᒃ

"ᐊᐱᖃᑦᑕᖏᓐᓂᖅᓴᖅᐹᓘᓕᖅᑐᒍᑦᓕ
ᐃᓖ ᐊᐳᑕᐅᙱᓂᖅᓴᐅᖃᑦᑕᓕᕐᑦᑖ
ᖃᓐᓂᓗᐊᖃᑦᑕᖏᒻᒪᑦᑖ ᖃᑦᓗᓈᑦᑕ
ᓄᓇᖓᑦ ᐊᐳᑎᐅᓂᖅᓵᓗᕙᓚᖅᑐᑦ
ᒫᙶᓂ."

—ᑕᐅᑭ ᖄᐱᒃ

"In the lakes [the ice] should become very thick. It doesn't get thick as it should be. I think it's getting warm from underneath the ground."

—Peterloosie Qappik

"[The permafrost] is thawing. It's visible; it seems to be sinking a little [...] because it's thawing underneath. You can pretty well see it."

—Peterloosie Qappik

"There is hardly any snow left. [...] The only people getting snow are the people down south."

—Taukie Qappik

Nunavut / Pangnirtung

ᐃᓅᐃᑦ ᐱᖁᑯᓯᑐᖃᖓᑦᑕ ᐊᒃᑐᖅᑕᐅᓂᖓ

“ᐃᒫᓪᓕ ᑕᒪᒃᑯᐊ ᓇᑦᑏᑦ ᕿᓯᖏ ᐱᐅᓯᓇᕋᖅᑐᖅᐹᓘᓕᖅᑐᑦ ᒦᓐᓇᓕ ᓯᑯᐃᓕᖅᓴᕋᐃᓗᐊᒍᑦ ᒥᖅᑯᖏᑦ ᐱᐅᓯᒌᖅᑎᓐᓇᒋᑦ ᓯᑯᐃᓕᖅᐸᓕᕐᒪᑦ ᐆᑦᑑᒐᓚᒍᓐᓇᐃᖀᖓᒥᒃ ᒥᖅᑯᖏᑦ ᐱᐅᓯᓇᕋᕐᓂᖅᓴᐅᓕᖅᑐᑦ. ᒦᓐᓇᓕ ᒥᖅᑯᖏᓐ ᐱᐅᓯᒌᖅᑎᓐᓇᒋᑦ ᓯᑯᐃᓕᖅᐸᓕᕐᖓᑦ ᐆᑦᑑᒐᓚᒍᓐᓇᐃᖀᖓᒥᒃ ᒥᖅᑯᖏᑦ ᐱᐅᓯᓇᕋᕐᓂᖅᓴᐅᕙᓕᖅᑐᑦ. ᑭᓯᐊᓂ ᒥᖅᑯᖏᑦ ᐃᓂᑦᑎᐊᕋᐃᒻᒪᑕ ᒪᒥᖓ ᒪᔾᔭᒐᕐᓂᖅᓯᓲᖅ, ᑕᒪᔾᔭ ᖃᐅᔨᒪᓪᓚᓇᒃᑕᒃᑲ ᐊᑯᓂᐊᓗᒃ ᕿᓯᓕᓇᖅᑲᑦᑕᖅᓯᒪᒐᒪ. ᑭᔅᓯᐊᓂ ᒥᖅᑯᖓ ᐱᐅᓯᑦᑎᐊᕐᖓᑦ ᒪᔾᔭᒐᕐᓂᖅᑐᑯᓗᐃᑦ, ᒥᖅᑯᖓ ᐱᐅᓯᒌᓚᐅᖅᑎᓐᓇᒍ ᒪᔾᔭᒐᓱᑦᑐᒍ ᐊᕿᑦᑐᐊᓗᐃᑦ. [...] ᑕᒪᐃᑦᑐᑦᑕᐅᖅ.”

—ᐄᕕ ᐊᓂᕐᓂᓕᐊᖅ

“ᖃᐅᔨᒪᔭᒃᑯᓪᓕ ᐃᓚᖓᒍ ᐱᔭᕆᑦᑐᓂᖅᓴᐅᓕᖅᓯᒪᕚ. ᐱᔭᕆᐊᑐᓂᖅᓴᐅᓲᖑᔪᑦ ᓇᑦᑎᑯᓗᐃᑦ ᐃᒻᒪᖄ ᑕᐃᔅᓱᒪᓂᒃᑲᓐᓂᖅ ᕿᓯᖏᑦ ᓱᒃᑲᐃᓕᓂᐊᕐᒪᑕ. ᕿᓯᖏᑦ ᐊᓱᔾᔨᕐᓇᖅᑲᑦᑕᓕᕐᒪᑕ ᖁᓚᕐᓇᖏᓪᒪᑦ ᕿᓯᑐᖅᑲᖅᑲᑦᑕᕐᓂᐊᓕᖅᑐᖅ ᓯᑯᐃᖅᓴᕋᐃᓗᐊᕐᓂᖓᓄᑦ. ᐆᑦᑑᑯᑖᖅᑲᑦᑕᕐᓗᑎᒃ ᑭᓯᐊᓂ ᕿᓯᖏᑦ ᐊᓱᔾᔨᖅᓴᕋᐃᓐᓂᖅᓴᐅᓲᖑᓪᒪᑕ.”

—ᔭᐃᑯ ᐃᓱᓪᓗᑕᖅ

Impacts on Traditional Ways of Life

“Back when we used to have longer winters, [ringed seals’] coats would be nice and shiny, but these days the coats are really not good for anything because the fur just tends to fall off the seal. [...] It’s not really great to try to scrape out the blubber, because it’s really soft and you have to be delicate. It’s really hard to scrape it off all at once. [...] [Other seals] are just the same.”

—Evee Anilniliak

“To some extent it has become more difficult to hunt. [...] Specifically, ring seals have changed because of the lack of spring. [...] When the sun is out and [the ice is melting] too quickly, the sealskin is not going to form. The seal has to bask in the sun for a period of time before the skin starts to grow properly. Because of the lack of basking in the sun, their fur is moulting too quickly, and as a result [we] are not getting good sealskin.”

—Jaco Ishulutaq

“ᓂᒃᑯᕐᑲᕐᕕᒃᓴᐅᒍᓐᓃᖅᑐᖅ ᒪᑯᐊᓕ ᕐᑭᓚᒥᐊᓐᓗ ᐊᒧᓱᕈᓲᑦ ᑕᐃᒃᑯᐊ ᕐᑯᐃᓇᖕᓇᖅᑐᐹᓗᐃᑦ.”

—ᐄᕕ ᐊᓂᕐᓂᓕᐊᖅ

“ᐅᑭᐅᖅᑕᖅᑐᒥᐅᑕᐅᒐᑦᑕᑖ ᑖᒃᑯᐊ ᓱᕐᕋᑦᑐᐃᓐᓇᓕᒫᑦ.”

—ᔮᐃᑯ ᐃᓱᑦᓗᑕᖅ

“ᐃᒪᓐᓇᐃᑦᑑᓕᖕᒐᑦᓕ ᒫᓐᓇ ᐃᕐᑲᓪᓕᐊᕈᒻᒪᐃᓐᓇᐅᓪᓗᓂ ᐃᕐᑲᓪᓕᐊᕐᕕᔅᓴᕈᓪᓚᑎᖕᖏᒃᑲᓗᐊᕐᒪᓪᓕ ᑕᕝᕙ ᑕᓯᖅ ᐊᕝᕙᓗᐊ ᐃᒪᐅᓕᖅᑎᓪᓗᒍ ᑕᐃᒪ ᐅᐸᒐᒃᓴᐅᔫᔪᓐᓂᕐᒪᑦ ᐅᐸᕐᑲᑦᑕᓗᐊᕈᓐᓃᖅᑐᕐᑲᐃ ᑕᒪᓐᓇ ᐊᖅᑯᑎᖕᒐ ᐱᐅᒍᓐᓃᖅᑐᑯᓗᐊᐱᐅᓕᖅᑐ. ᑕᐃᒫᑦᓕ ᐊᓕᐊᓇᐃᑉᐸᔪᒐᓗᐊᖕᒐᑦ ᐃᕐᑲᓗᒐᓱᑦᑐᓂ ᐳᖅᑭᓪᓚᑎᑕᐃᓐᓇᑎᓪᓗᒍ ᑕᓯᖅ ᐃᓛ ᓯᑯᖓ, ᒫᓐᓇ ᐅᐸᔅᓯᓇᕈᓐᓂᖅᑐᖅ.”

—ᓕᐊ ᐊᒃᐸᓕᐊᓗᒃ

“You can’t hang meat to dry, because now it will be completely covered [in black flies]. I’m petrified to dry meat because of all the flies.”

—Evee Anilniliak

“As Inuit, our way of thinking has a lot to do with our environment. For people in the North, you know how the seasons work and your body has to be in sync with the land all the time. It’s true that your mind and body are not so much in sync with the land anymore.”

—Jaco Ishulutaq

“We are not even able to travel out to our traditional hunting camp in the spring. We usually start our spring camping, but we don’t do that anymore because the sea ice is going out much quicker. [...] As an avid fisherwoman, I like to ice fish. It is more difficult for people to go out to fish. [...] It’s the best fishing time of the year, when the ice has floated on the lake. But we can’t get to the lake by skidoo anymore because the snow has all melted.”

—Leah Akpalialuk

ᓄᓇᕕᒃ Nunavik

ᐅᒥᐅᔭᖅ
Umiujaq

ᐅᒥᐅᔮᖅ

ᐅᒥᐅᔮᖅ ᑕᓯᐅᔭᕐᔪᐊᕐᒦᑦᑐᖅ, ᓴᓂᑭᓗᐊᑉ ᕿᑭᖅᑕᖕᒋᑕ ᑲᓇᖕᓇᖕᒐᓂ (N 56.56°, W 76.55°; ᖁᑦᑎᖕᓂᖕᒐ 76 m). ᐅᑭᐅᖅᑕᖅᑐᕐᒥ ᐊᖅᐸᓯᒃᑐᓄᑦ ᐃᓚᒋᔭᖅ, ᓇᐹᖅᑐᖅᑲᖅᑐᑉ ᑭᒡᓕᐊᓂ ᐅᖅᐱᒐᓂᒃ ᐱᕈᖅᑐᖅᑲᖁᓂ ᓄᓇᖕᒐᑕᓗ ᐃᑭᐊᖕᒐ ᖁᐊᖑᐃᓐᓇᖅᐸᒐᓂ. ᐅᒥᐅᔭᒥᐅᑦ ᑭᒡᒍᑎᖕᒋᑦ ᐊᒃᑐᖅᑕᐅᓯᒪᓂᖕᒋᓐᓄᑦ ᐱᔾᔪᑎᖅᑲᖅᑐᒃᓴᐅᔪᑦ. ᐅᒥᐅᔮᖅ ᓄᓇᓕᖕᓂ ᓄᑖᖑᓂᖅᐸᐅᖕᒪᑦ, ᐱᙳᖅᑐᕕᓂᖅ 1986ᒥ. ᑖᒃᑯᐊᓕ ᑲᖏᖅᓱᔪᐊᖅ ᐊᒻᒪ ᑲᖏᖅᓱᐊᓗᔾᔪᐊᖅ ᓄᓇᓕᙳᕐᓂᕐᒪᑎᒃ 1960ᖕᒋᑦ ᐊᑐᖅᑎᓪᓗᒋᑦ. 1975ᒥ ᔦᐃᒻᔅ ᐸᐃ ᐊᒻᒪ ᑯᐸᐃᒃ ᐅᐊᖕᓇᖕᒐᑕ ᐊᖕᒋᕈᑎᖕᒋᑦ ᐊᐃᕙᒍᑕᐅᓕᖅᑎᓪᓗᒋᑦ, ᐃᓄᐃᑦ ᑰᔾᔪᐊᕋᐱᖕᒥᐅᑦ ᓄᒃᑎᖅᑕᐅᓚᐅᖅᓯᒪᖕᒪᑕ ᐅᒥᐅᔪᒧᑦ, ᑕᓯᐅᔭᕐᔪᐊᕐᒦᑦᑐᖅ ᐃᓅᓯᑐᖅᑲᖕᒋᑦ ᐸᐸᑦᑕᐅᑎᐊᕐᓂᐊᕐᒪᑕ ᓯᓚᒧᑦ ᐃᖅᑲᓗᖕᒋᑦ ᐆᒪᔪᖕᒋᓪᓗ ᐊᑦᑐᖅᑕᐅᖕᒋᓪᓗᑎᒃ ᑕᐃᔅᓯᒪᓂ. ᐃᓄᐃᑦ ᓄᓈᓕᑦ ᐊᓯᐊᓂᑦ ᒪᓕᒃᓴᓕᓚᐅᖅᓯᒪᓐᓂᕐᒥᔪᑦ. 2011ᒥ, ᐃᓄᖕᒋᑦ 444ᖑᓕᖅᑐᕕᓃᑦ, ᑲᓇᑕᒥ ᓈᓴᐃᔩᑦ ᓈᓴᐃᓯᒪᓂᖕᒐᑦ ᒪᓕᒃᑐᒍ.

ᐅᑭᐅᒐᓴᖕᓂ ᐊᓂᒍᖅᑐᓂ, ᐅᒥᐅᔭᕐᒥᐅᑦ ᐅᔾᔨᕈᓱᒃᓯᒪᔪᑦ ᐊᔾᔩᙱᑦᑐᓂᒃ ᐊᕙᑎᒧᑦ ᐊᓯᔾᔨᖅᓯᒪᓕᖅᑐᓂᒃ. ᐊᐱᕆᔭᐅᒐᒥᒃ ᐱᕈᖅᑐᓄᑦ ᐊᓯᔾᔨᖅᓯᒪᓕᖅᑐᓂᒃ, ᐊᐱᖅᓱᖅᑕᐅᔪᑦ ᐅᕐᑲᕐᑲᑦᑕᓚᐅᖅᑐᑦ ᐱᕈᖅᑐᑦ ᐅᓄᕐᓂᖅᓴᐅᓕᕐᓂᖕᒋᓐᓂᒃ ᐊᖕᒋᓂᖅᓴᐅᓕᕐᓂᖕᒋᓐᓂᓪᓗ. ᐆᒃᑑᑎᒋᓗᒍ, ᐅᖅᐱᒐᐃᑦ, ᐅᖅᐱᐊᓛᓪᓗ

Umiujaq

Umiujaq is located on Hudson Bay's east shore, east of the Belcher Islands (N 56.56°, W 76.55°; elevation 76 m). It is part of the Low Arctic, at the edge of the treeline in the low shrub subzone with discontinuous permafrost. We believe that responses from Umiujaq residents may be influenced by important historical factors. Umiujaq is the youngest community in our sample, established in 1986. By comparison, Kangiqsujuaq and Kangiqsualujjuaq were established in the 1960s. In 1975, as part of negotiations of the James Bay and Northern Quebec Agreement, a number of Inuit from Kuujjuarapik were relocated to a new settlement called Umiujaq, near Richmond Gulf, to preserve their traditional lifestyle in an area where fish and game would not be threatened by hydroelectric developments at the time. People from other communities joined them. In 2011, the population reached 444 people, according to the Statistics Canada Census.

In the last decades, community members in Umiujaq have observed

ᐅᓄᕐᓂᖅᓴᐅᓕᕐᓂᖏᓐᓂᒃ, ᐅᖅᐱᑦᓗ ᑕᑭᓂᖅᓴᐅᕙᓕᕐᓂᖏᓐᓂᒃ, ᓄᓇ ᑕᐃᔅᓱᒪᓂᓂᑦ ᑐᖑᔪᕐᓂᖅᓴᐅᓕᖅᖢᓂ. ᐊᐱᖅᓱᖅᑕᐅᔪᑦ ᐅᓂᒃᑳᖃᑦᑕᓚᐅᖅᑐᑦ ᐱᕈᖅᑐᖃᕐᓂᖅᓴᐅᓕᕐᓂᖓᓄᑦ ᐱᓱᒋᐊᒃᓴᖅ ᑎᓴᒪᓕᒃᑯᑦᓗ ᐃᖏᕐᕋᔭᕆᐊᒃᓴᖅ ᐊᒃᓱᕈᕐᓇᕐᓂᖅᓴᐅᓕᕐᓂᖓᓂᒃ. ᐃᓚᖏᑦᓗ ᐅᖃᖅᓯᒪᑦᓗᑎᒃ ᐅᖅᐱᒐᕐᓄᑦ ᓄᓂᕙᒐᖃᖏᓐᓂᖅᓴᐅᓕᕐᓂᖓᓂᒃ ᐱᓗᐊᖅᑐᒥᒃ ᐸᐅᕐᖓᕐᓂᒃ, ᐸᐅᕐᖓᖃᙱᓐᓂᖅᓴᐅᓕᖅᖢᓂ. ᐃᓚᖏᑦ ᐱᕈᖅᑐᑦ, ᓲᕐᓗ ᑭᒍᑕᖏᕐᓇᑦ ᑭᖕᒥᖕᓇᐅᔭᑦᓗ ᓄᑕᐅᓂᕋᖅᑕᐅᑦᓗᑎᒃ ᐱᕈᖅᓯᒪᓂᖅᓴᐅᓕᕐᓂᕋᖅᑕᐅᑦᓗᑎᒡᓗ.

ᐊᖏᖃᑎᒌᒍᓐᓇᓚᐅᙱᑦᑐᑦ ᓂᕐᔪᑎᑦ ᐅᓄᕐᓂᖏᑕ ᐊᓯᔾᔨᖅᓯᒪᓕᕐᓂᖏᓐᓂᒃ. ᑭᓯᐊᓂ, ᐊᒃᕋᖃᕐᓂᖅᓴᐅᓕᕐᓂᕋᖅᑕᐅᑦᓗᓂ ᐅᒥᐅᔭᐅᑉ ᖃᓂᒋᔭᖓᓂ. ᓄᑖᑦ ᐆᒪᔪᑦ ᑕᑯᔭᐅᕙᓕᓚᐅᖅᑐᑦ, ᓲᕐᓗ ᑐᒃᑐᕙᑦ, ᐱᓗᐊᖅᑐᒥᒡᓗ ᓄᑖᑦ ᑎᖕᒥᐊᑦ, ᓲᕐᓗ ᖂᐱᓐ ᐊᒻᒪ ᖁᐸᓄᐊᑦ. ᐃᒥᖅᑯᑕᐃᓚᖅᑕᖃᐃᓐᓇᐅᔭᖅᓯᒪᔪᖅ, ᑭᓯᐊᓂ ᒫᓐᓇ ᑕᖅᑭᓕᒫᕐᒥᒃ ᑎᑭᓵᓕᕙᓕᖅᖢᑎᒃ, ᔫᓂᒥ ᔪᓚᐃᒥᐅᔪᓐᓃᖅᑐᖅ. ᐅᔾᔨᕆᔭᐅᓯᒪᔪᑦ ᐊᓯᔾᔨᖅᓯᒪᓃᑦ ᐊᐅᒃᑐᖅᐸᒃᑐᓂᒃ ᖁᐱᕐᕈᓂᒃ ᐊᔾᔨᒌᙱᒻᒪᕆᓚᐅᖅᑐᑦ, ᓄᑖᓂᒡᓗ ᑕᕐᕋᓕᑭᔅᓂᒃ ᑕᑯᕙᓕᕐᓂᕋᖅᖢᑎᒃ.

ᐊᑕᐅᓯᖅ ᐊᐱᖅᓱᖅᑕᐅᔪᖅ ᐅᖃᓚᐅᖅᑐᖅ ᐊᓯᔾᔨᖅᓯᒪᓕᕐᓂᖏᓐᓂᒃ ᓯᓚᐅᑉ ᐊᓯᔾᔨᖅᑕᕐᓂᖓᓄᑦ; "ᐅᐱᕐᖓᒃᓵᙳᖅᓵᓕᕙᓕᖅᑐᖅ, ᐊᑯᓂᐅᓂᖅᓴᖅ ᐊᐅᔭᐅᕙᓕᖅᖢᓂ, ᐊᑯᓂᐅᓂᖅᓴᖅ ᐅᑭᐊᒃᓵᖑᑦᓗᓂ, ᐅᑭᐅᕐᓗ ᓇᐃᓐᓂᖅᓴᐅᓕᖅᖢᓂ."

numerous changes to their environment. When asked about changes in vegetation, participants strongly asserted that plants are becoming more abundant and larger. For example, shrub species, such as dwarf birch, are increasing in abundance, and willows are growing taller, making the landscape greener than in the past. Interviewees described that changes in shrub growth can make it more difficult to walk or ride ATVs on the tundra. Some also noted that shrub growth is affecting berry plants, especially blackberries/crowberries (*Empetrum nigrum*), which are decreasing in abundance. Some plant species, such as blueberry/bilberry (*Vaccinium uliginosum*) and squashberry/pimbina (*Viburnum edule*), were reported as being new or more abundant.

There was no consensus in the community over obvious changes in animal abundance. However, black bears are seen more frequently around Umiujaq. New animal species have been observed, like moose, but especially bird species, such as robins and sparrows. Arctic terns have always been present around the community, but they seem to arrive almost one month earlier, in June instead

ᐊᐱᖅᓱᖅᑕᐅᔪᑦ ᐊᖏᖅᑲᑎᒌᒍᑎᖅᑲᓚᐅᖅᑐᑦ ᐊᓄᓇᑐᓂᖅᓴᐅᕙᓕᖅᓂᖕᖓᓂᒃ, ᐊᓄᖄᓗᔪᖕᓂᖅᓴᐅᓂᖕᖓᓂᒃ, ᓴᖅᑭᑳᑦᑕᖕᓂᖅᓴᐅᕙᓕᖅᓂᖕᖏᓐᓂᓪᓗ. ᐃᒪᖅᑲᖅᓂᖅᓗ ᐅᔾᔨᕆᔭᐅᒻᒪᕆᒃᓯᒪᓪᓗᓂ, ᓲᖅᓗ ᑕᓲᑦ ᑰᓪᓗ ᐃᒃᑲᓐᓂᖅᓴᐅᓕᖅᓂᖕᖏᓐᓂᒃ ᐊᖕᑖᑦ ᐃᒪᐃᓕᖅᓂᖕᖏᓐᓂᒃ. ᒪᖅᑯᖅᑲᑦᑕᖅᓂᖅᓴᐅᓂᖃᖅᑕᐅᓕᖅᓱᓂ ᐅᑭᐅᒃᑯᓪᓗ ᐊᐳᑎᑭᓐᓂᖅᓴᐅᓕᖅᓂᖃᖅᑕᐅᓪᓗᓂ. ᐊᐱᓇᓵᖅᓂᖅᓴᐅᕙᓕᖅᓱᓂ ᐊᓄᓇᓗᓪᓗ ᑎᒃᑕᐅᓂᖅᓴᐅᓕᖅᓱᑎᒃ, ᐊᐳᑎᑭᓐᓂᖅᓴᐅᓕᕈᑕᐅᓪᓗᓂ. ᐊᐱᖅᓱᖅᑕᐅᔪᑦ ᐅᔾᔨᕈᓱᒃᓯᒪᔪᑦ ᓯᓚ ᐆᓇᖅᓂᖅᓴᐅᓕᖅᓂᖕᖓᓄᑦ, ᐊᓄᓇᓂᖅᓴᐅᕙᓕᖅᓂᖕᖓᓄᑦ ᐃᒪᑭᓐᓂᖅᓴᐅᓕᖅᓂᖕᖓᓄᓪᓗ ᐱᕈᖅᑐᑦ, ᓲᖅᓗ ᑎᖕᖓᐅᔭᑦ ᐸᓂᖅᓂᖅᓴᐅᕙᓕᖅᓂᖕᖏᓐᓂᒃ ᑕᐃᓯᒪᓂᓂᑦ.

ᓯᑰᑉ ᖃᓄᐃᓐᓂᓇᓕᖅᑕᖕᖓ, ᐅᒥᐅᔭᖅᒥᐅᑦ ᐅᖅᑲᐅᓯᖅᑲᖅᑲᑦᑕᓚᐅᖅᑐᑦ ᓯᑯ ᓇᖏᐊᖅᓇᖅᓯᓕᖅᓂᖕᖓᓂᒃ ᓯᑯᓈᖅᐸᓕᖅᓂᖕᖓᓄᑦ ᓯᑯᐃᖅᓵᓕᕙᓕᖅᓂᖕᖓᓄᓪᓗ. ᐅᖅᑲᓚᐅᓇᓪᓗᑎᓪᓗ ᐊᐳᑎᑭᓐᓂᖅᓴᐅᓕᖅᓂᖕᖓᓄᑦ ᐊᐅᒃᓵᓕᕙᓕᖅᓂᖕᖓᓄᓪᓗ ᐅᑭᐅᒃᑯᑦ ᓄᒃᑕᖅᓂᖅ ᕿᓚᒻᒥᐅᓂᖅᓴᐅᕙᓕᖅᑐᖅ.

ᐊᐱᖅᓱᖅᑕᐅᔪᑦ ᐅᖅᑲᐅᓯᖅᑲᖅᑲᑦᑕᓚᐅᖅᑐᒥᔪᑦ ᖁᐊᖑᐃᓐᓇᖅᐸᒃᑐᑦ ᐊᐅᒃᐸᓕᖅᓂᖕᖏᓐᓄᑦ ᓯᓪᓴᖅᓗᓕᖅᓂᖕᖓᓂᒃ ᐃᕗᓂᖅᓴᐅᕙᓕᖅᓂᖕᖓᓂᓪᓗ. ᖁᐊᖑᐃᓐᓇᖅᐸᒃᑐᖅ ᐊᐅᒃᐸᓕᖅᓂᖕᖓᓄᑦ ᓯᓪᓴᒥ ᓯᐅᖅᑲᑦ ᐊᑖᓂ ᓂᓚᒃᑕᖅᑕᐅᕙᒃᑐᑦ

of July. Observed changes in the abundance of biting insects varied greatly, and new butterfly species were seen.

One interviewee summarized the changes in the timing of seasons: "We now have very early springs, long summers, long falls, and short winters." Participants largely agreed that the wind was stronger, more frequent, and prone to changing direction rapidly. Hydrology changes were also very much noted, such as lakes and rivers becoming shallower and even drying out. As for precipitation, rain tends to be more abundant and snow less abundant. Snow also arrives later and is rapidly blown away by the wind, which has a negative impact on the snow cover. Interviewees observed that a warmer growing season, more wind, and lower water levels are causing the plants, including lichens, to dry more than before.

In terms of sea ice conditions, Umiujamiut talked more about their impact on travel safety than the timing of freeze-up and breakup. They mentioned that a reduced snow cover melts faster and shortens the winter travelling season.

Participants also talked about how the thawing of

ᐱᑕᕐᑲᕐᒃᐸᒍᓐᓃᕐᓂᖕᒋᓐᓂᒃ,
ᐃᒪᑦᑎᐊᕙᖕᒥᒃ
ᐃᒻᒥᐅᒐᒃᓴᕐᑲᕈᓐᓃᕐᒃᓲᑎᒃ.

permafrost has increased beach erosion and hummock formation around the community. Another consequence of permafrost thawing is that people can no longer gather the ice once found under beach sand, thus depriving them of a source of good, clear drinking water.

ᐃᓐᓇᐃᑦ ᐊᐱᖅᑯᓯᖅᑕᐅᔪᑦ
Elders Interviewed

ᓕᓯ ᑯᕉ
Lizzie
Crow

ᓯᐊᓯ
ᓇᓗᒃᑐᖅ
Siasi
Naluktuk

ᓯᐅᕋ ᑯᕉ
Sarah
Crow

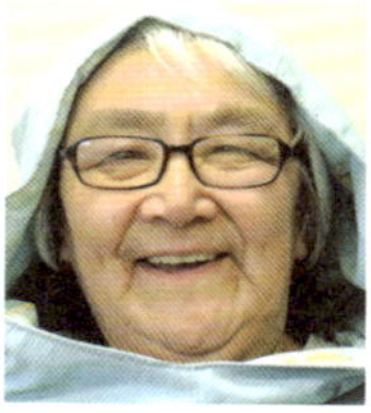

ᕙᐃᐅᓚ
ᓇᐹᖅᑐᖅ
Viola
Napartuk

ᑳᑎᓖᓐ
ᐃᓄᒃᐸᒃ
Kathleen
Inukpuk

ᓴᑭᐊᓯ
ᓂᕕᐊᖅᓯ
Zackiasie
Niviaxie

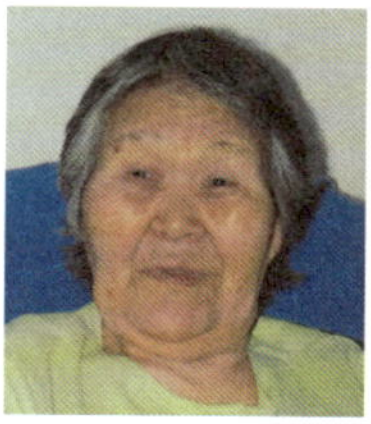

ᒫᑕ
ᖃᓴᓪᓗᐊᖅ
Martha
Kasudluak

ᒨᓯᓯ
ᓄᕙᓕᖖᒐᖅ
Moses
Novalinga

ᐅᐃᓕ
ᖁᒫᓗᒃ
Willie
Kumarluk

ᔪᓯᐊ ᓵᓚ
Joshua
Sala

ᑕᐃᓇ ᑐᑲᓗᒃ
Dinah
Tookalook

ᓵᓕ ᑐᒃᑐ
Charlie
Tooktoo

ᓕᓯ ᑐᑲᓗᒃ
Lizzie
Tookalook

ᓘᑲᓯ ᑐᒃᑐ
Lucassie
Tooktoo

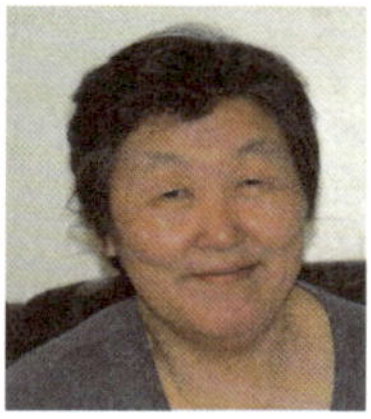

ᐋᓚᓯ ᑐᒃᑐ
Alice N.
Tooktoo

ᑭᓕᐅᕋ ᑐᒥᒃ
Clara
Tumic

ᐋᓚᓯ ᑐᒃᑐ
ᐊᖏᔪᒃᑎᖅ
Alice
Tooktoo Sr.

ᐆᓂᔅ ᑐᒥᒃ
Ernest
Tumic

ᐱᓕ ᑐᒃᑐ
Billy
Tooktoo

ᐸᐅᕐᖓᐃᑦ

“ᑕᑉᐹᓂ
ᑭᖕᒥᖕᓇᓕᒻᒪᓇᐊᓘᕙᓚᐅᖅᓯᒪᔪᖅ,
ᐊᕐᖔᒍᓄᑦ ᐱᖓᓯᓄᑦ
ᑕᑉᐸᐅᙵᐅᒥᐊᕐᑲᑦᑕᖅᓯᒪᔪᖕᒐ, [...]
ᑭᓱᕐᑲᐳᓐᓃᖅᑐᖅ. ᐊᑕᐅᓯᕐᒥᒡᓘᓐᓃᑦ.
ᐱᑕᓕᐊᓘᕙᓚᐅᖅᓯᒪᔫᒐᓗᐊᖅ. [...]
ᖃᐅᔨᒪᙱᑦᑐᖕᒐ ᖃᓄᐃᒻᒪᖔᑦ.”

—ᑭᓚᐅᕋ ᑐᒥᒃ

“ᐃᓚᖏᑦ ᑕᓰᑦ ᐃᒪᒃᓱᔪᑦ, ᐃᓚᖏᑦᑐ
ᐃᒪᖅᑲᙱᑦᑎᐊᒻᒪᓇᓕᖅᑐᑦ. [...]
ᑭᒍᑕᖏᕐᓇᖅᑲᙱᓐᓂᖅᓴᐅᓕᖅᑐᖅ.
ᓄᓇ ᑐᖑᔪᖅᐸᓚᐅᖅᓯᒪᔪᖅ, ᒫᓐᓇ
ᑭᒍᑕᖏᕐᓇᑦ ᐱᕈᙱᓐᓂᖅᓴᐅᕙᓕᖅᑐᑦ.”

—ᑕᐃᓇ ᑐᑲᓗᒃ

“ᑭᖕᒥᖕᓇᑦ ᒥᑭᓐᓂᖅᓴᐅᕙᓕᖅᑐᑦ
ᐃᓛᓐᓂ, ᐃᓛᓐᓂᒃᑯᑦᑐ ᐱᕈᖅᐸᒐᑎᒃ,
ᑕᐃᔅᓱᒪᓂ ᓇᓂᑐᐃᓐᓇᖅ
ᐱᕈᖅᐸᓚᐅᖅᓯᒪᔪᑦ ᐊᕐᖔᒍᑕᒫᑦ.”

—ᕙᐃᐅᓚ ᓇᐹᖅᑐᖅ

Berries

“There used to be a lot of cloudberries up north, but I tried to go there for the past three years, [...] there’s nothing. Not even one. There used to be a lot. [...] I don’t know what happened.”

—Clara Tumic

“Some lakes have gone down, and some lakes have no water now. [...] There are less berries. It used to be all blue on the ground, but now there are less berries growing.”

—Dinah Tookalook

“Cloudberries grow small in some seasons, and sometimes they don’t grow, but before they used to bloom every year.”

—Viola Napartuk

ᐱᕈᖅᑐᑦ ᐊᓯᖏᑦ

“ᐅᖅᐱᓕᒻᒪᕆᐊᓘᓕᖅᑐᖅ. ᐱᕈᖅᑐᒻᒪᕆᐊᓘᓕᖅᑐᑦ ᑭᒍᑕᖏᕐᓇᑦ ᐸᐅᕐᖓᓪᓗ ᐱᕈᕐᕕᒋᕙᓚᐅᖅᑕᖏᓐᓂ. ᒫᓐᓇ ᓇᓂᑐᐃᓐᓈᓗᒃ ᐱᕈᓕᖅᑐᑦ, ᐊᒡᓛᑦ ᐊᖅᑯᑎᑦ ᖃᓂᒋᔭᖏᓐᓂ ᑕᒫᓂ. ᑕᒪᐅᙱᓵᖅᑐᕕᓂᐅᓪᓗᑕ ᑭᒍᑕᖏᕐᓇᖅᐸᓚᐅᖅᑐᒋᑦ ᐊᖅᐱᖅᑕᖅᐸᒃᖢᑕᓗ, ᒫᓐᓇ ᐅᖅᐲᓐᓈᓘᓕᖅᑐᑦ.”

—ᐆᓂᔅ ᑐᒥᒃ

“ᐱᕈᖅᑐᑦ ᐱᕈᖅᑐᒻᒪᕆᐊᓘᓕᖅᑐᑦ ᐊᖏᓂᖅᓴᐅᓕᖅᖢᑎᒃ. ᐊᕐᕌᒍᑦ ᑕᓪᓕᒪᑦ ᖁᓕᓪᓗ ᐊᓂᒍᖅᑐᓂ ᐱᕈᖅᑐᑦ ᐊᓯᔾᔨᖅᓯᒪᔪᒻᒪᕆᐅᓕᖅᑐᑦ. [...] ᐊᕚᓚᕿᐊᓕᒻᒪᕆᐅᓕᖅᑐᖅ ᓯᓈᓂ. ᐃᓴᕈᑕᖏᑦ ᐱᕈᖅᓯᒪᓂᖅᓴᐅᓕᖅᑐᑦ, ᐊᕐᕌᒍᓂᒡᓗ ᐊᓂᒍᔩᖅᑑᖕᓂᒃ ᑕᑭᓂᖅᓴᐅᓕᖅᖢᑎᒃ. [...] ᑕᓯᐅᔭᖅ ᖃᓂᒋᔭᖓᓂ, 1950 ᓯᕗᓕᖏᓐᓂ ᓄᓇᖃᓚᐅᖅᑐᖅ ᐱᓱᒡᕕᒃᓴᑦᑎᐊᕙᐅᔪᓂᒃ, ᒫᓐᓇ ᐊᔪᕐᓇᕐᓂᖅᓴᐅᓕᖅᑐᖅ ᐅᑭᐅᑦ 30, 40-ᓗ ᐊᓂᒍᖅᑎᓪᓗᒋᑦ. ᐱᓱᒡᕕᒋᒋᐊᖏᑦ ᐊᔪᕐᓇᕐᓂᖅᓴᐅᓕᖅᑐᑦ ᐱᕈᖅᑐᐊᓗᖕᓄᑦ ᐃᓴᕈᑕᖏᓐᓄᓪᓗ.”

—ᐅᐃᓕ ᖁᒫᓗᒃ

Other Plants

“There are a lot of *urpiit* [willows]. They are taking over places where blueberries and blackberries used to grow. They are actually growing all over the place nowadays, even near the road here. When we first came here we were able to pick blueberries and *arpiit* [cloudberries/bakeapples] where all the bushes are now.”

—Ernest Tumic

“The plants are growing; all of them are bigger. Since five or ten years ago there is a big change in plants. [...] There is an abundance of *avaalaqiat* [birches] along the coast. The branches have grown more, and they have grown taller in the past couple of years now. [...] Around the area of Richmond Gulf, before the 1950s there were some areas you could walk easily on, but now it's harder compared to 30 or 40 years ago. It's difficult to walk through [those areas] because of plants and branches.”

—Willie Kumarluk

ᐆᒪᔪᑦ

"ᖁᐱᕐᕈᖃᕐᓂᖅᓴᐅᓕᖅᑐᖅ. ᐊᖏᔪᐊᓗᐃᑦ ᕿᕐᓂᖅᑐᑦ. ᑕᒫᓂ ᐱᑕᖃᓗᐊᖅᐸᓚᐅᖅᓯᒪᙱᑦᑐᖅ, ᐊᕐᕌᒍᑕᒫᑦ ᐅᓄᖅᓯᕙᓪᓕᐊᑐᐃᓐᓇᖅᑐᑦ. ᐅᑭᐅᖅᑕᖅᑐᕐᒥᐅᓄᑦ ᖁᐃᓇᒋᔭᐅᒻᒪᕆᒃᑐᑦ. ᓲᖃᐃᒻᒪ ᑕᑯᓚᐅᖅᓯᒪᙱᓐᓇᒥᒃ ᑕᐃᒪᐃᑦᑐᓂᒃ."

—ᐆᓂᔅ ᑐᒥᒃ

"ᑰᒃ ᑰᖏᓐᓂᖅᓴᐅᓕᖅᐸᑦ, ᐃᖃᓗᐃᑦ ᒪᔪᐊᕈᓐᓇᐃᓪᓕᓂᐊᖅᑐᑦ. ᖃᐅᔨᒪᔭᖃᖅᐳᖓ ᐃᖃᓗᖃᕐᓂᖅᓴᐅᓚᐅᖅᓯᒪᔪᒥᒃ ᑰᑉ ᖃᓂᒋᔭᖕᖓᓐᓂ. ᒫᓐᓇ ᐃᖃᓗᖃᖕᖏᓐᓂᖅᓴᐅᓕᖅᑐᖅ, ᑰᒃ ᑰᖏᓐᓂᖅᓴᐅᓕᕐᒪᑦ."

—ᒨᓯᓯ ᓄᕙᓕᙵᖅ

"ᑕᑯᖃᑦᑕᖅᓯᒪᓕᖅᑐᖓ ᕿᕐᓂᖅᑕᓂᒃ ᑎᖕᒥᐊᕐᓂᒃ. [...] ᐱᑕᖃᖅᐸᓚᐅᖅᓯᒪᙱᑦᑐᑦ ᑕᒫᓂ. ᕉᐱᓐ ᐊᔾᔨᒋᙱᑕᖕᒋᑦ. ᒪᕐᕉᖕᓂᒃ ᐊᔾᔩᙱᑦᑐᓂᒃ ᑎᖕᒥᐊᓂᒃ ᑕᑯᕙᓚᐅᖅᓯᒪᙱᑕᓐᓂᒃ ᑕᑯᕙᓕᖅᑐᖓ. ᕿᕐᓂᖅᑕᑦ ᑲᔪᑦᓗ."

—ᓕᓯ ᑐᑲᓗᒃ

"ᒥᓗᒋᐊᑦ ᖁᑦᑎᒃᑐᒧᐊᖅᐸᓪᓕᐊᓕᖅᑐᑦ ᐅᑭᐅᖅᑕᖅᑐᒥ. ᒥᓗᒋᐊᖃᓗᐊᓚᐅᖅᓯᒪᙱᑦᑐᓂ ᒥᓗᒋᐊᖃᕐᓂᖅᓴᐅᕙᓕᖅᑐᖅ."

—ᒨᓯᓯ ᓄᕙᓕᙵᖅ

Animals

"There are more bugs now. The big black ones. There hardly used to be any here, but there are more and more every year. People in the North are deathly afraid of them. They have never seen them before; that's why."

—Ernest Tumic

"If there is less running water in the river, [the fish] cannot go upstream. I know one place where there used to be more fish in the river. Now there are less, because that river is not strong enough anymore."

—Moses Novalinga

"I've seen some blackbirds. [...] We haven't had them before. They are different from the robins. I've seen two species that I had never seen before. The black ones and those brown ones."

—Lizzie Tookalook

"The *milugiat* [black flies] are going farther north. There are more *milugiat* at some places where there used to be few."

—Moses Novalinga

“ᖁᐱᕐᕈᖃᓗᐊᖏᓐᓂᖅᓴᐅᓕᖅᑐᖅ ᑕᐃᔅᓱᒪᓂᓂᑦ. [...] ᑕᐃᔅᓱᒪᓂ ᐅᓄᖅᑐᒻᒪᕆᐊᓘᕙᓚᐅᕐᒪᑕ ᐳᔪᐊᓘᔮᖅᕗᑎᒃ. [...] ᕿᒃᑐᕆᐊᑦ ᐊᓇᖏᑦᑐ ᐅᓄᖏᓐᓂᖅᓴᐅᓕᖅᑐᑦ, ᐊᒻᓛᑦ ᐃᖃᓗᖃᖏᓐᓂᖅᓴᐅᓕᖅᑐᖅ, ᐃᖃᓗᐃᑦ ᖁᐱᕐᕈᓂᒃ ᓂᕆᕙᖕᒪᑕ.”

—ᕙᐃᐅᓚ ᓇᐹᖅᑐᖅ

“[Insects] have become less abundant than before. [...] Before there used to be so many it looked like they were smoke. [...] [There are fewer] mosquitoes and black flies now, and there are even less fish, because the fish eat the bugs.”

—Viola Napartuk

ᐅᑭᐅᒥ ᐊᓯᔾᔨᖅᑕᕐᓂᖅ

“ᐱᕈᖅᓴᑎᑦᑐᖓ [...] ᐊᐱᖃᑦᑕᓚᐅᖅᓯᒪᔪᖅ ᓯᑎᐱᕆᒥ, ᐊᑯᓂᐅᓂᖅᓴᕐᓗ ᐅᑭᐅᖑᕙᒃᕗᓂ, ᒪᐃᒥᓗ ᐊᐅᒃᐸᑦᓕᐊᓕᖅᐸᒃᕗᓂ. ᒪᐃ ᓄᕐᖑᐊᓂ ᓯᑯᐃᖅᐸᑦᓕᐊᖃᑦᑕᓚᐅᖅᑐᖅ, ᔫᓂᒥᓗ ᐃᒫᓂ ᓯᑯᖃᒐᓛᒃᐸᒃᕗᓂ. ᔪᓚᐃᒥ ᐱᕈᖅᑐᑦ ᐱᕈᖅᐸᑦᓕᐊᖃᑦᑕᓚᐅᖅᑐᑦ, ᔪᓚᐃᒥᓗ ᓄᓇ ᑐᖑᔪᕐᓂᖅᓴᐅᕙᒃᕗᓂ. ᐋᒡᒋᓯᒥᓗ ᓱᓕ ᓄᓇ ᑐᖑᔪᖅᐸᒃᕗᓂ. ᓯᑎᐱᕆᒥ ᐅᑭᐊᒃᓵᖑᖅᐸᓚᐅᖅᑐᖅ, [...] ᐅᑐᐱᕆᒥᓗ ᓄᓇ ᑕᖅᓴᕆᒃᓯᑦᑎᐊᖅᐸᒃᕗᓂ. ᑕᒪᒃᑯᐊ ᐃᖅᑲᐅᒪᔭᒃᑲ, ᑕᐃᒪᐃᑦᑐᓐᓃᖅᑐᖅ. [...] ᐅᐱᕐᖓᒃᓵᓕᕙᓕᖅᑐᖅ, ᐊᐅᔭᐅᑯᑖᖕᓂᖅᓴᐅᕙᓕᖅᕗᓂ, ᐅᑭᐊᒃᓵᖑᑯᑖᖕᓂᖅᓴᐅᑦᑐᓂ, ᐅᑭᐅᑭᓐᓂᖅᓴᐅᓕᖅᕗᓂᓗ. ᑕᐃᒪᓐᓇ ᑕᑯᓯᒪᔭᕋ.”

—ᐋᓚᓯ ᑐᒃᑐ

Seasons

“When I was growing up [...] it started to snow in September, and we had a long winter, and in May everything started to melt. By the end of May the ice started to break, and in June we had floating ice. In July a lot of plants [started] to grow, and there was more green in July. It was still green in August. September it started to turn to autumn, [...] and in October it became more colourful. I remember that, but it's not like that anymore. [...] We have an early spring, a long summer, a long fall, and a short winter. That's how I see it now.”

—Alice N. Tooktoo

“ᓯᑯᖅᑲᖅᐸᓚᐅᖅᓯᒪᔪᖅ ᐃᑉᐱᐊᕐᔪᕐᒥ ᓯᑎᐱᕆᒥ, ᑭᓯᐊᓂ ᖁᕕᐊᓱᒻᕕᐅᑉ ᓯᕗᓂᑦᑎᐊᖓᓂ ᓯᑯᕋᓕᖅᑐᖅ. ᐊᓯᔾᔨᖅᓯᒪᕐᔪᐊᖅᑐᖅ.”

—ᕓᐃᐅᓚ ᓇᐹᖅᑐᖅ

“ᐊᐱᓇᕌᕐᓂᖅᓴᐅᕙᓕᖅᑐᖅ. ᐃᖅᑲᐅᒪᕗᖓ ᖃᒧᑕᐅᔭᕋ ᐅᑐᐱᕆᒥ ᐊᑐᓕᖅᐸᓚᐅᖅᑕᕋ. ᒫᓐᓇ ᑎᓯᐱᕆᖖᒍᖅᓲᔾᓂᐊᖅᑲᖅᐸᓕᖅᑐᖓ. [...] ᐅᑭᐅᕐᓇᕌᖅᑐᒻᒪᕆᐅᕙᓕᖅᑐᖅ. ᐅᓪᓗᐃᑦ ᐃᓚᖓᓐᓂ ᓯᑯᓯᒪᖖᒋᑦᑐᒥ ᖁᕕᐊᓱᒻᕕᒃᓯᐅᕈᒫᓕᖅᑐᒍᑦ. ᑕᒪᑐᒪᓂ ᓯᑯᑕᐃᓐᓇᓚᐅᖅᑐᖅ ᐱᓇᓱᐊᕈᓯᓕᒫᖅ ᖁᕕᐊᓱᒻᕕᖖᒍᓚᐅᖅᑎᓐᓇᒍ, [...] ᑭᖑᕙᓯᒃᑐᒻᒪᕆᐅᓕᖅᑐᖅ. ᖁᐊᖅᐸᓚᐅᖅᓯᒪᔪᖅ ᓄᕕᐱᕆᐅᑉ ᑭᖑᓂᐊᓂᐅᖖᒋᑦᑐᖅ.”

—ᐆᓂᔅ ᑐᒥᒃ

“ᐅᐱᕐᖔᖅ ᐊᓯᔾᔨᖅᓯᒪᓕᖅᑐᖅ. ᐱᐅᕙᓚᐅᖅᓯᒪᔪᖅ ᐅᖅᑰᓯᓕᖅᓱᓂᓗ ᐅᐱᕐᖔᒃᑯᑦ, ᑭᓯᐊᓂ ᑕᒪᑐᒪᓂ ᓂᒡᓚᓱᒃᑐᒻᒪᕆᐅᔪᖅ. ᐅᐱᕐᖔᒃᑯᑦ ᐊᐅᓪᓛᖅᐸᓚᐅᖅᑐᒍᑦ ᐊᖑᓇᓱᒋᐊᖅᓱᑕ, ᐅᓪᓗᐃᓪᓗ ᐃᓚᖏᓐᓂ ᓯᓚᑦᑎᐊᕙᒻᒪᕆᐅᕙᒃᓱᓂ [...] ᑭᓯᐊᓂ ᑕᒪᑐᒪᓂ ᓂᒡᓚᓱᒃᑐᒻᒪᕆᐅᔪᖅ.”

—ᓕᓯ ᑐᑲᓗᒃ

“ᐃᖅᑲᐅᒪᕗᖓ ᓂᒡᓚᓱᒃᑐᒻᒪᕆᐊᓘᕙᒃᑎᓪᓗᒍ, ᑮᓇᕐᒧᑦ ᑲᑭᒐᓛᖅᑰᔾᔭᓐᓇᒻᒪᕆᒃᓱᓂ. [...] ᑮᓇᒃᑯᑦ ᑲᑭᒐᓛᖅᑰᔾᔭᓐᓇᕈᓐᓃᖅᑐᖅ, ᐅᖅᑰᓂᖅᓴᐅᓕᕐᒪᑦ.”

—ᕓᐃᐅᓚ ᓇᐹᖅᑐᖅ

“There used to be ice on the bay in September, but now it becomes icy just before Christmas. That’s a pretty big change.”

—Viola Napartuk

“[Snow is] coming a lot later now. I remember I used to get my skidoo out in October. Nowadays we wait until December. [...] Winter comes a lot later. A lot later. One of these days we are going to have an iceless Christmas. [This year] the ice froze in the bay one week before Christmas, [...] and that’s late. It used to freeze at least in November.”

—Ernest Tumic

“Spring has changed. It used to be nice and warm in springtime, but this year it was very cold. In the springtime we used to travel to go hunting, and some days it would be nice and warm, [...] but this spring is very cold.”

—Lizzie Tookalook

“I remember when it used to be very cold, [it would feel] like pins and needles on your face. [...] There are no more pins and needles on your face, because it has become warmer.”

—Viola Napartuk

Nunavik / Umiujaq

ᓄᓇᕕᒃ / ᐅᒥᐅᔪᖅ

“ᒪᕐᑯᕐᑲᑦᑕᖖᒋᓐᓂᖅᓴᐅᓕᖅᑐᖅ, ᐱᓗᐊᖅᑐᒥᒃ ᐅᑭᐊᒃᑯᑦ. ᐅᐱᖕᒑᒃᑯᑦ, ᒪᕐᑯᓗᐊᖅᐸᒍᓐᓃᕐᒥᔪᖅ. ᑕᐃᔅᓯᒪᓂᑐᑦ ᒪᕐᑯᒃᐸᒍᓐᓃᖅᑐᖅ.”

“I think there’s less rain, especially in the fall. In spring, there is also less rain. It doesn’t rain as much as it used to.”

—ᒨᓯᓯ ᓄᕙᓕᖖᒐᖅ

—Moses Novalinga

“ᑕᒪᑐᒪᓂ ᐊᕐᖂᒍᒥ ᐊᐳᑎᓕᒻᒪᕆᐊᓗᒃ. ᐊᕐᖂᓂ ᐊᐳᑎᖅᑲᓚᐅᖖᒋᑦᑎᐊᖅᑐᖅ, ᑭᓯᐊᓂ ᑕᒪᑐᒪᓂ ᐊᐳᑎᓕᒻᒪᕆᐊᓗᒃ ᐅᐱᖕᒐᕐᓇᓵᖅᑐᒻᒪᕆᐅᓚᐅᕐᒪᑦ. ᑕᒪᑐᒪᓂᓗ ᓂᒡᓚᓱᖕᓂᖅᓴᐅᓚᐅᖅᕗᓂ. ᓂᒡᓚᓱᒃᑐᒻᒪᕆᐅᓚᐅᖅᑐᖅ.”

“This year there is a lot of snow. Last year there was no snow at all, but there is a lot of snow this time because the spring didn’t come in for a long period. And it was colder this year. Much colder.”

—ᓯᐊᓯ ᓇᓗᒃᑐᖅ

—Siasi Naluktuk

ᓯᓚ/ᐊᓄᕆ

Climate/Weather

“ᐊᐳᑦ ᐊᐅᒃᓵᓕᓂᖅᓴᐅᕙᓕᖅᑐᖅ. ᓯᑯ ᐱᓗᐊᖅᑐᒥᒃ ᐊᐅᒃᓵᓕᓂᖅᓴᐅᕙᓕᖅᑐᖅ.”

“[The snow] melts faster now. Especially the ice.”

—ᓵᓕ ᑐᒃᑐ

—Charlie Tooktoo

“ᓄᓇᐅᑉ ᐃᑭᐊᖕᒐᓂ ᖁᐊᖕᒍᐃᓐᓇᖅᐸᒃᑐᑦ ᐊᐅᒃᐸᓪᓕᐊᓕᖅᑐᖅ. ᓄᓇᒥᒃ ᐊᒃᑐᐃᓂᕐᑲᖅᓯᒪᔪᖅ. ᓯᕐᑭᓂᕐᒥᑦ ᐊᐅᒃᐸᓪᓕᐊᖖᒋᑦᑐᖅ, ᑕᐃᒪᐃᓐᓇᓱᒋᖖᒋᓚᕋ. ᓄᓇᕐᔪᐊᑦᑎᓐᓂᖖᒑᕋᓱᒋᕙᕋ. ᓄᓇᒥᖖᒑᖅᑐᓂᒃ ᐲᔭᐃᕙᓪᓕᐊᓗᐊᓘᔪᒍᑦ ᓄᓇᐅᑉ ᐆᓇᕐᓂᖕᒐ ᓄᓇᐅᑉ ᖄᖕᒐᓄᑦ ᖃᓪᓕᕚᓪᓕᖅᓯᒪᓕᖅᑐᖅ.”

“[The permafrost] is melting. That affects the land. The melting is not coming from the sun, I don’t think. I think it’s from the earth instead. We are removing so much raw material from the earth that the heat is coming closer to the surface.”

—ᒨᓯᓯ ᓄᕙᓕᖖᒐᖅ

—Moses Novalinga

“ᐃᒪᖅ ᐃᓱᖅᓯᕗᓪᓚᐊᖅᑰᐴᔪᖅ.
ᐃᒪᕐᒦᑦᑐᑦ ᑕᑯᔭᕐᓂᙱᓐᓂᖅᓴᐅᓕᖅᑐᑦ.
[...] ᐃᓱᕐᓂᖅᓴᐅᓕᖅᑐᖅ.”

—ᐋᓕᓯ ᑐᒃᑐ ᐊᙱᔪᒃᑎᖅ

“ᑰᒐᓛᖕᒍᕗᓚᐅᖅᑐᑦ
ᓄᑕᖅᐅᑎᓪᓗᖓ ᐱᑕᖅᑲᕈᓐᓃᖅᑐᑦ
ᐃᒪᖅᑲᙱᓐᓂᖅᓴᐅᓕᕐᒪᑦ. ᓈᔅᑖᐳᑲ
ᖃᓂᒋᔮᖓᓂ ᓄᓇᖅᑲᖅᐸᓚᐅᖅᓯᒪᔪᖓ
[...] ᑕᐅᕙᓂ ᑰᒐᓛᒃᑕᖅᑲᓚᐅᖅᓯᒪᔪᖅ.
ᑭᓯᐊᓂ ᒫᓐᓇ ᑰᒐᓛᖅᑲᕈᓐᓃᖅᑐᖅ.
ᑕᓱᑦ, ᐊᙱᔪᑦ, ᐱᑕᖅᑲᐃᓐᓇᕐᓂᐊᖅᑐᑦ
ᐊᑯᓂᑲᓪᓚᒃ, ᑭᓯᐊᓂ ᒥᑭᓐᓂᖅᓴᑦ ᑕᓲᑦ
ᑕᓯᕈᐃᓪᓗ ᐃᒪᐃᖅᐸᓪᓕᐊᓕᖅᑐᑦ.”

—ᔪᓯᐊ ᓵᓚ

“ᐸᓂᕐᓂᖅᓴᐅᕙᓪᓕᐊᓕᖅᑐᖅ
ᑕᓯᕈᖅᑕᖅᑲᖅᐸᓚᐅᖅᑐᓂ, ᑕᓲᓪᓗ
ᐃᒪᐃᖅᐸᓪᓕᐊᓕᖅᒍᑎᒃ. ᑕᐃᔅᓱᒪᓂᑦ
ᐃᒪᖅᑲᖅᑎᒋᔪᓐᓃᖅᑐᖅ.”

—ᓕᓯ ᑯᕉ

“ᐱᓱᒡᕕᒋᔭᒃᓴᐅᓚᐅᖅᑐᓂ
ᓇᐹᖅᑐᓕᖕᓂᓗ
ᐸᓂᖅᓯᒪᓂᖅᓴᐅᔮᓕᖅᑐᑦ.
ᒪᓴᐅᓂᖅᓴᐅᕙᓚᐅᖅᑰᖅᑑᒐᓗᐊᖅ.
[...] ᒥᑭᑦᑑᑎᓪᓗᖓ
ᖁᐊᐃᑦᑖᕐᓇᖅᐸᓚᐅᖅᓯᒪᔪᖅ,
ᑭᓯᐊᓂ ᒫᓐᓇ ᐱᓱᒡᕕᒋᔭᕗᑦ ᐸᓂᒧᑦ
ᓃᖅᑯᖅᐸᓕᖅᑐᖅ.”

—ᐱᓕ ᑐᒃᑐ

“It looks like the water is getting more murky. It’s harder to see through the water now. [...] It’s more muddy.”

—Alice Tooktoo Sr.

“The creeks [that were here] when I was a child, they don’t exist anymore because there is less water. I used to live around Nastapoka River [...] and there was a creek. There are no more creeks. Those lakes, the big ones, are going to stay for a while, but some smaller lakes or ponds will become dry.”

—Joshua Sala

“It’s becoming drier where there used to be ponds, and the lakes have become drier. There is less water compared to what there used to be.”

—Lizzie Crow

“There are some places where you walk or places in trees that seem to be drier now. I think they used to be wetter. [...] When I was little [the lichens] used to be slippery, but now they crunch when you walk on them.”

—Billy Tooktoo

“ᐃᓚᖏᑦ ᑰᑦ ᐃᒪᑭᓐᓂᖅᓴᐅᓕᖅᑐᑦ.
ᐊᒻᒪ, ᑰᑦ ᐃᓚᖏᓐᓂ
ᐱᕈᖅᑐᖅᑲᕐᓂᖅᓴᐅᓕᖅᖁᓂ.
ᐱᕈᖅᑐᖅᑲᓕᖅᐸᓪᓕᐊᔪᖅ ᖃᑦᑐᔪᓂ.”

—ᒨᓯᓯ ᓄᕙᓕᖖᒐᖅ

“ᐊᓄᕆ ᐊᓯᔾᔨᖅᓯᒪᔪᒻᒪᕆᐅᓕᖅᑐᖅ.
ᐊᓄᕆᑐᓂᖅᓴᐅᕙᓕᖅᑐᖅ.
ᑕᐃᔅᓯᒪᓂ ᐊᓄᖂᓗᐊᖅᐸᓚᐅᖅᓯᒪᙱᑦᑐᖅ,
ᐊᓄᖂᖁᓗᐊᖂᖖᒐᓪᓗ ᐊᓄᖂᖅᑐᐊᓘᕙᒐᓂ.”

—ᑕᐃᓇ ᑐᑲᓗᒃ

“ᐊᕐᖂᒍᑕᓖᑦ
ᐊᓄᕆᑐᓕᖅᐸᓪᓕᐊᑐᐃᓐᓇᖅᑐᖅ.”

—ᓕᓯ ᑐᑲᓗᒃ

“ᐊᓯᔾᔨᖅᓯᒪᔪᒻᒪᕆᐅᓕᖅᑐᖅ,
ᐊᓄᕆ. ᖃᐅᔨᓯᒪᓕᖅᑐᖓ
ᐊᓄᕆᑐᓂᖅᓴᐅᓕᕐᓂᖕᒐᓂᒃ
ᐊᓄᖂᒐᔪᖕᓂᖅᓴᐅᓕᕐᓂᖕᒐᓂᓪᓗ.
ᖃᔭᕆᐊᑦ ᑭᓴᖅᓯᒪᙱᑉᐸᑕ
ᑎᒃᑕᐅᓇᔭᖅᑐᑦ. [...] ᓯᓪᔭᒥ
ᒪᑦᓕᖅᑐᐊᓘᕙᓕᖅᑐᖅ, ᒪᓖᑦᓗ
ᐊᖏᓂᖅᓴᐅᕙᓕᖅᖁᑎᒃ.”

—ᕙᐃᐅᓚ ᓇᐹᖅᑐᖅ

“Some rivers have less water. Also, I think in some rivers there is vegetation growing. It’s starting to grow more on the banks.”

—Moses Novalinga

“There’s a big change with the wind. The wind has become stronger. Long ago there would hardly be any wind, or if there was wind, it would not be a really strong wind.”

—Dinah Tookalook

“[The wind] is becoming stronger every year.”

—Lizzie Tookalook

“It has changed a lot, the wind. I have observed that the wind has become stronger and more frequent. If the canoes are not tied, they will probably get blown away. [...] I have seen a lot of waves on the shore, and the waves have become bigger.”

—Viola Napartuk

“ᒥᑭᑦᑑᑎᑦᓗᖓ ᓯᕿᓂᖅ ᑕᑯᔪᓐᓇᖅᐸᓚᐅᖅᑕᕋ ᑕᐅᑐᒍᓐᓃᕈᑎᒋᓇᒍ. ᐃᒻᒪᖄ ᕿᓚᒥᒻᒪᕆᑯᓗᒃ ᑕᑯᕙᓚᐅᕋᒃᑯ. ᒫᓐᓇ ᓯᕿᓂᖅ ᕿᕕᐊᕈᓐᓇᐃᑦᓕᔭᕋ ᑕᐅᑐᒍᓐᓃᔭᓐᓇᒪ. ᐋᓐᓂᕐᓇᕐᓂᖅᓴᐅᓕᖅᑐᖅ ᓯᕿᓂᖅ ᑕᑯᔭᕆᐊᒃᓴᖓ ᓄᑕᕋᐅᓚᐅᕐᓂᕐᓂᑦ. ᐆᓇᖅᓯᔪᒻᒪᕆᐅᓕᖅᓱᓂᓗ. [...] ᐅᖅᑰᓂᖅᓴᐅᕙᓕᖅᑐᖅ.”

—ᔪᓯᐊ ᓵᓚ

“ᐊᓄᖄᕐᓂᖅᓴᐅᕙᓕᖅᑐᖅ, ᐊᓄᕆᑐᕙᑦᓕᐊᑐᐃᓐᓇᕐᓂᐊᖅᓱᓂᓗ. [...] ᐊᓄᕆᒧᑦ ᐊᐳᑦ ᑎᒃᑕᐅᓂᐅᓴᕙᓕᖅᑐᖅ ᑕᐃᔅᓯᒪᓂᓂᑦ, ᐊᐳᑦ ᒪᓂᕋᕐᒦᑐᐃᓐᓇᖅᐸᓚᐅᖅᓯᒪᔪᖅ. [...] ᒫᓐᓇ ᐊᐳᑦ ᐊᓄᕆᒧᑦ ᑎᒃᑕᐅᕙᓕᖅᑐᖅ ᕿᒧᒡᔪᖅᑲᕐᓂᖅᓴᐅᓕᖅᓱᓂᓗ, ᐃᓛᓐᓂᒃᑯᑦᓗ ᐊᐳᑦ ᓄᓇᒥ ᐊᓄᕆᒧᑦ ᑎᒃᑕᐅᕙᒃᑐᖅ, ᐃᓛᓐᓂᒃᑯᑦᓗ ᐃᒡᓗᐊᓂᑦ ᐊᓄᕆᓕᖄᖕᒐᑦ ᐊᐳᑦ ᓯᖅᔭᒧᑦ ᓯᑯᒧᑦᓘᓐᓃᑦ ᑎᒃᑕᐅᕙᒃᑐᖅ. ᓴᙱᔪᐊᓗᖕᓂᒃ ᐊᓄᖄᖅᐸᓕᖅᑐᖅ.”

—ᐅᐃᓕ ᖁᒫᓗᒃ

“ᓄᑕᕋᐅᑦᓗᖓ, ᐊᓄᖄᓗᐊᖅᐸᓚᐅᙲᑦᑐᖅ ᒫᓐᓇᑐᑦ. ᓄᑕᕋᐅᑦᓗᖓ ᐃᒪᖅ ᖃᐅᑕᒫᑦ ᐅᖅᓯᐊᖅᐸᓚᐅᖅᑐᖅ, ᐃᒪᖅ ᑕᕐᕋᖅᑑᑎᐅᖅᔮᖅᐸᓚᐅᖅᑐᖅ. ᒫᓐᓇ ᐊᓄᖄᖅᑐᐊᓘᕙᓕᖅᑐᖅ.”

—ᔪᓯᐊ ᓵᓚ

“When I was a child, normally there were no winds like we have now. When I was a child the water would be calm almost every single day, and the water would be like a mirror. [...] Now it’s really windy.”

—Joshua Sala

“It has become more windy, and it’s going to get windier. [...] The wind blows away the snow compared to many years ago, when it just dropped on the ground. [...] Now the snow gets blown away and makes snowdrifts, and sometimes the wind blows away the snow on the land, and sometimes wind from another direction will blow it back to the shore or on the ice. We’ve been having strong winds.”

—Willie Kumarluk

“When I was little I used to be able to look at the sun without getting blinded. Maybe I used to look at it for a couple of seconds. Now I cannot look at the sun because it blinds me. It hurts more to look at the sun compared to when I was young. It really is getting hot. It’s getting warmer.”

—Joshua Sala

“ᓯᓚ ᐅᖅᑰᓯᕙᓪᓕᐊᓂᖁᖅᑕᐅᔪᖅ,
ᑭᓯᐊᓂ ᐅᑭᐅᖅᑕᖅᑐᕐᒥ ᑭᒡᓗᒨᖁᓱᕆᔭᖁ.
ᓂᒡᓚᖅᐸᓪᓕᐊᔪᖅ, ᐃᓱᒪᒋᔭᒃᑯᓪᓕ.
ᓯᕐᑭᓂᖅ ᐆᓇᕐᓂᖅᓴᐅᒐᓗᐊᖅ;
ᑮᓇᒃᑯᑦ ᐃᒃᐱᖕᓇᖅᑐᖅ,
ᐅᑭᐅᒃᑯᓪᓘᓐᓃᑦ, ᐊᐅᔭᒃᑯᓪᓗ
ᓯᕐᑭᓂᖅ ᐆᓇᕐᓂᖅᓴᐅᓕᖅᕈᓂ.
ᓂᒡᓚᓱᖕᓂᖅᓴᐅᓇᓱᕆᔭᖁ
ᓄᕗᔭᕐᑲᕐᓂᖅᓴᐅᕙᓕᕐᒪᑦ
ᐅᐱᕐᖕᓖᒃᑯᑦ. ᓄᕗᔭᑎᓐᓇᒍᓗ,
ᓯᓚ ᕐᑲᐅᒪᓂᖅᓴᐅᓕᖅᕈᓂ.
ᑐᙳᔪᓗᐊᙱᓐᓂᖅᓴᐅᓕᖅᑐᖅ.
ᓄᕗᔭᑎᓐᓇᒍ ᓯᓚ
ᑐᖑᔪᕐᓂᖅᓴᐅᕙᓚᐅᖅᑐᖅ.”

—ᒨᓯᓯ ᓄᕙᓕᙵᖅ

“They say there is global warming, but I think in the North it’s the other way around. It’s getting colder, I think. Well, the sun is warmer; you can feel it on your face, even in the wintertime, and in the summer the sun is a lot hotter. I think the reason it’s getting colder is that we have more clouds now in the spring. And on a clear day, the sky is lighter now. Lighter blue. It used to be darker blue on clear days.”

—Moses Novalinga

“ᑕᑯᓯᒪᓕᖅᑐᖓ ᓄᓇᐅᑉ
ᐃᑭᐊᖓ ᖁᐊᖑᐃᓐᓇᖅᐸᒃᑐᖅ
ᐊᒻᒨᒃᐸᓪᓕᐊᓂᖓᓂᒃ. [...]
ᓄᓇ ᐊᖅᐸᖅᐸᓪᓕᐊᓕᖅᑐᖅ,
ᐊᕐᖀᒍᑕᓖᑦ ᖁᐊᒃᑲᓐᓂᓚᐅᖅᕈᓂ,
ᐊᖅᐸᖅᐹᓪᓕᖅᐸᒃᑐᖅ.
ᐊᕐᖀᒍᑕᓖᑦ ᐊᖅᐸᖅᐸᓪᓕᐊᔪᖅ,
ᖁᒡᕕᖅᐸᓪᓕᐊᙱᑦᑐᖅ.”

—ᐅᐃᓕ ᖁᒫᓗᒃ

“I’ve seen some [permafrost] go down. [...] The ground has been going down, and each year it will freeze again, and then it gets lower. Every year it seems to go down, but it’s not going up.”

—Willie Kumarluk

ᐃᓄᐃᑦ ᐱᖅᑯᓯᑐᖅᑲᖓᑦᑕ ᐊᒃᑐᖅᑕᐅᓂᖓ

"ᐊᐳᑎᖅᑲᖖᒋᓐᓂᖅᓴᐅᓕᖅᑐᖅ. [...]ᐊᐱᖅᑯᓕᓚᐅᖅᓯᒪᓕᕋ ᐅᖅᐱᑦ ᐱᑑᖅᕙᖏᓐᓂ ᐊᕿᒡᒋᕋᓱᒍᒪᓪᓗᖓ, ᑭᓯᐊᓂ ᒫᓇ ᐅᖅᐲᑦ ᐊᐱᔭᐅᓕᒍᓐᓃᖅᑐᑦ."

—ᐆᓂᔅ ᑐᒥᒃ

"ᑕᐃᔅᓱᒪᓂ ᖃᒧᒃᓯᕈᓐᓇᖅᐸᓚᐅᖅᑐᒍᑦ ᔫᓂ ᓄᙳᐊᓄᑦ ᑎᑭᔾᔪᒍ, ᑭᓯᐊᓂ ᓯᑯ ᐊᐅᒃᓵᓕᓕᖅᑐᖅ ᐃᓛᓐᓂᒃᑯᑦ. [...] ᖃᒧᒃᓯᖅᑐᖅᑲᖅᐸᒍᓐᓃᖅᑐᖅ ᐅᐱᕐᖔᑉ ᓄᙳᐊᓂ; ᐊᐅᒃᓵᓕᓂᖅᓴᐅᓕᕐᒪᑦ. ᐃᖅᑲᐅᒪᒐᖓ ᐃᓚᒌᑎᒍᑦ ᐊᐳᒻᒥᒃ ᓂᓚᖕᓂᓗ ᓯᐅᖅᑲᑦ ᐊᑖᓂ ᓂᓚᒃᑕᖅᐸᓚᐅᕋᑦᑕ ᐃᒥᒐᒃᓴᑦᑎᓐᓂᒃ ᐅᑭᐅᒃᑯᑦ. ᑭᓯᐊᓂ ᓯᑯᖅᑲᖅᐸᒍᓐᓃᖅᑐᖅ."

—ᕓᐃᐅᓚ ᓇᐹᖅᑐᖅ

Impacts on Traditional Ways of Life

"[There is] a lot less snow nowadays. [...] I used to wish for certain periods of time when all the bushes were snowed over so I could go to places where I could hunt ptarmigan, but nowadays [the bushes] don't get covered in snow."

—Ernest Tumic

"Before, we used to go out on the dog sleds until the end of June, but now some seasons [the ice] melts down sooner. [...] There are no dog teams at this time of year [late spring]; it melts down quicker these years. I remember my family picking some snow or ice under the sand during the winter for water. But the ice has thawed out."

—Viola Napartuk

ᑲᖏᖅᓱᐊᓗᔾᔪᐊᖅ
Kangiqsualujjuaq

ᑲᖏᖅᓱᐊᓗᔾᔪᐊᖅ

ᑲᖏᖅᓱᐊᓗᔾᔪᐊᖅ ᑰᑉ ᖃᒪᓂᐊᓃᑦᑐᖅ ᐅᖓᕙᒥ (N 58.69°, W 65.94°; ᖁᑦᑎᖕᓂᖓ 10 m). ᐅᑭᐅᖅᑕᖅᑐᕐᒥ ᐊᖅᐸᓯᒃᑐᓄᑦ ᐃᓚᒋᔭᖅ ᓇᐹᖅᑐᖃᖅᑐᑦ ᖃᓂᒋᔭᖓᓂ, ᐅᖅᐱᒐᕐᓂᒃ ᐱᕈᖅᑐᖃᖅᓱᑎᒃ, ᖁᐊᖑᐃᓐᓇᖅᐸᖏᑦᑐᑦ ᓄᓇᐅᑉ ᐃᑭᐊᖓ ᓇᖅᓴᓚᖕᒦᕐᓱᑎᒃ, ᖃᖅᑲᓪᓗ ᐊᑎᖏᓐᓂ ᓄᓇᐅᑉ ᐃᑭᐊᖓ ᖁᐊᖑᐃᓐᓇᖅᐸᒃᓱᓂ. ᓄᓇᓕᙳᓚᐅᖅᓯᒪᔪᖅ 1962-ᒥ ᐃᓄᒋᐊᖕᓂᖃᖅᓱᓂᓗ 874-ᓂᒃ 2011-ᑲᓇᑕᒥ ᑭᒡᓕᓯᓂᐊᖅᑕᓂᑦ.

ᑲᖏᖅᓱᐊᓗᔾᔪᐊᕐᒥᐅᑦ ᐅᔾᔨᕈᓱᒃᓯᒪᔪᑦ ᐊᒥᓱᓂᒃ ᐊᕙᑎᒧᑦ ᐊᓯᔾᔨᖅᓯᒪᔪᓂᒃ, ᐱᓗᐊᖅᑐᒥ ᐱᕈᖅᑐᓂᒃ. ᐅᔾᔨᕆᔭᐅᓗᐊᖅᓯᒪᔪᑦ ᐱᕈᖅᑐᑦ ᐅᓄᕐᓂᖅᓴᐅᓕᕐᓂᖏᓐᓂᒃ ᐱᕈᖅᓵᓕᓂᖅᓴᐅᕙᓕᕐᓂᖏᓐᓂᒡᓗ. ᐊᒥᓲᓂᖅᓴᑦ ᐅᔾᔨᕈᓱᒃᓯᒪᔪᑦ ᐱᕈᖅᑐᓂᒃ ᐊᔾᔨᒌᙱᑦᑐᓂᒃ, ᓇᐹᖅᑐᑦ, ᐅᖅᐱᒐᑦ (ᓲᕐᓗ ᐅᖅᐱᐊᓛᕐ) ᐱᑕᖃᕐᓂᖅᓴᐅᓕᕐᓂᖏᓐᓂᒃ ᓯᕗᑐᔪᓂ ᓯᒡᔭᖅᐸᓯᖕᓂᓗ. ᐃᓚᖏᑦ ᐱᕈᖅᑐᑦ, ᓲᕐᓗ ᓇᐹᖅᑐᑦ, ᒫᓇ ᐱᑕᖃᕐᓂᖅᓴᐅᕙᓕᕐᓂᖏᓐᓂᒃ, ᓄᑖᓂᒡᓗ ᓴᖅᑭᑦᑐᖃᖅᐸᓕᕐᓂᖓᓂᒃ, ᓲᕐᓗ ᐸᐅᓐᓇ. ᓄᓂᕙᒃᑕᐅᕙᒃᑐᑦ ᓲᕐᓗ ᐊᖅᐲᑦ ᐱᑕᖃᕐᓂᖅᓴᐅᕙᓕᖅᓱᑎᒃ ᐱᕈᐊᓂᒃᓵᓕᕙᓕᖅᓱᑎᒡᓗ.

ᓂᕐᔪᑎᓂ ᐊᓯᔾᔨᖅᓯᒪᓕᕐᒥᔪᑦ ᐅᖃᐅᓯᐅᕙᓚᐅᕐᒥᔪᑦ, ᓲᕐᓗ ᐃᑦᓛᕈᓯᖃᕐᓂᖅᓴᐅᓕᕐᓂᖓ ᐊᒻᒪ ᐊᒃᓴᖃᕐᓂᖅᓴᐅᓕᕐᓂᖓ. ᐅᖃᖅᑐᖃᖅᓯᒪᓪᓗᓂᓗ ᓇᓄᐃᑦ ᐸᐅᕐᖓᖅᑐᖅᐸᓕᕐᓂᖏᓐᓂᒃ

Kangiqsualujjuaq

Kangiqsualujjuaq is located at the mouth of George River, 25 kilometres from Ungava Bay (N 58.69°, W 65.94°; elevation 10 m). It is part of the Low Arctic near the treeline, in the low shrub subzone. In this region, discontinuous permafrost lies under valleys, and continuous permafrost is found beneath plateaus and mountains. It was established in 1962 and its population was estimated at 874 people in the 2011 Statistics Canada Census.

Kangiqsualujjuamiut have observed many environmental changes, especially regarding vegetation. The most obvious changes were related to a general increase in plant abundance and earlier blooming times. A majority of people have observed changes in plant distribution, with trees (e.g., larch) and erect shrubs (e.g., dwarf birch) becoming more abundant and colonizing the hillsides and seashore. Some plant species, such as balsam poplar, are more common nowadays, and new species, such as tall fireweed, are being observed. In terms of berry plants, cloudberries/bakeapples seem to be more abundant and ripen earlier.

ᓇᐅᔭᐃᑦᑐ ᒪᓐᓂᖕᒋᓐᓂᒃ
ᓂᕆᕙᓕᕐᓂᖕᒋᓐᓂᒃ,
ᓯᑯᕐᑲᖕᒋᓐᓂᖅᓴᐅᓕᕐᓂᖕᒐᓄᑦ
ᓇᕐᑎᕋᓱᖕᒋᓐᓂᖅᓴᐅᓕᕐᓂᖕᒋᓐᓄᑦᑐ.
ᐊᖕᒋᖅᑲᑎᒌᑐᐊᓚᐅᙱᑦᑐᑦ
ᐊᓯᖕᒋᓐᓂᒃ ᐆᒪᔪᑦ ᒥᒃᓵᓄᑦ, ᓲᕐᓗ
ᑎᖕᒥᐊᑦ, ᐃᕐᑲᓗᐃᑦ, ᐊᐅᒃᑐᖅᑏᑦᑐ
ᕐᑯᐱᕐᕈᑦ. ᑭᓯᐊᓂ ᓄᑖᓂᒃ ᒥᑭᑦᑐᓂᒃ
ᑎᖕᒥᐊᓂᒃ, ᐃᒍᑦᑕᕐᓂᒃ ᑕᕐᕋᓕᑭᓵᓂᒡᓗ
ᑕᑯᕙᓕᕐᓂᕋᖅᑐᑦ, ᐱᕐᑲᓯᐅᔾᔪᒋᑦ
ᐸᒥᐅᑯᑖᓛᑦ ᐊᕕᙵᖅᐸᑦ.

ᐊᒥᓲᓂᖅᓴᑦ ᐊᐱᖅᓱᖅᑕᐅᔪᑦ
ᐃᒪᑭᓐᓂᖅᓴᐅᓕᕐᓂᕋᐃᓚᐅᖅᑐᑦ, ᐊᒻᒪ
ᓯᑯ ᓵᓐᓂᖅᓴᐅᓕᕐᓂᕋᖅᑕᐅᑦᑐᓂ
ᓯᑯᐃᖅᓵᓕᓂᖅᓴᐅᕙᓕᖅᔪᓂᓗ.
ᐊᐳᑎᑭᓐᓂᖅᓴᐅᓂᕋᖅᑕᐅᓕᖅᔪᓂ
ᐊᐅᒃᓵᓕᓂᖅᓴᐅᕙᓕᖅᔪᓂᓗ.
ᑕᒪᐃᑲᓴᑦ ᐃᓄᐃᑦ ᐅᖅᑲᓚᐅᖅᑐᑦ ᓯᓚ
ᐆᓇᕐᓂᖅᓴᐅᕙᓕᕐᓂᖕᒐᓂᒃ, ᓄᓇᐅᑦᑐ
ᐃᑭᐊᖕᒐ ᕐᑯᐊᖕᒍᐃᓐᓇᖅᐸᒃᑐᖅ
ᐊᐅᖕᓂᖅᓴᐅᕙᓕᕐᓂᖕᒐᓂᒃ.
ᓄᓇᒥ ᑎᓱᐊᖅᓯᒪᔪᑦ ᓇᐹᖅᑐᑦᑐ
ᐅᕐᕈᓯᒪᓕᖅᑐᑦ ᑕᑯᔭᐅᓂᖅᓴᐅᕙᓕᖅᑐᑦ
ᑰᖕᓂ ᓯᒡᔭᓂᑦᑐ.

ᐃᓚᖕᒋᑦ ᑲᖕᒋᖅᓱᐊᓗᔾᔪᐊᕐᒥᐅᑦ
ᐃᓱᒫᓗᒋᔭᖅᑲᖅᑐᑦ ᓂᕐᑭᒋᕙᒃᑕᒥᖕᓂᒃ
ᐃᓅᓯᕐᒥᓂ ᐊᑐᐃᓐᓇᖅᑕᒥᖕᓂᒃ,
ᔭᒐᐃᔪᒪᓇᑎᒃ ᐃᓕᖅᑯᓯᓇᒐᒥᔾᔪᒃ,
ᐊᐱᕆᓪᓗᑎᒃ "ᓱᓇᔭᖅᐱᑕ?"
ᐃᓚᖕᒋᑦ ᐅᐱᕐᖔᒃᑯᑦ ᐊᐅᓪᓛᕐᓂᖅ
ᐊᑲᐃᓪᓕᐅᕈᑕᐅᕙᓕᕐᓂᖕᒐᓂᒃ
ᐅᖅᑲᖅᐸᓚᐅᖅᑐᑦ
ᓯᑯᐃᖅᓵᓕᓂᖅᓴᐅᕙᓕᕐᓂᖕᒐᓄᑦ.

Many changes in mammal abundance were described, such as increases in sightings of porcupines and black bears. Some people reported that polar bears are now eating blackberries and seagull eggs, because the reduction in sea ice limits their access to seals. There was little consensus on other animal groups, such as birds, fish, and biting insects. However, new species of small birds, bumblebees, and butterflies have been noticed, as well as rats.

Lower water levels were reported by most interviewees, as well as sea ice becoming thinner and breaking up earlier. Snow seems to be less abundant than before and is melting down faster. In general, people felt that the temperature was warmer than in the past, having an impact on permafrost, which is thawing more deeply. Erosion signs and fallen trees are also being seen more frequently nowadays along riverbanks and the seashore.

Some Kangiqsualujjuamiut expressed their concerns about the country food they are so used to eating, hoping they will not lose that part of their culture, because "What would we do?" Some spring travelling and camping issues were also mentioned because of the faster melting of the ice.

ᐃᓐᓇᐃᑦ ᐊᐱᖅᓱᖅᑕᐅᔪᑦ
Elders Interviewed

ᐱᐊᑦᓯ
ᐋᓇᓈᖅ
Betsie
Annanack

ᓘᑲᓯ ᐃᑦᑐᖅ
Lucas A. Etok

ᐄᕙ ᐋᓇᓈᖅ
Eva
Annanack

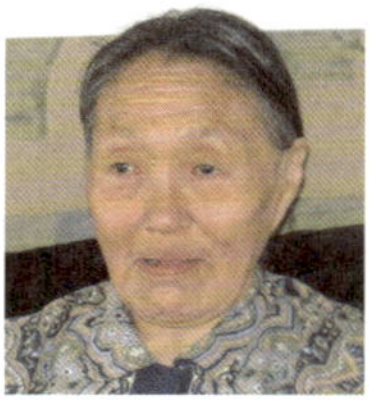

ᒥᐊᓕ ᐃᑦᑐᖅ
Mary Etok

ᔮᓂ ᔪᐊᔾᔨ
ᐋᓇᓈᖅ
Johnny George
Annanack

ᑏᕕ ᐃᑦᑐᖅ
Tivi Etok

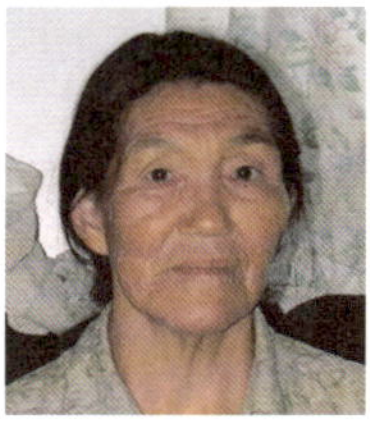

ᓯᐊᕋ ᐹᓴ
ᐋᓇᓈᖅ
Sarah Pasha
Annanack

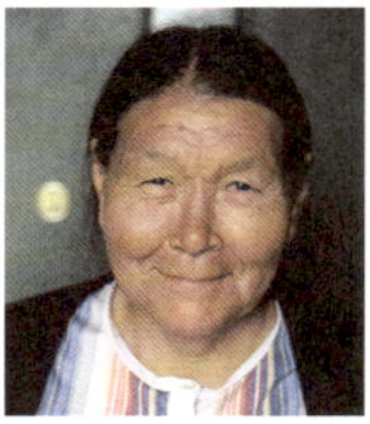

ᓲᓯ ᒧᐊᒐᓐ
Susie
Morgan

ᐅᐃᓕ
ᐃᒪᑦᓚᒃ
Willie
Emudluk

ᐸᐅᕐᖓᐃᑦ

“ᑕᒪᑐᒪᓂ ᐊᕐᖀᒍᒥ ᐊᓯᔾᔨᖅᓯᒪᔪᒻᒪᓇᐅᔪᖅ, ᓱᓕ ᐊᖅᐱᖕᓂᒃ ᑕᑯᓚᐅᙱᓐᓇᒪ. ᐱᕈᕐᓂᐊᕋᓗᐊᖅᐸᑦᑐᖅ.”

—ᐄᕙ ᐋᓇᓈᖅ

“ᑕᒪᐅᖓ ᑎᑭᓐᓇᑦᑕ ᐸᐅᕐᖓᕐᑲᖅᐸᓚᐅᖅᑐᖅ, ᑭᓯᐊᓂ ᒫᓐᓇ ᓯᐅᕋᐃᓐᓇᐅᓕᖅᑐᖅ.”

—ᐄᕙ ᐋᓇᓈᖅ

“ᐸᐅᕐᖓᑦ ᓇᓄᖕᒋᑦ ᐋᒡᒋᓯᒥ ᐱᕈᖅᐸᓚᐅᖅᑐᖅ. ᑕᒪᕐᒥᐊᓗᒃ ᐊᑕᐅᑦᑎᒃᑯᑦ ᐱᕈᖅᐸᓕᖅᑐᑦ. ᑭᒍᑕᖏᕐᓇᓪᓗ ᐊᐅᐸᔮᒃᓯᐊᓂᕇᖅᑐᑦ. ᓄᓇ ᓱᓕ ᑐᖑᔭᖅᓯᓪᓗᐊᙱᑦᑑᒐᓗᐊᖅ, ᑐᖑᔭᐊᓂᓕᓈᖅᑐᖅ.”

—ᓲᓯ ᒧᐊᒐᓐ

“ᐅᖅᐲᑦ ᐊᖏᔪᐊᓘᕙᓕᖅᑐᑦ. ᐊᒥᓱᐊᓗᖕᓂᒃ ᐅᖅᐱᖅᑲᓕᖅᑐᖅ, ᑕᐃᒪᐃᑉᐸᓚᐅᖅᓯᒪᙱᑦᑑᒐᓗᐊᖅ. ᐅᖅᐲᑦ ᒥᑭᓐᓂᖅᓴᐅᕙᓚᐅᖅᑐᑦ. ᐅᖅᐲᑦ ᒥᑭᓐᓂᖅᓴᐅᕙᓚᐅᖅᑐᑦ; ᒫᓐᓇ ᐊᖏᓕᕙᓪᓕᐊᑐᐃᓐᓇᖅᑐᑦ, ᐊᖏᔪᐊᓘᓪᓗᐊᙱᑦᑑᒐᓗᐊᑦ.”

—ᓲᓯ ᒧᐊᒐᓐ

“ᓄᓂᕙᒋᐊᖀᖓᑦᑕ ᐊᕐᖀᒍᑕᓖᑦ ᐱᓂᖅᓴᐅᕙᓪᓕᐊᑐᐃᓐᓇᖅᑐᒍᑦ, ᓱᓚ ᐅᖅᑰᓯᕙᓪᓕᐊᓕᕐᓂᖓᓄᑦ.”

—ᑏᕙ ᐃᑦᑐᖅ

Berries

“This year it’s changed a lot, because I haven’t seen the *aqpiit* [cloudberries/bakeapples] yet. I hope the berries will grow.”

—Eva Annanack

“When we arrived here there were berries here and there, but now only sand.”

—Eva Annanack

“[Green berries] were supposed to grow only in August. All of them, they grow all at once. And the blueberries are already pink. And the land is not supposed to be green yet, but it’s already green.”

—Susie Morgan

“The *uqpiit* [willows] are growing big. Huge amounts of *uqpiit*, and they weren’t like that before. *Uqpiit* were smaller; now they get bigger and bigger. They are not supposed to grow this big.”

—Susie Morgan

“When we pick berries we get more and more every year, because the weather is getting warmer.”

—Tivi Etok

ᐱᕈᖅᑐᑦ ᐊᓯᖏᑦ

“ᐱᕈᖅᑐᑦ ᐱᕈᖅᓵᓕᓂᖅᓴᐅᕙᓕᖅᑐᑦ ᑕᐃᔅᓱᒪᓂᓂᑦ. ᐊᖏᓕᕙᓪᓕᐊᑐᐃᓐᓇᕐᓂᐊᖅᖢᑎᒡᒍ.”

—ᔮᓂ ᔪᐊᔾᔨ ᐋᓇᓈᖅ

“ᐅᒥᐊᖅᑐᖅᖢᑕ ᓄᓇᐅᑉ ᑭᒡᓕᐊᒍᑦ ᑕᑯᕙᓕᖅᑐᒍᑦ ᓇᐹᖅᑐᓂᒃ ᐱᕈᓕᖅᑐᓂᒃ. ᐱᑕᖃᖅᐸᓚᐅᖅᓯᒪᙱᑦᑐᑦ, ᑭᓯᐊᓂ ᐱᕈᓕᖅᑐᑦ. ᑐᑯᓪᓗᒋᑦ ᐅᒃᐱᕐᓇᙱᑦᑐᑦ, ᐱᕈᓕᖅᑐᑦ. ᓇᐹᖅᑐᓂᒃ ᑕᑯᖅᑲᑦᑕᓕᓚᐅᖅᓯᒪᔪᒍᑦ 1985-ᒥ. ᓇᐹᖅᑐᑦ ᑕᑭᓂᖅᓴᐅᓕᖅᑐᑦ.”

—ᓘᑲᓯ ᐃᑦᑐᖅ

“ᐅᖅᐱᒐᐃᑦ ᐱᕈᖅᑐᐊᓘᕙᓕᖅᑐᑦ. ᓇᐹᖅᑐᓪᓗ ᐱᕈᕐᓂᖅᓴᐅᓕᖅᖢᑎᒃ—ᑕᐃᔅᓱᒪᓂᓂᑦ ᐱᕈᕐᓂᖅᓴᐅᓕᖅᑐᑦ.”

—ᓯᐊᕋ ᐹᓴ ᐋᓇᓈᖅ

“ᑕᑯᓯᒪᓕᖅᑐᖓ ᐱᕈᖅᑐᓂᒃ ᐊᓯᙵᓂ ᓄᓇᓂ ᑕᑯᓯᒪᔭᓐᓂᒃ. ᑕᒫᓂ ᓄᓇᑦᑎᓐᓂ ᑕᑯᓚᐅᖅᓯᒪᙱᑦᑕᒃᑲ. [...] ᐊᐅᓪᓛᕐᕕᒻᒥ, ᐅᖓᓯᒃᑐᒥ, ᓂᕆᕙᓚᐅᖅᑕᕗᑦ, ᑭᓯᐊᓂ ᒫᓂ ᐱᕈᖅᐸᓕᖅᑐᑦ.”

—ᓲᓯ ᒧᐊᒐᓐ

Other Plants

“The plants grow faster than they did before. They’re going to get bigger still.”

—Johnny George Annanack

“When we go along the coast we can see trees growing. They never used to be there, but they are growing. When I look there I don’t believe it, that they are growing. We started to see trees maybe in 1985. The trees are getting taller.”

—Lucas A. Etok

“The shrubs grow a lot. The trees grow more too—more than before.”

—Sarah Pasha Annanack

“I have seen a couple of plants that I’ve only seen in other communities. I’ve never seen them in this community. [...] At our cabin, somewhere far, we used to eat them, but now they grow here.”

—Susie Morgan

“ᐱᕈᖅᑐᑦ ᐊᓯᔾᔪᖅᓯᒪᙱᑦᑐᑦ,
ᑭᓯᐊᓂ ᓱᒃᑲᓂᖅᓴᒥ ᐱᕈᖅᐸᓕᖅᑐᑦ.
ᐊᖏᓂᖅᓴᐅᕙᓕᖅᑐᑦ. ᐊᕚᓚᕿᐊᑦ
ᐊᖏᓂᖅᓴᐅᓕᖅᑐᑦ. ᓄᓇᐅᓪᓗ
ᐃᑭᐊᙵᓂ ᕐᑯᐊᖑᐃᓐᓇᖅᐸᒃᑐᑦ
ᐊᐅᒃᐸᓪᓕᐊᓕᖅᓱᓂ. ᑭᓱᓕᒫᑦ
ᐊᖏᓕᕚᓪᓕᐊᓕᖅᑐᑦ.”

—ᑏᕕ ᐃᑦᑐᖅ

“[The plants] are the same, but they are growing faster. They grow bigger now. *Avaalaqiaq* [birch] is bigger now. The permafrost is thawing. Everything is getting big.”

—Tivi Etok

“ᑕᐃᒪᑐᖅ ᒪᕐᑯᖕᓂᐊᖅᐳᖅ
ᐱᕈᖅᑐᑦ ᐱᕈᑦᑎᐊᕐᓂᐊᕐᒪᑕ.
ᒪᕐᑯᕐᑲᑦᑕᙱᑦᑕᕚᖓᑦ
ᐸᓂᑐᐃᓐᓇᖅᐸᖕᒪᑕ.”

—ᐅᐃᓕ ᐃᒪᓪᓚᒃ

“I hope it will rain so the plants will grow. When it never rains, they will just burn.”

—Willie Emudluk

ᐆᒪᔪᑦ

Animals

“ᓇᓄᕐᓄᑦ ᓈᒻᒪᒍᓐᓃᖅᑐᖅ
ᓇᑦᑎᕐᑲᓗᐊᕈᓐᓃᕐᒪᑦ.
ᐊᖑᓇᓱᒋᐊᕚᖓᑦᑕ ᑕᑯᕙᓕᖅᑐᖓ
ᓇᓄᕐᒥᒃ ᐸᐅᕐᖓᖅᑐᖅᑐᓂᒃ.
ᓇᑦᑎᕐᑲᓗᐊᕈᓐᓃᖅᑐᖅ. [...] ᐃᒪᕐᒦᑦᑐᓂ
ᓯᑯᓃᑉᐸᓚᐅᖅᑐᑦ ᐅᑭᐅᓚᐅᖅᑎᓐᓇᒍ.
ᐃᒻᒪᕐᑳ ᓯᑯᐃᑦ ᐊᐅᒃᐸᓕᕐᒪᑕ
ᐊᐱᓚᐅᖅᑎᓐᓇᒍ. ᑕᐃᒪᐃᒻᒪᑦ
ᓄᓇᒧᐊᖅᐸᒃᑐᒃᓴᐅᓕᖅᑐᑦ.”

—ᓘᑲᓯ ᐃᑦᑐᖅ

“The polar bears have problems because there are not many seals around here anymore. When we go hunting I usually see the polar bears eating blackberries. Not many seals anymore. [...] They used to stay on the moving ice in the bay until the winter came. Maybe [moving ice] now melts before the snow. Could be that’s why they come to the land.”

—Lucas A. Etok

“ᐃᖅᑲᓗᐃᑦ
ᐊᖏᓪᓕᓵᓕᓂᖅᓴᐅᕙᓕᖅᑐᑦ.”

—ᐱᐊᑦᓯ ᐋᓇᓈᖅ

“Fish become bigger sooner.”

—Betsie Annanack

“ᓇᓄᐃᑦ ᐊᖕᒋᓪᓕᕙᓪᓕᐊᑐᐃᓐᓇᖅᑐᑦ. [...]
ᕐᑭᒧᒃᓯᒃᑯᑦ ᓇᓄᕐᓂᒃ
ᑕᑯᕙᓚᐅᖅᓯᒪᔪᖕᒐ, ᒫᓐᓇ
ᐊᖕᒋᓂᖅᓴᐅᓕᖅᑐᑦ. ᐊᕐᕌᒍᑕᒫᕐᓗ
ᓇᓄᕐᑲᕐᓂᖅᓴᐅᕙᓪᓕᐊᖕᒋᓐᓇᖅᓱᓂ.”

—ᔮᓂ ᔪᐊᔾᔨ ᐋᓇᓈᖅ

“ᐅᑭᐊᒃᑯᑦ ᑎᖕᒥᐊᓕᐊᓘᕙᓚᐅᖅᓯᒪᔪᖅ
ᓯᒡᓴᓃᑉᐸᒃᑐᓂᒃ. ᑕᑯᕙᒍᓐᓃᖅᑕᕗᑦ.
ᕐᑲᔅᓲᓐᓇᑯᓗᖕᓂᒃ ᑭᓯᐊᓂ.”

—ᓘᑲᓯ ᐃᑦᑐᖅ

“ᑐᒃᑐᕐᑲᓗᐊᕈᓐᓃᖅᑐᖅ
ᐊᕐᑭᒡᒋᕐᑲᓗᐊᕈᓐᓃᖅᓱᓂᓗ. [...]
ᑎᖕᒥᐊᕐᑲᕐᓂᖅᓴᐅᓕᖅᑐᖅ. ᓄᑖᓂᒃ
ᑎᖕᒥᐊᓂᒃ ᑕᑯᓚᐅᖅᓯᒪᙱᑕᑦᑎᓐᓂᒃ
ᑎᑭᑦᑐᕐᑲᖅᐸᓕᖅᑐᖅ. ᐃᓚᖕᒋᑦ
ᕐᑭᕐᓂᖅᑕᐅᓪᓗᑎᒃ ᓱᒃᑲᓪᓗᑎᒃ
ᑎᖕᒋᕙᒃᑐᑦ. ᑎᖕᒥᐊᖕᒍᕙᒃᑐᑐᑦ
ᐊᖕᒋᑎᒋᓪᓗᑎᒃ. [...] ᐊᒥᓱᓂᒃ
ᑕᑯᓯᒪᕗᖕᒐ ᑕᕆᐅᑉ ᓯᓈᓂ.”

—ᓴᐊᕋ ᐹᓴ ᐋᓇᓈᖅ

“ᒪᕐᕉᖕᓂᒃ ᓄᑖᓂᒃ ᑎᖕᒥᐊᓂᒃ
ᑕᑯᓯᒪᓕᖅᑐᖕᒐ. ᕐᑭᕐᓂᖅᑕᓂᒃ ᐊᒻᒪ
ᑕᖅᓴᓇᒃᑐᓂᒃ ᐊᓪᓚᖅᑲᐅᖅᑑᔭᖅᑐᓂᒃ.
ᒥᑭᑦᑐᐃᓐᓇᐅᓪᓗᑎᒃ. ᑎᖕᒥᐊᓪᓗ
ᑐᓵᖕᒋᓐᓇᖅᐸᓚᐅᖅᑕᕗᑦ
ᑐᓴᕐᓇᓗᐊᖅᐸᒍᓐᓃᖅᑐᑦ.
[...] ᓇᓄᐃᓪᓗ ᓄᓇᓕᖕᓄᑦ
ᑎᑭᖅᑲᑦᑕᓪᓗᐊᙱᑦᑑᒐᓗᐊᑦ.
ᑎᑭᖄᖅᐸᓚᐅᖅᓯᒪᙱᑦᑐᑦ.
ᑎᑭᖄᕐᓂᖅᓴᐅᕙᓕᖅᑐᖅ. [...]
ᓇᓂᑐᐃᓐᓈᓘᓕᖅᑐᑦ.”

—ᓰᓯ ᒧᐊᒐᓐ

“*Nanuit* [polar bears] are getting bigger and bigger. [...]On the dogsled I used to see the *nanuit*, but now they are bigger. There are more every year.”

—Johnny George Annanack

“In fall there used to be a lot [of seashore birds]. We don’t see many now. Only a few.”

—Lucas A. Etok

“We see less caribou and ptarmigan. [...] [There are] more birds. There are new species of birds that have never been seen. One is black and flies fast. It’s the same size as the other birds here. [...] I have seen many along the side of the sea.”

—Sarah Pasha Annanack

“I have seen two new birds. One is black and the other has some kind of colours and designs. Both are small birds. And the birds we used to hear a lot, I don’t hear them much. [...] Bears shouldn’t be hanging around town. It was uncommon. Now it is more common. [...] They are everywhere.”

—Susie Morgan

Nunavik / Kangiqsualujjuaq

“ᓴᖅᑭᓵᓕᕋᓕᖅᑐᑦ, ᑭᑦᑐᕆᐊᑦ ᒥᓗᒋᐊᓪᓗ ᐊᑕᐅᑦᑎᒃᑯᑦ ᓴᖅᑭᑉᐸᓕᖅᑐᑦ. ᑭᑦᑐᕆᐊᑦ ᓯᕗᓪᓕᐅᕋᓚᐅᖅᑐᑦ ᒥᓗᒋᐊᓪᓗ ᓴᖅᑭᑉᐸᒃᓯᑎᒃ ᐋᒡᒋᓯᒥ. ᒫᓐᓇ ᐊᑕᐅᑦᑎᒃᑯᑦ ᓴᖅᑭᑉᐸᓕᖅᑐᑦ.”

—ᑏᕕ ᐃᑦᑐᖅ

“ᐃᖃᓗᖃᕐᓂᖅᓴᐅᕆᓕᖅᑐᖅ. ᑲᖑᐃᑦ, ᑐᒃᑐᐃᑦ, ᐃᖃᓗᐃᑦ. [...] ᐃᓛᓐᓂᒃᑯᑦ ᐋᓐᓂᐊᓕᕈᑕᐅᕙᒃᑐᑦ. [...] ᓯᕿᓂᖅ ᓴᙱᔪᐊᓘᓕᖅᑐᖅ. ᐊᓪᓛᑦ ᓂᕐᔪᑎᑦ ᐋᓐᓂᐊᖃᓕᖅᐸᓕᖅᑐᑦ ᓯᕿᓂᕐᒥᑦ, ᐄᖑᔪᒃᓴᐅᔪᖅ. [...] ᐃᒻᒪᖄ ᒪᖓᑮᑦ ᑎᑭᑦᑐᒫᓕᖅᑐᑦ ᐊᓕᐸᐃᓪᓗ!”

—ᑏᕕ ᐃᑦᑐᖅ

“ᓄᓇ ᓂᓚᖃᓗᐊᓚᐅᕐᒪᑦ ᑐᒃᑐᑦ ᓴᓗᒃᑐᒻᒪᕆᐊᓗᐃᑦ, ᓂᕆᔭᒃᓴᖃᓗᐊᓚᐅᙱᒻᒪᑕ.”

—ᐅᐃᓕ ᐃᒪᓪᓚᒃ

“ᐅᔾᔮᕈᓯᓕᖅᑐᖓ ᑎᖕᒥᐊᑦ ᐅᑭᐅᓕᒫᖅ ᑕᒫᓃᑉᐸᓕᕐᒪᑕ, ᐅᑭᐅᖑᓗᐊᒃᑯᑦ ᐅᐱᕐᖔᒃᑯᓪᓗ.”

—ᐅᐃᓕ ᐃᒪᓪᓚᒃ

“They are coming earlier now, and *kitturiat* [mosquitoes] and *milugiat* [black flies] come out at the same time. We used to get *kitturiat* first, then *milugiat* in August. Now they come together.”

—Tivi Etok

“There are more fish coming. There are geese, caribou, fish. [...] They sometimes cause sickness. [...] The sun is real strong. Even the animals are getting sick from the sun, maybe. [...] The monkeys might even come! Elephants, too!”

—Tivi Etok

“Since there was too much ice on the ground, the caribou were very skinny because there was not enough grass.”

—Willie Emudluk

“I notice that the birds stay all year, even when it’s still winter and spring.”

—Willie Emudluk

“ᓯᑯᐃᕌᖕᒐᑦ ᐃᖃᓗᐃᑦ ᐊᑕᐅᑦᑎᒃᑯᑦ ᑎᓯᐊᕙᒃᑐᑦ, ᒥᑭᑦᑐᑦ ᐊᖏᔪᑦᑐ ᐃᖃᓗᐃᑦ. [...] ᑕᐃᔅᓯᒪᓂ ᐊᖏᔪᑦ ᐃᖃᓗᐃᑦ ᒥᑭᑦᑐᓄᑦ ᐊᑕᐅᑦᑎᑯᑦ ᑎᓯᐊᕙᓚᐅᖕᖏᑦᑐᑦ. ᒫᓇ ᐊᑕᐅᑦᑎᒃᑰᖅᐸᓕᖅᑐᑦ. ᑕᐃᒪᐃᓪᓗᐊᖕᖏᑦᑑᒐᓗᐊᖅ, ᐊᖏᓂᖅᓴᑦ ᐃᖃᓗᐃᑦ ᑕᓯᕐᒦᖕᖏᓐᓇᓪᓗᐊᕈᓗᐊᕐᒪᑕ.”

—ᐅᐃᓕ ᐃᒧᓪᓚᒃ

“When the ice breaks, the fish come down together, small and big chars. [...] It was uncommon for the big fish to go down with the small fish. Now they are floating down together. They are not supposed to, because the big fish should stay in the lakes.”

—Willie Emudluk

ᐅᑭᐅᒥ ᐊᓯᔾᔨᖅᑕᕐᓂᖅ

Seasons

“ᑲᖏᖅᓱᐊᓗᔾᔪᐊᑉ ᑰᖓ ᖁᐊᓚᐅᖕᖏᑦᑐᖅ ᑕᒪᑐᒪᓂ ᐅᑭᐅᕐᒥ, ᑭᓯᐊᓂ ᑎᓯᐱᓇᒥ ᓯᑯ ᓵᑦᑐᒻᒪᕆᐅᓚᐅᖅᑐᖅ. ᐅᐱᕐᖕᒑᒃᑯᑦ ᑲᖏᖅᓱᐊᓗᔾᔪᐊᑉ ᑰᖓ ᓯᑯᐃᖅᓵᓂᓚᐅᖅᑐᖅ.”

—ᐄᕙ ᐋᓇᓈᖅ

“George River didn’t freeze [this winter], but in December the ice was thin. In the spring George River broke very early.”

—Eva Annanack

“ᖃᓂᓚᐅᖕᖏᑦᑐᖅ ᑕᒪᑐᒪᓂ ᐅᑭᐅᕐᒥ, ᖃᓂᕋᓗᐊᕌᖕᒐᑦᑐ ᐊᐅᑲᐅᑎᒋᕙᒃᑐᓂ ᓄᓇᒧᑦ ᐊᒃᑐᑐᐊᕋᒥ. [...] ᐊᐱᑦᓚᓇᓚᐅᖕᖏᑦᑐᔪᑦ. ᑕᐃᒪᐃᒻᒪᑦ ᓄᓂᕙᒃᓴᑦ ᐱᕈᖅᓵᓕᓚᐅᖅᑐᑦ, ᐊᐳᑎᖃᑦᑎᐊᓚᐅᖕᖏᓐᓇᑦᑕ.”

—ᔫᓂ ᔪᐊᔾᔨ ᐋᓇᓈᖅ

“It did snow [this winter], but when the snow touched the ground it was already melting. [...] We didn’t get any regular snow. This is why the berries grew faster, because we didn’t have enough snow.”

—Johnny George Annanack

“ᐊᐳᑎᖃᕐᓇᖅᐸᓚᐅᖅᑐᒍᑦ ᐊᐅᔭᓕᒫᖅ. ᓯᑯ ᐊᐅᓚᐅᖅᓯᒪᓇᓂ. ᐃᓛᓐᓂᒃᑯᑦ ᐊᐅᓚᓛᒃᐸᒃᑲᓗᐊᖅᑯᓂ. [...] ᐱᑕᖃᓐᓇᐅᔭᖅᐸᓚᐅᖅᑐᖅ, ᒫᔾᔨᒥᑦ ᔫᓐᓃᑦ, ᐊᐅᔭᒃᑯᔾ ᔫᓐᓃᑦ, ᐅᑭᐅᕈᓗᐊᕐᒪᑦ ᐊᓂᐅᒐᖃᓐᓇᖅᐸᓚᐅᖅᑐᖅ.”

—ᔫᑲᓯ ᐃᑦᑐᖅ

“The snow would stay here all summer. The ice never melted. Sometimes it melted a little bit. [...] It was here all the time, even in March, in summer, even when the [next] winter came.”

—Lucas A. Etok

ᓄᓇᕕᒃ / ᑲᖏᖅᓱᐊᓗᔾᔪᐊᖅ

“ᐊᐅᑦᑖᖀᖕᒐᑦᑕ ᐊᑦᑕᕐᓇᕐᓂᖅᓴᐅᓕᖅᑐᖅ. ᓯᑯ ᓱᖃᒃᑳᖕᒐᑦ ᐊᓄᕆᒧᑦ ᑎᒃᑕᐅᕙᒃᑐᑦ. ᔫᓂᒥ ᓯᑯᖃᕈᓐᓃᖅᐸᒃᑐᖅ. ᓯᑯᐃᖅᓵᓕᓂᖅᓴᐅᕙᓕᖅᑐᖅ ᐆᓇᕐᓂᖕᒐᓗ ᑕᐃᔅᓱᒪᓂᑑᕈᓐᓃᖅᖢᓂ. [...] ᑕᐃᔅᓱᒪᓂ ᓂᒡᓚᓱᖕᓂᖅᓴᐅᕙᓚᐅᖅᑐᖅ.”

—ᒥᐊᓕ ᐃᑦᑐᖅ

“ᐊᐳᑎᖃᙱᓐᓂᖅᓴᐅᓕᖅᑐᖅ. ᓯᑯ ᐃᔾᔪᙱᓐᓂᖅᓴᐅᓕᖅᖢᓂ ᐊᐳᑦᑐ ᐊᐅᓵᓕᓂᖅᓴᐅᕙᓕᖅᖢᓂ. ᐃᓚᖕᒋᑦ ᐃᑳᓇᐊᖕᒋᑦ ᐊᑦᑕᕐᓇᓕᖅᑐᑦ. ᐅᑭᐅᖃᑦᑕᕈᓐᓃᑖᑐᐃᓐᓇᓕᖅᑐᒍᑦ.”

—ᓯᐊᕋ ᐹᓴ ᐋᓇᓈᖅ

“When we go out it’s more dangerous. The ice on the bay cracks and leaves with the wind. In June, there is no ice left. The ice melts earlier now and the temperature is not the same as before. [...] Before it was colder.”

—Mary Etok

“There is less snow. The ice is less thick and the snow melts fast. There are areas that are dangerous to pass. Eventually one day we will not see winter.”

—Sarah Pasha Annanack

ᓯᓚ/ᐊᓄᕆ

“ᐅᐱᕐᖕᒑᒃᑯᑦ ᒪᕐᑯᒐᔪᓚᐅᖅᓯᒪᔪᖅ. [...] ᑕᒪᑐᒪᓂ ᐊᕐᕌᒍᒥ ᒪᕐᑯᓗᐊᓚᐅᙱᑦᑐᖅ. [...] ᑕᐃᔅᓱᒪᓂ ᒪᕐᑯᖃᑦᑕᕐᓂᖅᓴᐅᓚᐅᖅᑐᖅ, ᐅᑭᐊᒃᓵᒃᑯᑦᓘᓐᓃᑦ. ᐃᑖᓐᓂ ᐃᒪᖃᕐᓂᖅᓴᐅᓕᖅᐸᒃᖢᓂ. ᒫᓇᒃᑯᑦ ᐅᑭᐊᒃᓵᒃᑯᑦ ᒪᕐᑯᓗᔪᒍᓐᓃᖅᑐᖅ, ᐃᒪᖃᓗᐊᕈᓐᓃᖅᑐᖅ.”

—ᓘᑲᓯ ᐃᑦᑐᖅ

“ᐊᐅᔭᖅ ᐅᖅᑰᔪᒻᒪᑎᐊᓗᒃ. ᒪᓐᑐᕆᐊᒥᑦ ᐅᖅᑰᓂᖅᓴᐅᒻᒪᕆᒃᑐᖅ. ᓱᓕᔫᖅ. ᐊᓯᔾᔨᓕᖅᑐᖅ, ᑭᐳᓕᖅᑐᖅ. ᐊᐳᑎᓕᒫᕗᑦ ᒪᓐᑐᕆᐊᒧᐊᖕᒑᓚᐅᖅᑐᖅ.”

—ᓲᓯ ᒧᐊᒐᓐ

Climate/Weather

“It used to rain often in the spring. [...] This year we barely had rain. [...] Before it rained more, even in the fall. Sometimes the water level would go up. These years we barely have rain in the fall, so not much water.”

—Lucas A. Etok

“This summer is very hot. It’s even hotter than in Montreal. It’s true. It’s turning, switching. All our snow went to Montreal.”

—Susie Morgan

ᐃᓄᐃᑦ ᐱᖅᑯᓯᑐᖃᖓᑦᑕ ᐊᒃᑐᖅᑕᐅᓂᖓ

"ᐅᑭᐅᒃᑯᑦ ᐅᑭᐊᒃᑯᓪᓗ, ᓯᑯᖓ ᐊᔾᔨᒋᔭᓐᓃᖅᑕᖓ. ᖃᐅᔨᒋᐊᕆᐊᖃᖅᑕᑦ. ᑕᐃᔅᓯᒪᓂ, ᕿᒧᒃᓯᒃᑯᑦ, ᕿᒻᒦᑦ ᖃᐅᔨᒪᕙᓚᐅᖅᑐᑦ ᓯᑯ ᐊᑐᒐᒃᓴᐅᖕᒪᖔᑦ. ᖃᒧᑕᐅᔭᓄᑦ ᖃᐅᔨᔭᒃᓴᐅᙱᑦᑐᖅ."

—ᐱᐊᑦᓯ ᐋᓇᓈᖅ

"ᑎᓂᓯᒪᑎᓪᓗᒍ ᒪᐅᖓ ᒪᑦᑎᑦᑕᐅᓯᖅᕈᑕ ᐊᒥᓱᐊᓗᖕᓂᒃ ᐃᖃᓗᒃᐸᓚᐅᖅᑐᒍᑦ, ᒫᓐᓇ ᐊᔾᔨᒋᔭᓐᓃᖅᑕᖓ. ᐃᒻᒪᖄ ᖃᒧᑕᐅᔭᖃᓗᐊᓕᕐᒪᑕ. ᑲᐸᑎᓯᒪᖅᑰᕐᑎᒍᑦ ᕿᒫᑎᓯᒪᖅᑰᓕᕐᑎᒍᑦ."

—ᓘᑲᓯ ᐃᑦᑐᖅ

"ᐊᖑᓇᓱᖕᓂᖅ ᐊᒃᑐᖅᑕᐅᓯᒪᓕᕐᒥᔪᖅ. ᒫᓂ ᑐᒃᑐᓕᐊᖅᐸᓚᐅᓯᒪᔪᒍᑦ, ᓇᑦᑎᓕᒻᒪᕆᐅᓚᐅᖅᓯᒪᑦᓗᓂᓗ ᑕᕆᐅᕐᒥ. ᐊᕐᑭᒡᒋᑦ ᑕᒪᐅᙵᖅᐸᓚᐅᖅᓯᒪᖕᒥᔪᑦ ᐃᓛᓐᓂᒃᑯᑦ; ᑕᒪᐅᙵᖅᑐᖃᓗᐊᖅᐸᒍᓐᓃᖅᑐᖅ. [...] ᐊᒥᓱᓂᒃ ᓇᑦᑎᕐᓂᒃ ᑕᑯᕙᓚᐅᖅᓯᒪᔪᖓ ᓯᑯᒥ. ᖃᒧᑕᐅᔭᒃᑯᑦ ᓯᑯᓕᐊᕐᓗᐊᕐᒪ ᓇᑦᑎᕐᓂᒃ ᑕᑯᓗᐊᖅᐸᒍᓐᓃᖅᑐᖓ. [...] ᐅᖅᑰᓂᖓᓅᓇᓱᒋᔭᕋ. ᐅᑭᐅᖅᑕᖅᑐᖅ ᐊᓯᔾᔨᕐᓂᐊᓕᖅᑐᖅ."

—ᓘᑲᓯ ᐃᑦᑐᖅ

Impacts on Traditional Ways of Life

"In the winter and fall, the ice is not the same. You have to check. Before, with dogsleds, the dogs knew if the ice was safe. With skidoos, no."

—Betsie Annanack

"At high tide we would put out our nets here and we would catch lots of fish, but it's not the same now. Maybe [because] there's so many machines going around. We scared them off."

—Lucas A. Etok

"Hunting is affected too. We used to have caribou here, and lots of seals in the ocean there. Ptarmigan used to come here sometimes. Sometimes nothing; sometimes they come; sometimes they barely come. [...] I used to see lots of seals on the ice. When I go there by skidoo I barely see any seal. [...] I think it's from the heat. Everything is going to change. The Arctic is going to change."

—Lucas A. Etok

“ᐅᐱᕐᖕᒑᒃᑯᑦ ᓯᑯᕐᑲᕐᒃᐸᒍᓐᓃᕐᒃᑐᕐᒃ. [...] ᓯᑯᑦ ᑕᓯᕐᓂ ᑲᖕᒋᕐᒃᓱᐊᓗᔾᔪᐊᑉ ᑰᖕᒐ-ᒥᓪᓗ ᐊᒃᑐᕐᒃᑕᐅᓯᒪᓂᕐᒃᑐᑦ, ᓇᓂᓕᒫᕐᒃ. ᑭᓯᓕᒫᑦ ᐊᐅᓚᐅᕐᒃᑐᑦ. ᐅᑭᐅᒃᑯᑦ ᐊᐳᑎᕐᑲᖖᒋᓐᓂᕐᒃᓴᐅᓂᕐᒃᑐᕐᒃ. ᐃᕐᑲᓗᖕᓂᒃ ᐊᒃᑐᕐᒃᑯᓴᐅᔾᔪᐊᕐᒃᑎᑦᑎᒋᐊᓚᐅᕋᓗᐊᕐᒪᑕ, ᐊᐅᓪᓚᕈᓐᓇᓚᐅᖖᒋᑦᑐᑦ ᑭᓯᓕᒫᑦ ᐊᐅᒃᓯᒪᓕᓚᐅᕐᒪᑕ.”

—ᑏᕕ ᐃᑦᑐᕐᒃ

“ᑕᒪᑐᒪᓂ ᐅᑭᐅᒥ ᑕᓰᑦ ᓄᓇᒥ ᐃᔾᔪᕐᒃᓯᓚᐅᖖᒋᑦᑐᑦ. [...] ᐃᕐᑲᓪᓕᐊᕐᑲᑦᑕᕈᓐᓃᖏᐊᕐᑲᓚᐅᕐᒃᑐᒍᑦ ᐊᑦᑕᕐᓇᓗᐊᓕᓚᐅᕐᒪᑦ.”

—ᐅᐃᓕ ᐃᒪᓪᓚᒃ

“In spring there was no more ice. [...] It affected the ice in the lakes and George River, everywhere. Everything has melted. In the winter there was less snow. There was a fishing contest, but they could not go because everything melted.”

—Tivi Etok

“This winter the ice on the tundra lakes was not very thick. [...] We had to stop going fishing because it was too dangerous.”

—Willie Emudluk

ᑲᖏᖅᓱᔪᐊᖅ
Kangiqsujuaq

ᑲᖏᖅᓱᔪᐊᖅ

ᑲᖏᖅᓱᔪᐊᖅ 10 ᑭᓛᒥᑐ ᐅᖓᓯᖕᓂᕐᑲᖅᑐᖅ ᕐᑭᑭᖅᑖᓘᑉ ᑯᐸᐃᑦᑐ ᐃᑭᕋᓴᖓᓐᓂᑦ (N 61.60°, W 71.96°; ᕐᑯᑦᑎᖕᓂᖓ 13 m). ᐅᑭᐅᖅᑕᖅᑐᕐᒥ ᐊᖅᐸᓯᒃᑐᓄᑦ ᐃᓚᒋᔭᖅ, ᐅᖅᐱᐊᓛᓂᒃ ᐱᕈᖅᑐᕐᑲᖅᑐᓂ ᓄᓇᐅᑦᑐ ᐃᑭᐊᖓ ᕐᑯᐊᖑᐃᓐᓇᖅᑐᓂ. 1960-ᒥ ᓄᓇᓕᙱᓚᐅᖅᓯᒪᔪᖅ 696-ᓂᒡᓗ ᐃᓄᒋᐊᖕᓂᕋᖅᑕᐅᑦᑐᓂ 2011-ᒥ ᑲᓇᑕᒥ ᑭᒡᓕᓯᓂᐊᖅᑕᓂᑦ.

ᑲᖏᖅᓱᔪᐊᕐᒥ, ᐊᖏᕐᑲᑎᒌᒻᒪᕆᓚᐅᖅᑐᑦ ᐅᖅᐱᒐᑦ ᐅᓄᕐᓂᖅᓴᐅᓕᕐᓂᖏᓐᓂᒃ, ᐱᑐᐊᖅᑐᒥ ᐅᖅᐲᑦ, ᑕᑭᓂᖅᓴᐅᕙᓕᖅᑐᑎᒡᓗ. ᐃᓚᖏᑦ ᐊᐱᖅᓱᖅᑕᐅᔪᑦ ᐅᔾᔨᕈᓯᓚᐅᖅᑐᑦ ᓄᓂᕙᒃᑕᐅᕙᒃᑐᑦ ᐅᓄᕐᓂᖅᓴᐅᓕᕐᓂᖏᓐᓂᒃ, ᑕᐃᒪᐃᓐᓂᕋᖅᑕᐅᑦᑐᓂᓗ ᐊᐳᑎᕐᑲᕐᓂᖅᓴᐅᓕᕐᓂᖓᓄᑦ. ᐸᐅᓐᓃᑐᐃᑦ ᐊᒥᓱᕈᒃᑲᓐᓂᖅᓯᒪᔪᑦ.

ᐊᒥᓱᑦ ᐊᓯᔾᔨᖅᓯᒪᓕᖅᑐᑦ ᐅᔾᔨᕆᔭᐅᔪᑦ ᐆᒪᔪᑦ ᐅᓄᕐᓂᖏᓐᓂᒃ—ᐆᒃᑑᑎᒋᓗᒍ ᑐᒃᑐᑭᓐᓂᖅᓴᐅᓕᖅᑐᖅ, ᓇᑦᑎᑭᓐᓂᖅᓴᐅᓕᖅᑐᖅ, ᓇᓄᐃᑦᑐ ᑕᑯᔭᐅᒐᔪᖕᓂᖅᓴᐅᓕᖅᑐᑎᒃ ᓄᓇᓕᖕᓂ. ᐊᒻᒪᓗ, ᓄᑖᓂᒃ ᐆᒪᔪᓂᒃ ᑎᑭᑦᑐᕐᑲᖅᐸᓕᕐᓂᖓᓂᒃ, ᓲᕐᓗ ᐊᒃᓱᐃᑦ, ᐅᒥᖕᒪᐃᑦ, ᐃᓛᕈᓰᑦ, ᕌᐱᓐ (ᐃᕙᕝᕕᒃᑐᑎᒃ), ᕐᑭᕐᓂᑦ ᕐᑯᒡᔪᑦᑐ ᒥᑏᑦ, ᐊᒻᒪ ᑭᑦᑐᓰᑦ ᓄᑖᑦᑐ ᐃᒍᑦᑕᐃᑦ.

ᐊᒥᓱᑦ ᓯᓚᒧᑦ ᐊᓯᔾᔨᖅᓯᒪᔪᑦ ᐊᒻᒪ ᐃᒪᕐᑲᕐᓂᖅ ᐅᔾᔨᕆᔭᐅᓯᒪᔪᑦ ᑲᖏᖅᓱᔪᐊᕐᒥ. ᐊᐱᖅᓱᖅᑕᐅᔪᑦ ᐊᖏᕐᑲᑎᒌᓚᐅᖅᑐᑦ

Kangiqsujuaq

Kangiqsujuaq is located on the southeast shore of Wakeham Bay, 10 kilometres from Hudson Strait (N 61.60°, W 71.96°; elevation 13 m). It is part of the Low Arctic, with erect dwarf-shrub tundra vegetation and continuous permafrost. It was established in 1960 and its population was estimated at 696 people in the 2011 Statistics Canada Census.

In Kangiqsujuaq, there was strong consensus that shrubs are becoming more abundant, particularly willows, which are also growing taller. Some participants observed increases in berry abundance, and attributed this to more snow. Tall fireweed has recently become more abundant.

Many changes were noticed in mammal abundance—for example, declining caribou stocks, fewer seals, and polar bears being more frequently sighted around the community. In addition, new animal species have been seen in this area, such as black bears, muskoxen, porcupines, robins (laying eggs), and a black-and-white duck, as well as dragonflies and new bumblebee species.

ᐊᓄᓇᑐᓂᖅᓯᐅᕙᓕᕐᓂᖕᒪᓂᒃ ᐊᓄᖀᒪᔪᖕᓂᖅᓯᐅᓕᕐᓂᖕᒪᓂᒡᓗ. ᐊᓯᖕᒋᑦ ᐅᔾᔨᕆᔭᐅᔪᑦ ᓯᑯ ᑕᓯᕐᓂᓗ ᓯᑯᐃᖅᓵᓕᕙᓕᕐᓂᖕᒪᓂᒃ, ᓯᑯ ᓵᓐᓂᖅᓯᐅᓕᕐᓂᖕᒪᓂᒃ, ᐊᐳᑎᑭᓐᓂᖅᓯᐅᓕᕐᓂᖕᒪᓂᒃ, ᐃᒪᑭᓐᓂᖅᓯᐅᓕᕐᓂᖕᒪᓂᒃ, ᓯᓚᐅᑦᑐ ᖃᓄᐃᓐᓂᖕᒪ ᐊᔾᔨᒌᖖᒋᓐᓂᖅᓯᐅᕙᓕᕐᓂᖕᒋᓐᓂᒃ. ᐊᐅᔭᖅ ᐊᒃᑐᖅᑕᐅᓂᖅᑲᕐᓂᖅᐸᐅᓂᕋᖅᑕᐅᔪᖅ, ᓂᒡᓚᔪᖕᓂᖅᓯᐅᓕᖅᓱᓂ ᐊᐅᔭᑯᑖᖕᒍᓂᖅᓯᐅᓕᖅᓱᓂᓗ. ᐊᐱᖅᓱᖅᑕᐅᔪᑦ ᐅᖃᓚᐅᕐᒥᔪᑦ ᒪᕐᑯᒐᔪᖕᓂᖅᓯᐅᓕᕐᓂᖕᒪᓂᒃ.

ᐊᐱᖅᓱᖅᑕᐅᔪᑦ ᐅᖃᓚᐅᖅᑐᑦ ᐊᖅᑯᑎᒋᔭᐅᕙᒃᑐᖅᑲᑦ ᐊᐅᓪᓛᕐᓂᕐᒧᑦ ᐊᖑᓇᓱᖕᓂᕐᒧᓪᓗ ᐊᒃᑐᖅᑕᐅᓯᒪᓂᖅᑲᕐᓂᖕᒋᓐᓂᒃ ᐊᔾᔨᒌᖖᒋᑦᑐᓄᑦ ᓯᓚᐅᑉ ᐊᓯᔾᔨᕐᓂᖕᒪᓄᑦ, ᐊᐅᓪᓛᕐᓂᖅ ᓄᓇᒃᑯᑦ ᐃᒪᒃᑯᓪᓗ ᐊᒃᓱᕈᕐᓇᕐᓂᖅᓯᐅᓕᕐᓂᖕᒪᓂᒃ. ᐊᓯᔾᔨᕐᓂᐅᔪᑦ ᐊᐳᑎᑭᓐᓂᖅᓯᐅᓕᕐᓂᖕᒪᓄᑦ, ᑕᓰᑦ ᓯᑯᐃᖅᓵᓕᕙᓕᕐᓂᖕᒋᓐᓄᑦ, ᐊᓄᓇᑐᓂᖅᓯᐅᓕᕐᓂᖕᒪᓄᑦ, ᐱᕈᖅᑐᓪᓗ ᐱᕈᕐᓂᖅᓯᐅᕙᓕᕐᓂᖕᒋᓐᓄᑦ.

Many changes in climate and hydrology have been observed in Kangiqsujuaq. Interviewees agreed that the wind is stronger and more frequent. Other observations included earlier sea ice and lake ice breakup, thinner sea ice, less snow accumulation, lower water levels, and greater variability in the weather. Summer is perceived as the most affected season, tending to be colder and longer. Participants also described that rainfall has increased.

Interview participants mentioned that traditional travel routes are being affected by several environmental changes, making it more difficult to travel on land and water. These changes include lack of snow, earlier lake ice melting, stronger winds, and increased shrub growth.

ᐃᓐᓇᐃᑦ ᐊᐱᖅᓱᖅᑕᐅᔪᑦ
Elders Interviewed

ᒥᐊᓕ ᐊᓄᒐᖅ
Mary
Anogak

ᐄᕙ ᐃᓕᒪᓴᐅᑦ
Eva
Ilimasaut

ᓕᕕ
ᐊᕐᓇᐃᑦᑐᖅ
Livi
Arnaituk

ᓕᓯ ᐃᕐᓂᖅ
Lizzie
Irniq

ᓇᑉᐹᓗᒃ
ᐊᕐᓇᐃᑦᑐᖅ
Nappaaluk
Arnaituk

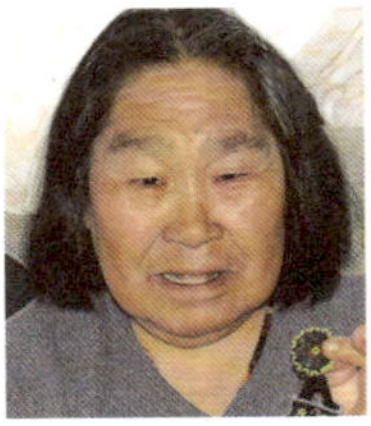

ᒥᐊᓕ
ᑭᐊᑕᐃᓐᓇᖅ
Mary
Kiatainaq

ᔨᐊᓯᑲ
ᐊᕐᖓᖅ
Jessica
Arngaq

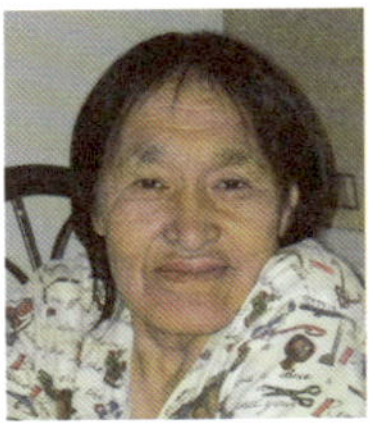

ᐋᓚᓯ ᑯᓂᐊᖅ
Alasie Koneak

ᒥᓂ ᐃᑎᑦᓗᐃ
Minnie
Etidloie

ᔫᓯᐱ ᓇᑉᐹᓗᒃ
Juusipi
Nappaaluk

ᓘᑲᓯ ᓇᑉᐹᓗᒃ
Lukasi
Nappaaluk

ᐋᓂ ᑎᖅᑎᓗᒃ
Annie
Tertiluk

ᓈᓚ ᓇᑉᐹᓗᒃ
Naalak
Nappaaluk

ᒫᒃ ᑎᖅᑎᓗᒃ
Mark
Tertiluk

ᐱᑦᓯᐅᓛᖅ
ᐱᖑᐊᖅᑐᖅ
Pitsiulaq
Pinguatuq

ᒫᑕ ᑐᓂᒃ
Maata
Tuniq

ᐊᖁᔮᖅ ᕿᓰᖅ
Aqujaq Qisiiq

ᐃᓄᓗᒃ ᕿᓰᖅ
Inuluk
Qisiiq

ᐸᐅᖅᖓᐃᑦ

“ᒪᓕᑦᓱᐊᒐᐃᑦ ᐱᑕᖅᑲᐹᓐᓃᑲᐅᑎᒋᕙᓕᖅᑐᑦ. ᐊᑯᓂᐊᓗᒃ ᐱᐅᖅᑲᑦᑕᓚᐅᖅᓯᒪᔪᑦ. [...] ᓄᑕᐅᑎᓪᓗᒋᑦ ᐱᐅᒻᒪᑕ, ᑭᓯᐊᓂ ᖃᔅᓰᓐᓇᓄᑦ ᐱᓇᓱᐊᕈᓯᕐᓄᑦ ᓄᑕᐅᕙᓕᖅᑐᑦ. ᐊᑯᓂᒃᑲᓐᓂᖅ ᓄᑕᐅᕙᓚᐅᖅᑑᒐᓗᐊᑦ.”

—ᐊᖁᔭᖅ ᕿᓰᖅ

“ᓄᓂᕙᒐᑦ ᐊᖏᔪᐊᓘᓕᖅᑐᑦ. ᑕᐃᔅᓱᒪᓂ ᒥᑭᑦᑐᑯᓘᕙᓚᐅᖅᓯᒪᔪᑦ, ᑭᓯᐊᓂ ᐊᐅᔭᖅ ᐃᒪᐃᓕᖅᑎᓪᓗᒍ ᐊᖏᓕᕙᓪᓕᐊᑐᐃᓐᓇᖅᑐᑦ, ᑕᒪᕐᒥᒃ. ᐊᖅᐲᑦ, ᐸᐅᖅᖓᐃᑦ, ᑭᒻᒥᓐᓇᐃᑦ, ᑭᒍᑕᖏᕐᓇᐃᓪᓗ. ᑭᒍᑕᖏᕐᓇᐃᑦ ᓄᓇᓕᖕᓂ ᐱᑕᖅᑲᕐᓂᖅᓴᐅᔪᑦ. ᒪᒃᑯᒃᑎᓪᓗᖓ ᒥᑭᒃᑐᓂᒃ ᑭᒍᑕᖏᕐᓇᕐᑯᑎᖅᑲᖅᐸᓚᐅᖅᑐᖅ, ᒫᓐᓇ ᓄᓂᕙᒋᐊᕌᖓᒪ ᐊᖏᔪᐊᓘᓕᖅᑐᑦ.”

—ᓕᓯ ᐃᕐᓂᖅ

Berries

“*Malitsuagait* [seabeach sandwort] are gone right away. They used to be good for a long time. [...] When they’re fresh they’re good, but they’re only fresh a few weeks. Usually they stay fresh for a longer time.”

—Aqujaq Qisiiq

“The berries are getting very big. They used to be small a long time ago, but this time of year they are getting bigger and bigger, all four of them. *Arpiit* [cloudberries/ bakeapples], *paurngait* [blackberries/crowberries], *kimminait* [cranberries], *paurngaqutiit* [blackberry plants]. *Kigutangirnait* [blueberries] are more common around the village. When I was younger there were small berry plants, but when I go now they are very big.”

—Lizzie Irniq

ᐱᕈᖅᑐᑦ ᐊᓯᖏᑦ

"ᐅᐱᕐᙶᑕᒫᑦ ᐊᐅᔭᖅᑕᒫᓪᓘᓐᓃᑦ ᓄᑖᓂᒃ ᐱᕈᖅᑐᓂᒃ ᑕᑯᕙᓕᖅᑐᒍᑦ. [...] ᑐᖑᔪᕐᓂᖅᓴᐅᓪᓗᑎᒃ ᓄᑕᐅᙱᓐᓂᖅᓴᓂᒃ. ᑕᑯᓯᒪᓕᖅᑕᒃᑲ ᓇᓂᓕᒫᑲᓴᒃ. ᐃᒐᔾᔪᑎᒋᕙᒃᑕᕗᑦ. ᓄᑖᓂᒃ ᐱᕈᖅᑐᓂᒃ ᑕᑯᕙᓕᖅᑐᒍᑦ, ᐊᒥᓲᓂᖅᓴᓂᒃ ᐱᕈᖅᑐᓂᒃ ᐱᕈᖅᑐᖅᑲᖅᐸᓕᖅᑐᖅ."

—ᔫᓯᐱ ᓇᑉᐹᓗᒃ

"ᐱᕈᖅᑐᑦ ᐱᕈᖅᓵᓕᓂᖅᓴᐅᕙᓕᖅᑐᑦ ᐊᐳᑦ ᐊᐅᒃᓵᓕᓂᖅᓴᐅᕙᓕᕐᒪᑦ."

—ᔫᓯᐱ ᓇᑉᐹᓗᒃ

Other Plants

"Every spring or summer we see new plants. [...] The colour is greener than the older ones. I've seen them almost everywhere. We use them for fires. We see new plants, more plants starting to grow."

—Juusipi Nappaaluk

"The plants are starting to grow earlier because the snow melts faster."

—Juusipi Nappaaluk

ᐆᒪᔪᑦ

"ᓄᑖᒥᒃ ᑎᖕᒥᐊᒥᒃ ᐊᖏᔪᒥᒃ ᐱᑕᖅᑲᓕᖅᑐᖅ. ᖃᓄᖅ ᑕᐃᓂᐊᕆᐊᖓ ᓇᓗᔭᕋ. ᓄᓇᓕᖕᒥ ᑕᑯᓚᐅᖅᓯᒪᙱᓐᓇᒃᑯ. ᒪᕐᕉᖕᓂᒃ ᐊᖏᔪᐊᓘᖕᓂᒃ ᑎᖕᒥᐊᖕᓂᒃ ᑕᑯᓚᐅᖅᑐᖓ, ᐊᖏᔪᑦ ᕿᒻᒥᑐᑦ ᐊᖏᑎᒋᓪᓗᓂ. ᑲᔫᓚᐅᖅᑑᒃ, ᑕᑭᔪᐊᓗᖕᒥᒃ ᖁᖑᓯᖅᑲᖅᖢᑎᒃ. [...] ᒥᑎᐅᔮᖅᑐᖅ, ᑭᓯᐊᓂ ᒥᑎᐅᙱᑦᑐᖅ. ᐃᒻᒪᖃ ᒥᑎᐅᓗᐊᖅ ᖃᑦᓗᓈᑦ ᓄᓇᖓᓂᑦ ᐅᑭᐅᖅᑕᖅᑐᒧᑦ ᑎᑭᓯᒪᓕᖅᑐᖅ."

—ᐊᖁᔭᖅ ᕿᓰᖅ

Animals

"We have a new big bird. I don't know what to call it. I never saw it before in town. I saw two big birds, almost the same size as a big dog. They were brown, with a long, long neck. [...] It looks like a duck, but it's not a duck. Maybe it's a duck from the south that moved to the North."

—Aqujaq Qisiiq

"I used to see the birds in the springtime, but I'm not seeing them anymore."

—Eva Ilimasaut

“ᑎᖕᒥᐊᓂᒃ ᐅᐱᕐᖔᒃᑯᑦ ᑕᑯᕙᓚᐅᖅᓯᒪᔪᖓ, ᑭᓯᐊᓂ ᑕᑯᕙᒍᓐᓃᖅᑕᒃᑲ.”

—ᐄᕙ ᐃᓕᒪᓴᐅᑦ

“ᑕᑯᖅᑲᑦᑕᖅᓯᒪᔪᖓ ᐃᒍᑦᑕᕐᓂᒃ ᐱᖓᓐᓇᕐᒥᑦ ᐊᓄᕆᓚᐅᖅᑎᑦᓱᒍ. ᑕᒫᓂ ᐃᒍᑦᑕᐅᕙᒃᑐᑦ ᐊᔾᔨᒋᙱᑕᖕᒋᓐᓂᒃ. ᑭᑦᑐᕆᐊᑦ ᒥᓗᒋᐊᑦᓗ ᓴᖅᑭᓵᓕᕙᓕᖅᑐᑦ, ᔫᓂ ᓄᙳᐊᓂ, ᑭᓯᐊᓂ ᑕᒪᑐᒪᓂ ᑭᑦᑐᕆᐊᕐᒥ ᑕᑯᓚᐅᖅᑐᖓ ᒪᐃᒥ. ᐱᖅᑲᑎᒐ ᑕᑯᓚᐅᕐᓂᕋᓚᐅᖅᑐᖅ ᐄᑉᐳᕈᒥ. [...] ᐃᒻᒪᖄ ᐆᒻᒪᖅᓵᓕᕙᓕᕐᒪᑕ.”

—ᔫᓯᐱ ᓇᑉᐹᓗᒃ

“ᑰᑦ ᐃᒃᑲᓗᐊᓕᖅᐸᑦᓕᐊᔪᑦ. ᐃᖅᑲᓗᐃᑦ ᑐᖁᕙᓕᖅᑐᑦ. ᒪᕐᕉᒃ ᑰᒃ ᖃᐅᔨᒪᔮᒃᑲ ᐃᐱᐅᕋᖅ ᐃᖅᑲᓗᕕᒡᓗ, ᐃᖅᑲᓗᐃᑦ ᑐᖁᕙᒃᑐᑦ ᐅᑭᐅᒃᑯᑦ. ᐃᒪᖅ ᐃᖅᑲᖓᓄᑦ ᖁᐊᓚᐅᕐᒪᑦ. ᐃᒪᖓ ᑕᐅᓇᓂ ᐃᒪᖅᑲᕈᓐᓃᖅᐸᑦᓕᐊᓕᖅᑐᖅ.”

—ᔫᓯᐱ ᓇᑉᐹᓗᒃ

“ᓇᓄᐃᑦ ᓯᑯᒥ ᓂᕿᒃᓴᖅᓯᐅᖅᐸᒃᑑᒐᓗᐊᑦ, ᑭᓯᐊᓂ ᓯᑯᐃᖅᓵᓕᓗᐊᖅᐸᓕᖅᑐᖅ. ᐅᑦᓗᒥ ᓇᓄᐃᑦ ᐊᐅᑦᓛᓄᑦ ᑎᑭᕌᖅᐸᓕᖅᑐᑦ ᓂᕿᒃᓴᖅᓯᐅᖅᓱᓂᑎᒃ.”

—ᓘᑲᓯ ᓇᑉᐹᓗᒃ

“I’ve seen *igutsait* [bumblebees] after the wind blew from the west. Different than the ones we have here. *Kitturiat* [mosquitoes] and *milugiat* [black flies] are starting to fly sooner, at the end of June, but this year I saw a mosquito in May. My friend told me he saw one in April. [...] Maybe they are starting to hatch earlier.”

—Juusipi Nappaaluk

“The rivers are getting too shallow. Fish die. I know two rivers, Ipiuaq and Iqaluvik, where some fish died in the winter. The water froze to the bottom. The water down there is drying up.”

—Juusipi Nappaaluk

“The polar bears usually hunt on the ice, but the ice is gone so early. Today the polar bears come into camps to look for food.”

—Lukasi Nappaaluk

Nunavik / Kangiqsujuaq

“ᓱᓕ ᓇᐅᔭᖃᖅᐸᒃᑐᖅ ᔮᓄᐊᕆᒥ ᕕᕈᐊᕆᒥᓪᓗ. ᑳᒃᑐᐊᓘᕙᒃᑐᑦ. [...] ᓇᐅᔭᓄᑦ ᐊᔪᕐᓇᓕᖅᑐᖅ. ᐊᒻᒪ, ᐱᓇᓱᐊᕈᓲᒃ ᒪᕐᕉᓚᐅᖅᑑᒃ, ᐊᐅᓪᓛᖅᓯᒪᖕᕕᓐᓃᑎᓪᓗᖓ, ᒥᑎᕐᒥᒃ ᑕᑯᓚᐅᖅᓯᒪᙱᑕᓐᓂᒃ ᑕᑯᓚᐅᖅᑐᖓ. ᕿᕐᓂᖅᑑᓚᐅᖅᑐᖅ. ᓯᕗᓪᓕᖅᐹᒥ ᐅᑭᐅᖅᑕᖅᑐᒥ ᑕᑯᓚᐅᖅᑕᕋ. [...] ᕿᕐᓂᖅᑐᒥᒃ ᒥᑎᕐᒥᒃ ᑕᑯᓚᐅᖅᓯᒪᔪᖓ ᓯᓕᒦᑎᓪᓗᖓ, ᑕᐃᒪᐃᑦᑐᒥᒃ ᑕᑯᖅᑰᓚᐅᖅᑐᖓ ᐱᓇᓱᐊᕈᓯᐅᓚᐅᖅᑑᒃ ᒪᕐᕉᖕᓂ. ᓱᒃᑲᐃᑦᑐᒥᒃ ᐅᑭᐅᖅᑕᖅᑐᒧᑦ ᑎᑭᐸᑦᓕᐊᖅᑰᔨᓕᖅᑐᑦ.”

—ᓘᑲᓯ ᓇᑉᐹᓗᒃ

“ᕿᓚᓗᒐᖃᙱᓐᓂᖅᓴᐅᔪᖅ ᑕᒪᑐᒪᓂ ᐊᕐᕌᒍᒥ. [...] ᓯᑯᖃᓗᐊᖅᐸᑦ ᑕᓯᐅᔭᕐᔪᐊᑉ ᐃᑭᕋᖓᓂ, ᓯᑯᓂᒃ ᓄᓈᖅᑐᖃᖃᑦᑕᕐᓂᖅᓴᐅᔪᖅ. ᑕᒪᑐᒪᓂ ᓯᑯᐃᖅᓵᓕᓗᐊᓚᐅᖅᑐᖅ, ᓯᑯᐃᑦ ᑕᑉᐸᐅᖓᖅᑐᐃᓐᓇᐅᓚᖅᑐᑦ. ᐃᒃᐱᐊᕐᔪᒃ ᓯᑯᖃᓗᐊᕌᖓᑦ ᓄᓇᒃᑯᑦ ᐊᐅᓪᓛᖅᐸᒃᑐᑦ. ᓯᑯᖃᙱᑦᑕᕌᖓᑦ ᐃᒪᒃᑰᖅᐸᒃᑐᑦ.”

—ᓘᑲᓯ ᓇᑉᐹᓗᒃ

“ᐊᒥᓱᐊᓗᖕᓂᒃ ᓂᕐᔪᑎᖃᓕᖅᑐᖅ! [...] ᐱᓗᐊᖅᑐᒥᒃ ᑎᖕᒥᐊᑦ. ᐊᕐᕌᒍᑕᒫᑦ ᑎᑭᐸᒃᑐᑦ ᖃᑉᓗᓈᑦ ᓄᓇᖓᓂᑦ. [...] ᐃᓚᖏᑦ ᐊᖏᔪᐊᓗᖕᓂᒃ ᓂᐊᖃᖅᖢᑎᒃ ᑕᑭᔪᓂᓪᓗ ᖁᖓᓯᖃᖅᖢᑎᒃ. [...] ᕿᕐᓂᖅᑐᓂᓪᓗ ᐊᒃᖤᓂᒃ ᑕᑯᖃᑦᑕᓕᕐᒥᔪᑦ.”

—ᒫᒃ ᑎᖅᑎᓗᒃ

“We still see seagulls in January and February. They are hungry. [...] For the seagull it's difficult. Also, two weeks ago, when I was at the camp, I saw a duck that I had never seen before. It was a black duck. First time I've seen it in the North. [...] I saw a black duck in Chile, and I think I saw it up here two weeks ago. It seems to be coming north slowly.”

—Lukasi Nappaaluk

“We had much less beluga this year. [...] If there is too much ice in Hudson Strait, we get more coming to the shore. [This year] we lost the ice too early, so they travelled right up north. When the bay is full of ice, they travel around the land. When there is no ice, they just travel straight [through].”

—Lukasi Nappaaluk

“Lots [of new animals]! [...] Especially birds. Every year they come from the south. [...] Some have big legs and a long neck. [...] We're starting to have black bears, too.”

—Mark Tertiluk

“ᑐᒃᑐᑦ ᐊᒪᕐᒃᑯᑦᑐ
ᐱᑕᕐᑲᑐᐊᓚᐅᕐᒃᓯᒪᙱᑦᑐᑦ.
ᒫᓐᓇ ᐊᒪᕐᒃᑯᑦ ᑐᒃᑐᐃᑦᑐ
ᐅᓄᕐᓂᕐᒃᓴᐅᓕᕐᒃᑐᑦ. ᐊᒻᓛᑦ ᓇᓄᐃᑦ
ᒥᑏᑦᑐ ᐅᓄᕐᓂᕐᒃᓴᐅᓕᕐᒃᑐᑦ.
ᐃᕐᑲᓗᓕᒻᒪᕆᐅᓕᕐᒥᔦᕐᒃ,
ᑭᓯᐊᓂ ᐃᕐᑲᓗᐃᑦ ᑐᕐᑯᖁᓕᕐᒃᑐᑦ
ᐅᔪᖁᖕᓂᐊᕐᒃᐸᒃᑐᓄᑦ.”

—ᒥᐊᓕ ᐊᓄᒐᕐᒃ

“There were fewer caribou and wolf a long time ago. Now there are more wolf and caribou. Even the *nanuk* [polar bear] and *mitiq* [eider duck], too. There's a lot of fish, too, but a lot of fish are dying because of the mining.”

—Mary Anogak

“ᑐᒃᑐᑦ ᑎᑭᒃᑳᖕᒐᑕ ᑭᑦᑐᕆᐊᑦ
ᐊᑕᐅᑦᑎᒃᑯᑦ ᓴᕐᒃᑭᑉᐸᒃᑐᑦ.
ᒫᓐᓇ ᑐᒃᑐᑭᓐᓂᕐᒃᓴᐅᓕᕐᒃᑐᕐᒃ
ᑭᑦᑐᕆᐊᑭᓐᓂᕐᒃᓴᐅᓕᕐᒃᖢᓂᓗ.”

—ᐱᑦᓯᐅᓛᕐᒃ ᐱᖖᒍᐊᕐᒃᑐᕐᒃ

“When the caribou arrive, the mosquitoes arrive at the same time. Now there are less caribou, less mosquitoes.”

—Pitsiulaq Pinguatuq

“ᑕᓰᑦ ᐃᓚᖕᒋᓐᓂ ᐃᕐᑲᓗᖕᓂᒃ
ᑕᑯᕙᓕᕐᒃᑐᖕᒐ ᑕᑯᓚᐅᕐᒃᓯᒪᙱᑕᓐᓂᒃ.
ᐊᐅᐸᕐᒃᑐᓂᒃ ᑕᕐᒃᓴᑦᓛᔦᓂᒃ. [...]
ᐃᕐᑲᓗᑭᓐᓂᕐᒃᓴᐅᓕᕐᒃᖢᓂᓗ.”

—ᐱᑦᓯᐅᓛᕐᒃ ᐱᖖᒍᐊᕐᒃᑐᕐᒃ

“In some lakes I have seen a fish that I have never seen before. With red spots. [...] There is less Arctic char.”

—Pitsiulaq Pinguatuq

“ᓇᓄᐃᑦ ᑳᒃᑐᐊᓘᓕᕐᒃᑐᑦ. ᒥᑏᑦ
ᒪᓐᓂᖕᒋᓐᓂᒃ ᓂᕆᕙᓕᕐᒃᑐᑦ.
ᓯᑯᐃᕐᒃᓵᓕᓗᐊᕐᒃᐸᓕᕐᒪᑦ. [...] ᒥᑏᑦ
ᒪᓐᓂᖕᒋᑦ [...] ᐱᑕᕐᑲᕈᓐᓃᕐᒃᑐᑦ
ᓇᓄᕐᓄᑦ ᓂᕆᔭᐅᔦᐃᓐᓇᐅᕙᓕᕐᒪᑕ.”

—ᐱᑦᓯᐅᓛᕐᒃ ᐱᖖᒍᐊᕐᒃᑐᕐᒃ

“Polar bears are very hungry. They eat all the *mitiq*'s eggs. The ice is melting too fast. [...] The eggs of the *mitiq* [...] are all gone because the polar bears ate them all.”

—Pitsiulaq Pinguatuq

“ᑐᒃᑐᐃᑦ ᓴᓄᒃᑐᐊᓘᓕᕐᒃᑐᑦ
ᐊᑯᓂᐊᓗᒃ ᐱᓱᒃᐸᓕᕐᒪᑕ
ᓂᕆᔭᒃᓴᕐᒃᓯᐊᕈᓐᓇᑦᑎᐊᖁᑎᒃ.
ᒥᑭᑦᑐᑯᓘᓕᕐᒃᑐᑦ ᓴᓄᒃᑐᑯᓘᓕᕐᒃᖢᑎᒡᓗ.
ᑕᐃᒪᐃᑉᐸᓚᐅᕐᒃᓯᒪᙱᑦᑑᒐᓗᐊᑦ.”

—ᐱᑦᓯᐅᓛᕐᒃ ᐱᖖᒍᐊᕐᒃᑐᕐᒃ

“The caribou are very skinny because they walk a long time and they hardly find [any] food. They are very small and skinny. [They didn't] used to be like that.”

—Pitsiulaq Pinguatuq

ᐅᑭᐅᒥ ᐊᓯᔾᔨᖅᑕᕐᓂᖅ

“ᐅᑐᐱᕆᒥ ᑕᓰᑦ ᖁᐊᓚᐅᙱᑦᑐᑦ. […] ᓄᓇᐅᑉ ᐃᓗᐊ ᐆᓇᕐᓂᖅᓴᐅᓚᐅᕐᒪᑦ. ᓯᓚ ᐅᖅᑰᓂᖅᓴᐅᕙᓕᖅᑐᖅ ᓄᓇᒥᑦ, ᓯᓚᒥᐅᙱᑦᑐᖅ.”

—ᐊᖁᔭᖅ ᕿᓰᖅ

“ᑕᐃᔅᓯᒪᓂ, ᐅᑐᐱᕆᒥ ᖃᒧᑕᐅᔭᒃᑯᑦ ᐊᐅᓪᓛᖅᐸᓚᐅᖅᑐᑦ, ᑭᓯᐊᓂ ᐅᑐᐱᕆᒃᑯᑦ ᓄᓇᓕᐊᖅᐸᒍᓐᓃᖅᑐᑦ. ᑕᐃᒪᐃᑦᑐᓐᓇᖅᐸᓚᐅᖅᓯᒪᒐᓗᐊᖅᑐᑦ.”

—ᐊᖁᔭᖅ ᕿᓰᖅ

“ᐅᑭᐊᖑᓚᐅᖅᑐᒥ ᐊᐳᑎᖃᙱᓗᐊᓚᐅᖅᑐᖅ. ᐊᑯᓂᐊᓗᒃ ᑕᐃᒪᓐᓇ ᐊᐳᑖᕐᓇᓵᓚᐅᖅᑐᖅ. […] ᐃᓄᐃᑦ ᐊᖑᓇᓯᒋᐊᕈᒪᓚᐅᖅᑑᒐᓗᐊᑦ ᑭᓯᐊᓂ ᐊᐳᑎᖃᙱᓗᐊᓚᐅᖅᑐᖅ.”

—ᐄᕙ ᐃᓕᒪᓴᐅᑦ

“ᓯᓚᐅᑉ ᓂᒡᓚᓯᖕᓂᖓ ᐅᑭᐅᒃᑯᑦ, ᐊᔾᔨᒋᔪᓐᓃᖅᑕᖓ. […] ᑕᐃᔅᓯᒪᓂᓂᑦ ᓂᒡᓚᓯᖕᓂᖅᓴᐅᔮᓕᖅᑐᖅ. ᐃᓄᐃᑦ ᐅᖃᖅᐸᒃᑐᑦ ᐅᖅᑰᓯᕙᓪᓕᐊᓕᕐᓂᖓᓂᒃ ᑭᓯᐊᓂ ᑕᐃᒪᐃᙱᑦᑐᖅ, ᑕᐃᔅᓯᒪᓂᓂᑦ ᓂᒡᓚᓯᖕᓂᖅᓴᐅᓕᕐᒪᑦ. ᓯᓚᐅᑉ ᐆᓇᕐᓂᖓ ᐃᒃᐱᖕᓇᕈᓐᓃᖅᑐᖅ—ᐱᓗᐊᕈᓐᓃᖅᑐᖅ. ᓯᕿᓐᓂᕋᓗᐊᖅᑎᓪᓗᒍ ᐆᓇᖅᐸᒍᓐᓃᖅᑐᖅ. […] ᑕᒪᔾᔭ ᐊᓯᔾᔨᖅᓯᒪᓂᐅᔪᖅ ᓯᓚᐅᑉ ᓂᒡᓚᓯᖕᓂᖓᓄᑦ. ᓂᒡᓚᓯᓕᖅᑐᖅ, ᓯᑯ ᐊᐳᓪᓗ ᐊᐅᒃᓵᓕᓂᖅᓴᐅᕙᓕᖅᓱᑎᒃ.”

—ᔫᓯᐱ ᓇᐹᓗᒃ

Seasons

“In October the lakes were not frozen. […] The reason is because the weather was warmer under the ground. The warmer temperatures are coming from the ground, not the air.”

—Aqujaq Qisiiq

“In the past, in October we would go on the land by snowmobile, but we never go out on the land in October anymore. We used to be able to do that.”

—Aqujaq Qisiiq

“There was not enough snow last fall. We didn’t have any snow for a long time. […] People wanted to go hunting, but there was not enough snow.”

—Eva Ilimasaut

“The feeling of the cold when it’s winter, it’s not the same anymore. […] It seems to be cooler than before. Some people say it’s warming up, but it’s not, because it’s colder than before. You don’t feel warm air—not too much. The sun is shining, but it’s not really hot. […] That’s the change in the feeling of the temperature.

“ᐊᐱᓵᓕᓚᐅᖅᑐᖅ, ᓯᑎᐱᕆ ᐱᒋᐊᕐᓂᖓᓂ ᐊᐱᑦᑐᓂ, ᐃᓚᖏᑦᑐ ᓄᓂᕙᒃᓴᑦ ᐱᕈᑦᑎᐊᓚᐅᕋᑎᒃ. ᐊᐳᑦ ᐊᐅᖕᒪᑦ ᐊᐱᒃᑲᓂᓚᐅᙱᑦᑐᖅ ᔮᓐᓄᐊᕆ ᕿᑎᐊᓄᑦ ᑎᑭᕐᓗᒍ. […] ᓂᒡᓚᓱᓐᓂᖅᓴᐅᓚᐅᖅᑐᖅ ᐊᐳᑎᖃᓚᐅᙱᒻᒪᑦ.”

—ᓘᑲᓯ ᓇᑉᐹᓗᒃ

“ᐊᐱᕙᒃᑰᒐᓗᐊᖅ ᓯᑎᐱᕆ ᓄᙳᐊᓂ ᐅᑐᐱᕆᒥᑦ ᔫᓐᓂᑦ, ᑭᓯᐊᓂ ᒫᓐᓇ ᐊᐱᕙᓕᖅᑐᖅ ᓄᕕᐱᕆᒥ ᑎᓯᐱᕆᒥᑦ ᔫᓐᓃᑦ. ᐊᓪᓛᑦ ᐊᐱᒐᓗᐊᖅᐸᑦ ᓯᑎᐱᕆᒥ ᐅᑐᐱᕆᒥᑦ ᔫᓐᓃᑦ ᐊᓄᕆᒧᑦ ᑎᒃᑕᐅᑐᐃᓐᓇᓲᖑᓕᖅᑐᑦ. ᐊᐳᑎᖃᐃᓐᓇᐸᙱᑦᑐᖅ. ᓂᒡᓚᓱᓐᓂᖅᓴᐅᓕᖅᑐᖅ, ᐅᔾᔨᕆᓯᒪᔭᕋ, ᓂᒡᓚᖅᐸᓕᖅᑐᖅ ᑎᓯᐱᕆᒥ ᑭᓯᐊᓂ. […] ᓂᒡᓚᖅᓵᓕᓂᖅᓴᐅᕙᓚᐅᖅᑰᒐᓗᐊᖅ.”

—ᐋᓂ ᑎᖅᑎᓗᒃ

“ᐃᒃᐱᐊᕐᔪᖕᒥ ᓯᑯ ᐊᐅᒃᓵᓕᕙᓕᖅᑐᖅ. ᓱᒃᑲᓗᐊᖅᑐᒥᒃ.”

—ᒥᐊᓕ ᑭᐊᑕᐃᓐᓇᖅ

It’s cold, and the ice and snow melt fast.”

—Juusipi Nappaaluk

“We had early snow, arriving in the beginning of September, and some of the berries didn’t grow very well. The snow melted and didn’t come back until the middle of January. […] It was much colder than usual because there was no snow.”

—Lukasi Nappaaluk

“Normally snow would come at the end of September or October, but now [it comes in] November or December. Even now, if we get it at the end of September or October, it keeps blowing away. It’s not going to stay. The cold weather, I have noticed, doesn’t come until late December. […] It used to come much earlier.”

—Annie Tertiluk

“The ice [on the bay] melts very early. Too fast.”

—Mary Kiatainaq

“ᑕᐃᔅᓱᒪᓂᑐᑦ ᖃᓐᓂᖅᐸᒍᓐᓃᖅᑐᖅ. ᑕᐃᔅᓱᒪᓂ ᐊᐅᓪᓚᕈᓐᓇᖅᓯᓵᓕᕙᓚᐅᕋᓗᐊᖅᑐᒍᑦ. [...] ᐅᑭᐅᓄᑦ ᖁᓕᓄᑦ ᐊᐳᑎᖅᑲᖏᓐᓂᖅᓴᐅᓯᒪᓕᖅᑐᖅ. ᑕᐃᔅᓱᒪᓂ ᖃᒧᑕᐅᔭᒃᑯᑦ ᐊᐅᓪᓚᕈᓐᓇᖅᐸᓚᐅᖅᑐᒍᑦ ᓄᕕᐱᕆᒥ—ᒫᓐᓇ ᑭᓯᐊᓂ ᑎᓯᐱᕆᒥ ᔮᓐᓄᐊᕆᒥᓪᓘᓐᓃᑦ.”

—ᐱᑦᓯᐅᓛᖅ ᐱᖑᐊᖅᑐᖅ

“It snows less than before. Before we could travel earlier. [...] For about 10 years [there has been less snow]. Before [we could travel by skidoo] in November—now December or even January.”

—Pitsiulaq Pinguatuq

ᓯᓚ/ᐊᓄᕆ

Climate/Weather

“ᑕᐃᔅᓱᒪᓂᓂᑦ ᐃᒪᑭᓐᓂᖅᓴᒻᒪᕆᐅᓕᖅᑐᖅ. ᐅᑭᐅᒃᑯᑦ ᐊᐳᑎᖅᑲᖏᓗᐊᓕᖅᒪᑦ ᐊᐅᔭᒃᑯᓪᓗ ᒪᖁᖅᑲᑦᑕᖏᓗᐊᓕᖅᔪᓂ. ᐊᒡᓛᑦ ᑕᓯᖅᔪᐊᖅ ᒥᑭᓪᓕᕙᓪᓕᐊᓕᖅᑐᖅ.”

—ᐊᖁᔭᖅ ᕿᓰᖅ

“The water is much, much lower than before. The reason is because we didn’t receive much snow or rain in the summer. Even the big lake is getting lower.”

—Aqujaq Qisiiq

“ᔮᓄᐊᕆᒥ ᖃᓐᓂᓗᐊᓚᐅᖏᑦᑐᖅ; ᐅᑐᐱᕆᒥᑦ ᒫᔅᓴᒧᑦ ᖃᓐᓂᓚᐅᖏᑦᑎᐊᖅᑐᖅ. ᓯᕗᓪᓕᖅᐹᒥ ᑕᐃᒪᓐᓇ ᐊᑯᓂᐅᑎᒋᔪᖅ ᖃᓐᓂᓚᐅᖏᑦᑐᖅ, ᑕᖅᑭᑦ ᑕᓪᓕᒪᑦ. ᑕᐃᒪᐃᑦᑐᖅ ᑕᑯᓚᐅᖅᓯᒪᖏᑦᑕᕋ.”

—ᐊᖁᔭᖅ ᕿᓰᖅ

“It didn’t snow much in January; from October up to March, we didn’t receive the snow at all. That’s the first time we didn’t get snow for such a long time, in five months. I’ve never seen that before.”

—Aqujaq Qisiiq

“ᓯᓚ ᐊᑯᓂᐊᓗᒃ
ᐊᓯᔾᔩᖅᑕᖅᐸᓚᐅᙱᑦᑐᖅ, ᐅᓪᓗᒐᓴᓄᑦ
ᐱᓇᓱᐊᕈᓯᕐᓄᑦ ᑑᓐᓃᑦ, ᑭᓯᐊᓂ ᒫᓐᓇ
ᐊᓯᔾᔩᑳᓪᓚᒃᐸᓕᖅᑐᖅ.”

—ᐄᕙ ᐃᓕᒪᓴᐅᑦ

“ᐊᓄᓇᑐᓂᖅᓴᐅᕙᓕᖅᑐᖅ
ᑕᐃᔅᓯᒪᓂᓂᑦ. ᖃᐅᔨᒪᓇᖅᐸᒍᓐᓃᖅᑐᖅ.
ᐅᓪᓗᒥ ᓯᓚᑦᑎᐊᕙᐅᔪᖅ ᑭᓯᐊᓂ
ᐃᑲᕐᕋᒥ ᓯᓚᓗᓕᕈᓐᓇᖅᑐᖅ.”

—ᔫᓯᐱ ᓇᑉᐹᓗᒃ

“ᒪᕐᑯᖃᑦᑕᕈᓐᓃᕋᓱᒋᕙᕋ,
ᑕᓯᐅᕙᓚᐅᖅᑐᑦ ᐃᒪᖃᕈᓐᓃᕐᒪᑕ.
[…] ᓄᑲᑉᐱᐊᙳᑎᓪᓗᖓ
ᐃᒪᖃᖅᐸᓚᐅᖅᑐᑦ, ᒫᓐᓇ
ᐃᒪᖃᕈᓐᓃᖅᑐᑦ. ᒪᕐᑯᓚᐅᖅᑎᓪᓗᖑ
ᐃᒪᖃᓕᖅᐸᒃᑑᒐᓗᐊᑦ, ᑭᓯᐊᓂ
ᐅᓪᓗᐃᑦ ᖃᔅᓰᓐᓇᐃᑦ ᐊᓂᒍᖅᑎᓪᓗᒋᑦ
ᐃᒪᖃᕈᓐᓃᖅᐸᒃᑐᑦ. ᐃᒪᖓ
ᓅᓇᒧᐊᖅᐸᓕᖅᑐᖅ ᐊᑖ
ᖁᐊᙳᔭᓐᓂᕐᒪᑦ. ᖃᐅᔨᒪᔭᖓ ᐃᓚᖏᑦ
ᑰᑦ ᐃᒃᑲᒡᓕᓯᒪᓕᕐᓂᖏᓐᓂᒃ.”

—ᔫᓯᐱ ᓇᑉᐹᓗᒃ

“ᐊᑖᑕᒐ ᐊᑖᑕᑦᑎᐊᕆᓚᐅᖅᑕᖁᓗ
ᖃᐅᔨᒪᕙᓚᐅᖅᑑᒃ ᓯᓚ
ᖃᓄᐃᓪᓘᓂᐊᕐᒪᖔᑦ ᓯᓚᒥᒃ
ᕿᒥᕐᕈᓪᓗᑎᒃ. […] ᒫᓐᓇ ᓯᓚ
ᓱᒃᑲᔪᐊᓗᒻᒥᒃ ᐊᓯᔾᔨᓕᖅᑐᖅ.
ᒫᓐᓇ ᓇᓗᓇᖅᐸᓕᖅᑐᖅ ᓯᓚ
ᖃᓄᐃᓪᓘᓂᐊᕐᒪᖔᑦ.”

—ᓕᓯ ᐃᕐᓂᖅ

“It used to be that the weather would be stable for a long time, days and weeks, but now it changes quickly.”

—Eva Ilimasaut

“The wind is getting stronger than before. It doesn’t give warning anymore. Today is a nice day, but in one hour it could be really bad.”

—Juusipi Nappaaluk

“I don’t think it’s raining more, because I see dry lakes. […] When I was a boy there was water, but now they are dry. After the rain they will be back. It will fill with water after the rain, but after a few days it dries very fast. The water is leaking out because the bottom isn’t frozen. I know some rivers, too, are getting shallow.”

—Juusipi Nappaaluk

“My father and grandfather used to say what the weather was going to be like by looking outside. […] Now the weather is changing very fast. Now it’s hard to say what the weather will be like.”

—Lizzie Irniq

Nunavik / Kangiqsujuaq

“ᐊᐅᔭᖅ ᒫᖁᖅᑲᑦᑕᖅᑐᐊᓘᓚᐅᖅᑐᖅ.
[…] ᓄᓇᒧᑦ ᐱᐅᔪᓐᓃᕈᑕᐅᓚᐅᖅᑐᖅ.
ᓇᓂᑐᐃᓐᓈᓗᒃ ᐃᒪᖅᑲᓕᓚᐅᖅᑐᖅ.”

—ᓘᑲᓯ ᓇᑉᐹᓗᒃ

“ᐃᒻᒪᖄ 6 7-ᓘᓐᓃᑦ ᐊᖅᑮᔪᓚᐅᖅᑐᑦ
ᐊᐳᑦ ᐊᐅᒃᐸᑦᑕᐊᖑᔭᖅᐸᓚᐅᖅᑐᖅ.
ᐅᐱᕐᖔᒃᑯᑦ ᐊᐳᑦ ᐅᓐᓄᐊᒃᑯᑦ
ᖁᐊᖅᓯᐊᓇᕆᓚᐅᖅᑕᕗᑦ […]
ᑎᓯᒃᑎᑦᑎᐊᖅᑐᐊᓘᕆᓚᐅᖅᑐᖅ.
ᐃᖏᕐᕋᒡᒐᒃᓴᑦᑎᐊᕆᐅᓪᓗᓂ,
ᖃᒧᑏᒃ ᖃᒧᑕᐅᔭᓪᓗ
ᐃᖏᕐᕋᑦᑎᐊᕈᓐᓇᖅᕗᓂᑎᒃ ᒫᐅᔭᐅᓇᓂ
ᐊᐳᒻᒥ, ᑭᓯᐊᓂ ᖃᖓᑦᑎᐋᔅᓂᑦ
ᑕᐃᒪᐃᑦᑐᓐᓃᖅᑐᖅ.”

—ᐋᓂ ᑎᖅᑎᓗᒃ

“ᐊᓄᓇ ᐱᖕᖓᓐᓂᒐᔪᖕᓂᖅᓴᐅᓕᖅᑐᖅ
ᑕᐃᓯᒪᓂᓂᑦ. ᐊᒻᒪ
ᓂᒡᒋᒐᔪᓕᖅᕗᓂ. ᑕᐃᓯᒪᓂᓂᑦ
ᐊᓂᓇᑐᓂᖅᓴᐅᓕᖅᐸᑦᑕᐊᑐᐃᓐᓇᖅᑐᖅ.”

—ᓇᑉᐹᓗᒃ ᐊᕐᓇᐃᑦᑐᖅ

“ᑕᒪᑐᒪᓂ ᐊᓄᓇᔪᐊᓘᖅᑲᑦᑕᓚᐅᖅᑐᖅ,
ᖃᓐᓂᖅᐸᒃᕗᓂ, ᒫᖁᒃᐸᒃᕗᓂ
ᓇᑕᖅᑯᕐᓇᖅᐸᒃᕗᓂᓗ ᐊᑕᐅᑦᑎᒃᑯᑦ.”

—ᓕᕕ ᐊᕐᓇᐃᑦᑐᖅ

“ᐊᐳᑎᖃᕐᓂᖅᓴᐅᕆᓕᖅᑐᖅ
ᐅᐱᕐᖔᒃᑯᑦ. ᐅᖃᖅᑐᖅᑲᖅᐸᒃᑑᒐᓗᐊᖅ
ᓄᓇᕐᔪᐊᕐᒥ ᐅᖅᑰᓯᕆᑦᑕᐊᑕᕐᓂᖕᒐᓂᒃ,
ᑭᓯᐊᓂ ᐅᒃᐱᓇᙱᑕᕋ ᓂᓚᓯᒃᑐᐊᖕᒪᑦ.
ᓂᓚᓯᐃᓐᓇᐅᔭᖅᑐᖅ.
ᐅᖅᑰᓯᕆᑦᑕᐊᙱᑦᑐᖅ.”

—ᓕᕕ ᐊᕐᓇᐃᑦᑐᖅ

“We had so much rain this summer. […] It caused a lot of damage to the land. Lots of water everywhere.”

—Lukasi Nappaaluk

“Maybe six or seven years ago, [the snow] used to melt gradually. In the springtime we would wait for the snow to freeze at night […] and it would become really nice and hard. This is when we wanted to travel, because the sled and skidoo could move fast and not [get stuck] in the snow, but since a few years ago now we don’t have that.”

—Annie Tertiluk

“[The wind] blows often from the north, more than before. Also more wind from the south. Stronger and stronger than before.”

—Nappaaluk Arnaituk

“[The wind] comes from the north—northeast wind. It is not usual. It used to be from the south. […] [This year] it was very windy, snowing, raining, and hailing all at the same time.”

—Livi Arnaituk

“ᐅᔭᖅᑲᓂᒃ ᑎᓯᐊᔪᓂᒃ ᑕᑯᐃᓐᓇᐅᔭᓕᖅᑐᖓ. [...] ᑕᐃᒪᐃᒐᔪᙱᖅᓴᐅᓕᖅᑐᖅ ᐊᐅᒃᓵᓕᓗᐊᖅᐸᓕᕐᒪᑦ ᓄᓇ, ᐊᐳᑦ.”

—ᐱᑦᓯᐅᓛᖅ ᐱᖑᐊᖅᑐᖅ

“There is more snow when the spring comes. They say it’s getting warmer in the world, but I don’t believe that because it’s cold. It’s cold all the time. It’s not getting warm.”

—Livi Arnaituk

“I see rock slides all the time. [...] [It’s happening] more now because it melts too fast on the earth, the snow.”

—Pitsiulaq Pinguatuq

ᐃᓄᐃᑦ ᐱᖅᑯᓯᑐᖅᑲᖓᑦᑕ ᐊᒃᑐᖅᑕᐅᓂᖓ

“ᑐᒃᑐᑦ ᐊᐅᓪᓚᖅᓵᓕᕋᓕᖅᑐᑦ ᑎᑭᓐᓇᓵᖅᐸᒃᓱᑎᒡᓗ. ᑎᑭᑉᐸᓪᓕᐊᓕᖅᐸᒃᑐᑦ ᒫᓐᓇᒃᑯᑦ [ᔪᓚᐃᒥ] ᒪᐃᒥᐅᙳᒋᑦᑐᖅ, ᑕᖅᑮᖕᓄᒃ ᒪᖅᑮᖕᓄᒃ ᑭᖑᕙᓯᒃᓱᑎᒃ. ᓄᓇᕗᖕᒍᐊᖅᐸᓕᖅᑐᑦ ᑕᖅᑭᓕᒫᖅ ᓯᕗᕙᖅᓱᑎᒃ. [...] ᑕᒫᓃᔅᓰᔭᒃᑲᓐᓂᕐᓂᐊᖅᐳᑦᑐᑦ [ᑕᒪᑐᒪᓂ] ᑐᒃᑐᖅᑲᑦᑕᕐᓂᐊᕐᑦᑕ ᖁᐊᖅᑎᑦᓗᒋᑦ ᐅᑭᐅᒧᑦ ᓂᕿᒃᓴᑦᑎᓐᓂᒃ.”

—ᓘᑲᓯ ᓇᑉᐹᓗᒃ

Impacts on Traditional Ways of Life

“The caribou go out early and come late. They start to arrive now [in July] instead of May, two months late. They go out to the land one month early. [...] I hope they stay here a bit longer [this year] so we can catch the caribou and freeze it and use it for the winter.”

—Lukasi Nappaaluk

“ᐃᓄᐃᑦ ᓄᓇᖅᐸᒃᑳᖕᒐᑕ,
ᐅᑎᕐᓯᒋᐊᒃᓴᖅ
ᐱᒐᓇᕐᓂᖅᓴᐅᓂᕋᖅᑕᐅᓕᖅᑐᖅ ᐊᐳᑦ
ᐊᐅᒃᓵᓕᓗᐊᖅᐸᓕᕐᒪᑦ.”

—ᓇᑉᐹᓗᒃ ᐊᕐᓇᐃᑦᑐᖅ

“ᐊᖑᓇᓱᒋᐊᒃᑲᐣᓂᕈᒪᔪᖕᒐ, ᐊᐳᑦ
ᐊᐅᒃᓵᓕᓗᐊᓚᐅᕐᒪᑦ [ᑕᒪᑐᒪᓂ
ᐅᑭᐅᕐᒥ]. ᐊᐳᑎᕐᑲᙱᓗᐊᖅᑐᖅ
ᐊᐅᓪᓛᕐᕕᒋᒋᐊᖕᒐ.”

—ᐱᑦᓯᐅᓛᖅ ᐱᙳᐊᖅᑐᖅ

“When the people go out on the tundra, they say it is more difficult to come back because the snow melts [too early].”

—Nappaaluk Arnaituk

“I want to go back out hunting, because the snow melted too fast [this winter]. Not enough snow to travel.”

—Pitsiulaq Pinguatuq

ᓄᓇᑦᓯᐊᕗᑦ Nunatsiavut

ᓇᐃᓂ
Nain

ᓇᐃᓂ

ᓇᐃᓂ ᓛᐸᑐᐊᕇᒦᑦᑐᖅ (N 56°33, W 61.41°) ᖁᑦᑎᖕᓂᖅᐸᐅᓪᓗᓂᓗ ᓄᓇᑦᓯᐊᕗᒻᒥ ᓄᓇᓕᖕᓂᒃ, 200 ᑭᓛᒥᑐᓂᒃ ᐅᖓᓯᖕᓂᖅᑲᖅᑐᖅ ᓇᐹᖅᑐᖅᑲᕐᕕᖕᒥᑦ. ᓄᓇᒥ ᐱᕈᖅᑐᑦ ᐊᔾᔨᒌᒃᐸᖖᒋᑦᑐᑦ ᓄᓇᐅᑉ ᖅᑲᕐᕿᓯᖕᓂᖓ ᒪᓕᒡᓗᒍ, ᓇᐹᖅᑐᖅᑲᖅᓱᓂ ᐅᖅᐱᒐᖅᑲᖅᓱᓂᓗ, ᓄᓇᐅᓪᓗ ᐃᑭᐊᖓ ᖁᐊᖑᐃᓐᓇᖅᐸᒐᓂ. ᓄᓇᓕᖖᒍᖅᑎᑕᐅᓚᐅᖅᓯᒪᔪᖅ ᒧᕆᕕᐊᓐ ᑐᒃᓯᐊᕐᕕᓕᕆᔨᒃᑯᓐᓄᑦ 1771-ᒥ, ᓄᓇᓕᖕᓂ ᐃᓄᒋᐊᖕᓂᖅᐸᐅᓕᖅᑐᖅ ᓄᓇᑦᓯᐊᕗᒻᒥ, ᐃᓄᒋᐊᖕᓂᖅᑲᖅᓱᓂ 1,188-ᓂᒃ 2011-ᒥ ᑲᓇᑕᒥ ᑭᒡᓕᓯᓂᐊᖅᑕᓂᑦ.

ᐊᕐᕌᒍᓂ ᐊᓂᒎᓴᖅᑐᓂ, ᓇᐃᓂᒥᐅᑦ ᐅᔾᔨᕈᓱᒃᓯᒪᔪᑦ ᐊᒥᓱᓂᒃ ᓄᓇᒥ ᑕᕆᐅᕐᒥᓗ ᐊᓯᔾᔨᖅᓯᒪᓕᖅᑐᓂᒃ - ᐊᓯᔾᔨᕐᓂᖕᒋᓐᓄᑦᓗ ᐃᓅᓯᕐᒥᓄᑦ ᐃᓕᖅᑯᓯᕐᒥᓄᑦ ᐊᒃᑐᖅᑕᐅᓯᒪᓂᖅᑲᖅᓱᑎᒃ. ᐊᐱᕆᔭᐅᒐᒥᒃ ᐱᕈᖅᑐᑦ ᐊᓯᔾᔨᖅᓯᒪᓕᕐᓂᖕᒋᓐᓂᒃ, ᑕᒪᕐᒥᑲᓴᑦᑎᐊᑦ ᐊᐱᖅᓱᖅᑕᐅᔪᑦ ᐅᕐᑲᓚᐅᖅᑐᖅ ᐱᕈᖅᑐᖅᑲᕐᓂᖅᓴᐅᓕᕐᓂᖓᓂᒃ. ᐱᓗᐊᖅᑐᒥᒃ ᓇᐹᖅᑐᖅᑲᕐᓂᖅᓴᐅᓕᕐᓂᖓᓂᒃ ᐅᖅᐱᒐᕐᓂᖅᓴᐅᓕᕐᓂᖓᓂᒃ (ᐅᖅᐱᖓᓂᒃ ᐅᖅᐱᐊᓛᓂᒡᓗ), ᐱᕈᖅᓵᓕᓂᖅᓴᐅᕙᓕᕐᓂᖅᑕᐅᓪᓗᑎᒃ ᑕᑭᓂᖅᓴᐅᕙᓕᖅᓱᑎᓪᓗ. ᐃᓚᖓᓂ ᐅᖅᐱᒐᐃᑦ ᐊᖅᑯᑎᕿᔭᐅᕙᒃᑐᓂ ᐱᕈᖅᐸᓕᕐᓂᖕᒋᓐᓄᑦ ᐊᖅᑯᑕᐅᕙᒍᓐᓃᕐᓂᖕᒋᓐᓂᒃ. ᑕᒪᕐᒥᑲᓴᒃ ᐊᐱᖅᓱᖅᑕᑦ ᐅᔾᔨᕈᓱᒃᓯᒪᔪᑦ ᓄᓂᕙᒃᑕᐅᕙᒃᑐᑦ ᐊᕐᕌᒍᑕᓖᑦ ᐅᓄᙱᓐᓂᖅᓴᐅᕙᑦᓕᐊᖕᒋᓐᓇᕐᓂᖕᒋᓐᓂᒃ;

Nain

Nain is located along the Labrador coast (N 56°33, W 61.41°) and is the northernmost permanent community in Nunatsiavut, approximately 200 kilometres south of the latitudinal treeline. The vegetation transitions from forest to shrub tundra with increasing elevation, and is underlain with discontinuous permafrost. It was established by the Moravian missionaries in 1771, and is now the largest communities in Nunatsiavut, with an estimated population of 1,188 people as of the 2011 Statistics Canada Census.

In recent years, community members in Nain have observed numerous changes in land and sea—changes that have affected their traditional ways of life. When asked about changes in vegetation, nearly all the participants stated that plants are more abundant now. This is especially true for trees (spruce and tamarack) and shrubs (willow and birch), which are said to be growing more rapidly and getting taller. In some cases, shrubs have overgrown traditional travel routes on the land. Nearly all interview participants observed an

ᑭᓯᐊᓂ ᖃᓄᐃᑦᑐᑦ ᓄᓂᕙᒃᑕᐅᕙᒃᑐᑦ ᐅᓄᕈᓐᓃᖅᐸᑦᓕᐊᖕᒪᖔᖏᑦ ᑭᒡᒍᑕᐅᔪᑦ ᐊᔾᔨᒌᖃᑦᑕᓚᐅᙱᑦᑐᑦ.

ᐃᓄᑐᖃᐃᑦ ᐊᐱᖅᓱᖅᑕᐅᔪᑦ ᓇᑉᐸᖏᓐᓂ ᐅᓄᕐᓂᖅᓴᑦ ᐅᔾᔨᕈᓱᓚᐅᖅᑐᑦ ᐊᖅᐲᑦ ᑭᒍᑕᖏᕐᓇᐃᑦᓗ ᐅᓄᖏᓐᓂᖅᓴᐅᓕᕐᓂᖏᓐᓂᒃ, ᐸᐅᕐᖓᐃᑦ ᐅᓄᕈᓐᓃᓕᕐᒥᔫᒐᓗᐊᑦ ᐊᓯᖏᓐᓂᒃ ᐱᙱᓐᓂᖅᓴᐅᑦᓗᑎᒃ. ᐊᒻᒪᓗ, ᐊᒥᓱᑦ ᐃᓄᑐᖃᐃᑦ ᐅᖃᓚᐅᖅᑐᑦ ᓄᓂᕙᒐᒃᓴᑦ ᒥᑭᓐᓂᖅᓴᐅᓕᕐᓂᖏᓐᓂᒃ, ᑎᐱᖏᑦᓗ ᐊᓯᔾᔨᓕᕐᓂᖏᓐᓂᒃ. ᐊᔾᔨᒌᙱᑦᑐᓂᒃ ᐃᓱᒪᒋᔭᖃᓚᐅᖅᑐᑦ ᓄᓂᕙᒐᒃᓴᑦ ᐱᕈᐊᓂᒃᐸᖕᓂᖏᑦ ᐊᓯᔾᔨᖅᓯᒪᓕᕐᒪᖔᑦ, ᑭᓯᐊᓂ ᐃᓚᖏᑦ ᐃᓄᑐᖃᐃᑦ ᐅᖃᓚᐅᖅᑐᑦ ᐱᕈᐊᓂᖕᓇᓵᕐᓂᖅᓴᐅᕙᓕᕐᓂᖏᓐᓂᒃ.

ᑕᒪᕐᒥᑲᓴᒃ ᐊᐱᖅᓱᖅᑕᐅᔪᑦ ᐊᖏᖃᑎᒌᓚᐅᖅᑐᑦ ᐆᒪᔪᑦ ᐅᓄᕐᓂᖏᑦ ᐊᓯᔾᔨᓕᕐᓂᖏᓐᓂᒃ. ᐱᓗᐊᖅᑐᒥᒃ, ᐅᔾᔨᕈᓱᒃᓯᒪᔪᑦ ᑐᒃᑐᐃᑦ ᓇᑦᑏᑦᓗ ᐅᓄᙱᓐᓂᖅᓴᐅᓕᕐᓂᖏᓐᓂᒃ ᐱᐅᖏᓐᓂᖅᓴᐅᓕᕐᓂᖏᑦᓗ. ᐆᒃᑑᑎᒋᓗᒍ, ᐃᓚᖏᑦ ᐃᓄᑐᖃᐃᑦ ᐅᖃᖅᓯᒪᔪᑦ ᑐᒃᑐᐃᑦ ᐸᑎᖏᑦ ᑕᖏᑭᓐᓂᖅᓴᐅᓕᕐᓂᖏᓐᓂᒃ, ᐊᒡᓛᑦ ᐃᓚᖏᑦ ᑕᑯᓯᒪᔪᑦ ᐸᑎᕐᓂᒃ ᐊᐅᓕᖕᓂᒃ. ᐊᒃᖤᐃᑦ ᑕᑯᔭᐅᒐᔪᖕᓂᖅᓴᐅᓕᖅᑐᑦ ᓇᐃᓂᐅᑉ ᒥᒃᓯᖓᓂ ᓇᓄᐃᑦᓗ ᓄᓇᓕᖕᒧᑦ ᖃᒡᓕᓂᖅᓴᐅᕙᓕᕐᓂᖏᓐᓂᒃ ᐅᓄᕐᓂᖅᓴᐅᓕᖅᖢᑎᒡᓗ. ᐃᓚᖏᑦ ᐅᖃᓚᐅᕐᒥᔪᑦ ᓄᑖᓂᒃ ᑎᖕᒥᐊᖅᑕᖃᖅᐸᓕᕐᓂᖓᓂᒃ, ᓲᕐᓗ ᒥᑭᑦᑐᓂᒃ ᕿᕐᓂᓂᒃ ᑎᖕᒥᐊᓂᒃ. ᐊᓇᙲᑦ ᕿᒃᑐᕆᐊᑦᓗ ᐅᓄᕐᓂᖏᑦ ᐊᐱᖅᓱᖅᑕᐅᔪᑦ ᑭᒡᒍᑎᖏᓐᓂ ᐊᔾᔨᒌᒃᐸᓚᐅᙱᑦᑐᑦ.

ᑕᒪᕐᒥᑲᓴᒃ ᐊᐱᖅᓱᖅᑕᐅᔪᑦ ᐅᖃᓚᐅᖅᑐᑦ ᓯᓚᐅᑉ ᖃᓄᐃᓐᓂᖓ ᐊᔾᔨᒌᙱᓐᓂᖅᓴᐅᕙᓕᕐᓂᖓ

overall decrease in the number of berries produced each year; however, opinions on what berry species are declining varied among participants. More than half of Elders observed declines in cloudberries/bakeapples and blueberries, and to a lesser extent in blackberries/crowberries. Few Elders observed declines in redberries/cranberries. In addition, many Elders felt that berries are getting smaller, and that their taste is changing. There were mixed opinions on whether or not berry ripening times have changed, but some Elders observed that ripening is occurring later in the season.

Nearly all the interviewees agreed that mammal abundance is changing. In particular, they have observed decreases in the number and quality of caribou and seal. For example, some Elders observed that caribou marrow is more watery now, and some have even observed blood in the marrow. Black bears are also seen more frequently around Nain, and polar bears appear to be travelling closer to town and are more abundant along the north coast. Some people also noted new birds in the area, such as small black birds. Perceptions of changes in black fly and mosquito abundance varied greatly.

ᐊᓯᔾᔨᖅᓵᓕᓂᖅᓴᐅᕙᓕᕐᓂᖕᒐᓗ (ᐱᓗᐊᖅᑐᒥ ᐅᑭᐅᒃᑯᑦ). ᑕᒪᐃᑲᓴᒃ ᐊᐱᖅᓱᖅᑕᐅᔪᑦ ᐅᖃᓚᐅᖅᑐᑦ ᓂᒡᓚᓱᖕᓂᖅᓴᐅᕙᓕᕐᓂᖕᒐᓂᒃ, ᐊᒥᓱᑦᑐ ᐅᖃᓚᐅᖅᑯᑎᒃ ᐊᓄᕆ ᐊᓯᔾᔨᖅᓯᒪᓕᕐᓂᖕᒐᓂᒃ, ᓱᕐᓗ ᓇᑭᖖᒐᔾᔭᖅᐸᖕᓂᖕᒐ. ᐊᐱᕆᔭᐅᒐᒥᒃ ᒪᕐᑯᖅᑲᑦᑕᕐᓂᐅᐸ ᖃᓐᓂᖅᑲᑦᑕᕐᓂᐅᐸ ᒥᒃᓵᓄᑦ, ᐊᒥᓲᓂᖅᓴᑦ ᐃᓄᑐᖅᑲᐃᑦ ᐅᖃᓚᐅᖅᑐᑦ ᖃᓐᓂᕿᔪᖖᒋᓐᓂᖅᓴᐅᓕᕐᓂᖕᒐᓂᒃ ᐅᑭᐅᓂ ᐊᓂᔫᖅᓵᖅᑐᓂ, ᐊᒻᒪ ᒪᕐᑯᒐᔪᖕᓂᖅᓴᐅᓕᕐᓂᖕᒐᖕᒐᓂᒃ. ᓯᓚᒥ ᐊᓯᔾᔨᖅᓯᒪᔪᑦ ᐅᔾᔨᕆᔭᐅᓯᒪᖕᒥᔪᑦ ᓯᑰᐸ ᐊᓯᔾᔨᖅᓯᒪᓕᕐᓂᖕᒐᓄᑦ. ᐊᐱᖅᓱᖅᑕᐅᔪᑦ ᓇᑉᐸᑲᓴᖕᒋᑦ ᐅᖃᓚᐅᖅᑐᑦ ᐅᔾᔨᕈᓱᒃᓯᒪᓂᕐᒥᓂᒃ ᓯᑯᓇᓵᕐᓂᖅᓴᐅᕙᓕᕐᓂᖕᒐᓂᒃ ᓯᑯᐃᖅᓵᓕᕙᓕᕐᓂᖕᒐᓂᓪᓗ, ᐊᓯᖕᒋᑦ ᐊᓯᔾᔨᖅᓯᒪᔪᒥᒃ ᐅᔾᔨᕈᓱᖖᒋᓐᓂᖅᑯᑎᒃ. ᑭᓯᐊᓂ ᐊᒥᓲᓂᖅᓴᑦ ᐊᐱᖅᓱᖅᑕᐅᔪᑦ ᐅᖃᓚᐅᖅᑐᑦ ᑕᐃᔅᓱᒪᓂᓂᑦ ᓯᑯ ᓵᓐᓂᖅᓴᐅᓕᕐᓂᖕᒐᓂᒃ, ᓯᑰᑦᑐ ᐊᓯᔾᔨᖅᓯᒪᓕᕐᓂᖕᒐᓄᑦ ᐊᐅᑦᓵᕐᕕᐅᕙᒃᑐᑦ ᐊᖑᓇᓱᒡᕕᐅᕙᒃᑐᑦ ᐊᖅᑯᑎᖕᒋᑦ ᐊᓯᔾᔨᖅᓯᒪᔭᕆᐊᖅᑲᓕᕐᓂᖕᒋᓐᓂᒃ. ᐊᐱᕆᔭᐅᒐᒥᒃ ᑕᓲᑦ ᓯᑯᕙᖕᓂᖕᒋᓐᓂᒃ, ᐊᒥᓲᓂᖅᓴᑦ ᐅᖃᖅᓯᒪᔪᑦ ᐅᔾᔨᕈᓱᒃᓯᒪᖖᒋᓐᓂᖅᑯᑎᒃ ᓯᑯᕙᖕᓂᖕᒋᓐᓂᒃ ᓯᑯᐃᖅᐸᖕᓂᖕᒋᓐᓂᓪᓗ, ᐃᔾᔪᓂᖕᒐᓗ ᐊᓯᔾᔨᖅᓯᒪᓂᖅᖃᖅᓂᔾᔪᒃ. ᐊᒥᓱᑦ ᐃᓄᑐᖅᑲᑦ ᐅᔾᔨᕈᓱᒃᓯᒪᖕᒥᔪᑦ ᓄᓇᐅᑉ ᐃᑭᐊᖕᒐ ᕐᑯᐊᖑᐃᓐᓇᖅᐸᒃᑐᖅ ᐊᐅᒃᐸᓕᕐᓂᖕᒐᓂᒃ, ᐊᓪᓚᑦ ᐃᓪᓗᐃᑦ ᐅᕚᓕᕐᓂᖕᒋᓐᓂᒃ, ᐊᐱᖅᓱᖅᑕᐅᔪᑦᑐ ᓇᑉᐸᑦᑐᐊᖕᒋᑕ ᐅᖕᒐᑎᑦᑎᐊᖕᒐᓅᑦᑐᑦ ᐅᖃᖅᓯᒪᑦᑐᑎᒃ ᓄᓇ ᑎᓱᐊᕐᓂᖅᓴᐅᕙᓕᕐᓂᖕᒐᓂᒃ.

Nearly all the interviewees believed that the weather is becoming increasingly variable and that the timing and nature of seasons (especially winter) has changed. Most participants stated that it is colder now, and many also stated that some characteristics of the wind, such as direction, have changed. When asked about changes in precipitation, the majority of Elders indicated that it snows less in Nain now than in past years, and many have observed more rain. Changes in weather and climate are also concurrent with observed changes in ice conditions. Approximately half of the interview participants observed later sea ice freeze-up and earlier breakup, whereas the other half observed no change. Yet the majority of interviewees said that sea ice is thinner now than in the past, and that changes in ice conditions have affected travel routes. When asked about lake ice conditions, the majority of participants said they have not observed changes in the timing of freeze-up and breakup, nor in the ice thickness. Many Elders have also observed the effects of permafrost thawing, such as slumping houses, and just over half of interview participants have observed increased erosion in recent years.

ᐃᓐᓇᐃᑦ ᐊᐱᖅᓱᖅᑕᐅᔪᑦ
Elders Interviewed

ᑕᐃᔭᐅᔪᒪᙱᑦᑐᖅ
Anonymous

ᔪᓕᐊᔅ
ᐃᒃᑯᓯᖅ
Julius
Ikkusek

ᑯᕆᔅᑏᓐ ᐸᐃᑭ
Christine
Baikie

ᐃᓕᓴᐱ
ᐃᑦᑐᓪᓚᒃ
Elizabeth
Ittulak

ᐄᑐᐊᖅ
ᕗᓚᐅᕙᔅ
Edward
Flowers

ᓘᑲᔅ ᐃᑦᑐᓪᓚᒃ
Lucas
Ittulak

ᓗᐃᓴ
ᕗᓚᐅᕙᔅ
Louisa
Flowers

ᓴᐃᕋ ᐃᑦᑐᓚᒃ
Sarah
Ittulak

ᔩᐊᓯᑲ ᕙᐊᕐᑦ
Jessica
Ford

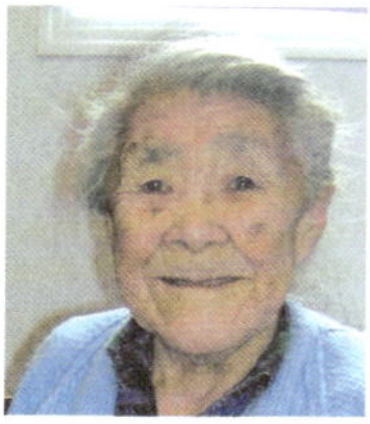

ᕕᕈᓇ
ᐃᑦᑐᓪᓚᒃ
Verona
Ittulak

ᔫᓐ ᔭᕋᕈᔅ
John
Jararuse

ᔪᓕᐊᔅ
ᒥᕐᑯᕋᑦᓱᒃ
Julius
Merkuratsuk

ᐋᓂ ᓛᒻᑉ
Annie Lampe

ᓇᐃᒥ
ᒥᕐᑯᕋᑦᓱᒃ
K. Naeme
Merkuratsuk

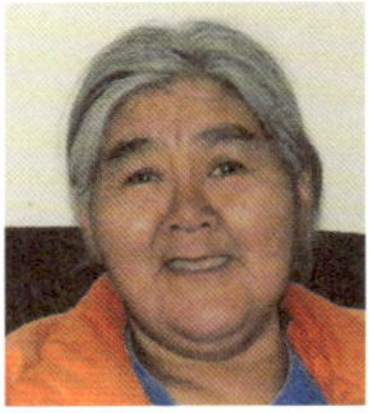

ᐋᓂ ᓕᑦ
Annie Lidd

ᒥᓂ
ᒥᕐᑯᕋᑦᓱᒃ
Minnie
Merkuratsuk

ᐃᓚᐃ
ᒥᕐᑯᕋᑦᓱᒃ
Eli
Merkuratsuk

ᒫᑕ
ᐅᒃᑯᐊᑦᓯᐊᖅ
Martha
Okkuatsiak

ᔭᐃᑯ
ᒥᕐᑯᕋᑦᓱᒃ
Jacko
Merkuratsuk

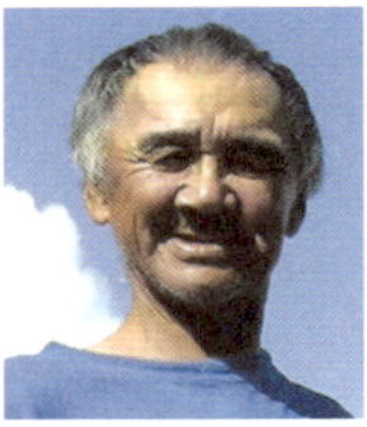

ᑎᒧᑎ ᑕᐅᓐᓕ
Timothy
Townley

ᕌᓐ ᕗᐊᑉ
Ron Webb

ᑳᑎ ᐅᐃᓐᑐᔅ
Katie Winters

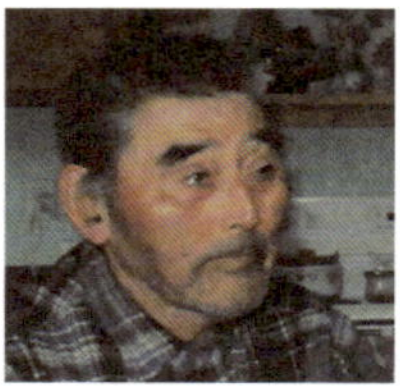

ᐅᐃᓕᐊᒻ ᐅᐃᓐᑐᔅ
William
Winters

ᐸᐅᕐᖓᐃᑦ

"ᐊᕐᕿᓂ ᓄᓂᕙᒐᒃᓴᑦ ᐊᔾᔨᐅᙱᒻᒪᓇᓚᐆᖅᑐᑦ. ᐊᕐᕿᓂ ᑕᐃᒪᐃᑦᑐᐊᓕᖅᑎᑦᑐᒍ ᑭᒍᑕᖏᕐᓇᖅᑕᓚᐅᖅᑐᖓ. ᑕᒪᑐᒪᓂ ᐱᕈᖅᐸᑦᑕᐊᓕᑕᐃᓐᓇᖅᑐᑦ; ᐱᕈᐊᓂᙱᑦᑐᑦ. [...] ᑭᒍᑕᖏᕐᓇᑦ ᑭᖑᕙᓯᒃᑐᑦ. ᐊᐅᐸᖅᑐᑦᑐ, ᐱᕈᖅᐸᑦᑕᐊᓕᑕᐃᓐᓇᖅᑐᑦ. [...] ᐊᐅᐸᖅᑐᑦ ᖃᓄᐃᑐᐊᙱᑦᑐᑦ ᑭᖑᑦᓕᐅᑦᑐᑎᒃ ᐱᕈᖅᐸᖕᒪᑕ. ᑭᓯᐊᓂ ᑭᒍᑕᖏᕐᓇᑦ ᑭᖑᕙᓯᒻᒪᓕᒃᑐᑦ, ᐊᖅᐲᑦᑐ. ᐃᒻᒪᖄ ᑕᒪᕐᒥᒃ ᓄᓂᕙᒐᒃᓴᑦ ᑭᖑᕙᓯᒃᑐᑦ."

—ᑯᓕᔅᑏᓐ ᐸᐃᑭ

"ᐊᕐᕿᒍᑕᒫᖑᔫᖅᑐᖅ [ᓄᓂᕙᒐᒃᓴᑦ ᐱᕈᐊᓂᒃᐸᖕᓂᖏᑦ] ᐊᔾᔨᒋᔦᓐᓃᖅᐸᑦᑕᐊᔭᖓ. ᐊᕐᕿᒍᑦ 20-30-ᑐ ᐊᓂᒍᖅᑐᑦ ᐱᕈᖅᐸᓚᐅᖅᑐᑦ ᓈᓪᒪᒃᓯᒐᐃᖕᒪᑦ ᐊᔾᔩᒃᑐᓂᒃ, ᒫᓐᓇ ᑭᖑᕙᓯᒃᐸᓕᖅᑐᑦ, ᐃᓛᓐᓂ ᐱᑦᑐᐊᕐᕚᖕᒥᓂ ᐱᕈᖅᐸᒃᓱᑎᒃ, ᐃᓛᓐᓂᑐ ᐱᕈᖅᓵᓕᕙᒃᓱᑎᒃ; ᖃᐅᔨᒪᓇᖅᐸᒍᓐᓃᖅᑐᖅ. ᖃᐅᔨᓴᐃᓐᓇᓇᐊᖅᑲᒻᒪᓇᓕᖅᑐᒍᑦ ᐊᕝᕗᒐᒃᓴᐅᓕᕐᒪᖔᑕ."

—ᑯᓕᔅᑏᓐ ᐸᐃᑭ

Berries

"I found a big difference in the berries last year. This time last year I was picking blueberries. This year there are only flowers; there are no berries forming. [...] But it's late for blueberries. And the redberries, very few blossomed this year, very few flowered. [...] The redberries are not so bad, because they're the last berries to grow. But the blueberries are definitely late, and [also the] bakeapples. Probably all the berries."

—Christine Baikie

"It seems like each year [the time the berries ripen] is a little bit different. Twenty or 30 years ago it was always the same time. Now it's late, or it's around the time it should be, or it's early; you never know now. Now you really have to keep checking if they're ripe or not."

—Christine Baikie

“ᖃᐅᔨᓕᑕᐃᓐᓇᓚᐅᖅᑐᖓ
ᓄᓂᕙᒐᒃᓴᑦ ᐱᕈᕐᓂᖏᑕ
ᐊᓯᔾᔨᖅᓯᒪᓕᕐᓂᖏᓐᓂᒃ ᑕᒪᐅᖓ
ᓅᓇᒪ, ᓇᐃᓂᒧᑦ. ᐊᒻᒪ ᓇᐃᓂᒥᑐᐊᖅ
ᖃᐅᔨᓕᓚᐅᖅᑐᖓ ᓯᓚᐅᑉ
ᐊᓯᔾᔨᖅᐸᑦᓕᐊᓂᖓᓂᒃ; ᑕᐃᒃᓱᒪᓂᓗ
ᐅᔾᔨᕆᓕᓚᐅᖅᑕᒃᑲ ᓄᓂᕙᒐᒃᓴᑦ
ᐊᓯᔾᔨᓕᕐᓂᖏᓐᓂᒃ. ᐊᒻᓛᑦ ᑕᖅᓴᖏᑦ
ᓄᓂᕙᒐᒃᓴᐃᑦ ᐊᓯᔾᔨᓕᖅᑐᑦ.”

—ᓴᐃᕋ ᐃᑦᑐᓚᒃ

“The only time I started to recognize the change in the berries was when I got here, to Nain. And it was only in Nain that I noticed the climate was slowly changing; that’s about the same time the berries started changing as well. Even the colours of the berries.”

—Sarah Ittulak

“ᓄᓂᕙᒋᐊᒃᓴᖅ [ᐊᖅᐱᓂᒃ]
ᐱᐅᓂᖅᐸᐅᓚᐅᖅᓯᒪᔪᖅ ᐋᒡᒋᓯ
ᕿᑎᖅᐸᓯᐊᓂ, ᑭᓯᐊᓂ ᒫᓐᓇ
ᓯᕿᓂᖅ ᐆᓇᓗᐊᖅᐸᓕᖅᑐᖅ,
ᓈᒻᒪᒃᓯᓚᐅᖅᑎᓐᓇᒋᑦ
ᐱᐅᔪᓐᓃᖅᐸᓕᖅᑐᑦ [...] ᓯᓚ
ᐅᖅᑰᓗᐊᓕᕐᓂᖓᓄᑦ.”

—ᔪᓕᐊᔅ ᐃᒃᑯᓯᖅ

“The best time [to pick bakeapples] was around mid-August, but sometimes the sun is too hot nowadays, and even before they ripen they get rotten [...] because the weather is too hot.”

—Julius Ikkusek

“ᐊᐅᔭᐅᓚᐅᖅᑐᒥ [2009]
ᐱᐅᓚᐅᖖᒋᑦᑐᐊᓗᒃ ᓄᓂᕙᒐᕐᓄᑦ,
ᐱᕈᓚᐅᖖᒋᑲᓴᓚᕆᒃᑐᑦ. ᐊᖅᐲᑦ
ᐊᕝᕙᒐᒃᓴᐅᓕᖅᐸᒃᑑᒐᓗᐊᑦ ᓯᑎᐱᕆᒥ,
ᑭᓯᐊᓂ ᖃᒐᓕᓃᓚᐅᖅᑐᑦ ᐋᒡᒋᓯᒥ. [...]
ᓯᓚᒧᑦ ᓄᓂᕙᒐᒃᓴᑦ ᐱᕈᖅᐸᒍᓐᓃᖅᑐᑦ.
ᐊᐅᔭᒃᑯᑦ ᓂᒡᓚᓱᖕᓂᖅᓴᐅᓕᖅᑐᖅ,
ᓂᒡᓚᖅᓵᓕᕙᓕᖅᔪᓂᓗ. [...]
ᒪᖁᖃᑦᑕᕐᓂᖅᓴᐅᔭᕌᖓᑦ
ᓯᕿᓐᓂᑦᑎᐊᖅᐸᒃᔪᓂᓗ
ᐱᕈᖅᑐᑦ ᐱᕈᖅᓵᓕᓂᖅᓴᐅᕙᒃᑐᑦ
ᒪᒪᕐᓂᖅᓴᐅᓪᓗᑎᒡᓗ.”

—ᒫᑕ ᐅᒃᑯᐊᑦᓯᐊᖅ

“Last summer [2009] was a really bad year for berries; they hardly grew. The bakeapples are usually ripe by September, but they were rotten in August. [...] The weather has to do with the berries not growing much anymore. It is colder now in the summers compared to before, a lot colder and earlier. [...] When there is more rain and more sunshine the plants tend to grow faster and tastier.”

—Martha Okkuatsiak

Nunatsiavut / Nain

“ᐃᖅᑲᐅᒪᔫᖓ ᔫᓂᒥ, ᓄᓂᕙᒐᑦᓴᑦ
ᐱᕈᖅᐸᓪᓕᐊᕙᓕᓚᐅᕐᓂᖕᒋᓐᓂᒃ
ᒪᒪᖅᑐᒻᒪᓇᐅᑦᑐᑎᒡᒎ, [...] ᑭᓯᐊᓂ
ᑭᖑᕙᓯᖕᓂᖅᓴᒻᒪᓇᐅᓕᖅᑐᑦ,
ᐃᓖᓐᓂ ᐊᕝᕙᒐᒃᓴᐅᖖᒋᒻᒪᓇᒃᑯᑎᒃ;
ᖃᒐᓕᖅᑯᑎᒃ
ᖁᐊᖅᓯᒪᓕᖅᑯᑎᒡᔫᓐᓃᖑᕙᓕᖅᑐᑦ.”

—ᕓᕉᓇ ᐃᑦᑐᓚᒃ

“ᑎᐱᖓ [ᑕᐃᔅᓯᒪᓂᑐᑦ]
ᐊᔾᔨᖖᒋᒻᒪᓇᓕᖅᑕᖓ
ᓄᑕᖃᐅᓚᐅᕐᓂᓐᓂᑦ. [...] ᒫᓐᓇ
ᑎᐱᖃᖅᑔᔭᕉᓐᓃᖅᑐᑦ.”

—ᕓᕉᓇ ᐃᑦᑐᓚᒃ

“ᓄᓂᕙᒐᒃᓴᖃᕐᓂᖅᓴᐅᓚᐅᖅᑐᖅ
ᖃᐅᔨᒪᔭᖃ. ᒫᓐᓇ ᐱᑕᓕᐅᒐᓗᐊᖅ,
ᑭᓯᐊᓂ ᐱᕈᖅᐸᓚᐅᕉᓯᕐᒦᑐᔾᔫᓐᓃᖅᑐᖅ
ᐃᓚᖏᓐᓂ ᓄᓇᓂ.”

—ᔮᓐ ᔭᖃᕉᔅ

“ᓄᓂᕙᒐᒃᓴᑦ ᐅᓄᕈᓐᓃᖅᐸᓪᓕᐊᓕᖅᑐᑦ,
ᐅᔾᔨᕈᓱᒃᓯᒪᓕᖅᑯᖓᒎ ᐊᕐᕌᒍᓂ
ᐊᓂᒍᕗᖅᑐᓂ ᑎᐱᖏᑦ
ᐊᔾᔨᖖᒋᓕᕐᓂᖓᓂᒃ; ᑕᐃᔅᓯᒪᓂᑐᑦ
ᒪᒪᖅᑎᒋᔫᓐᓃᖅᑐᑦ.”

—ᐃᓕᓴᐱ ᐃᑦᑐᓚᒃ

“I remember in June, that’s when berries were starting to grow and becoming very good, [...] but they’re getting ripe a lot later, and sometimes they don’t even get ripe; they’re burned out or frostbitten or something like that.”

—Verona Ittulak

“There’s a big difference in the taste [compared to] when I was younger. [...] They’re tasteless now.”

—Verona Ittulak

“I’m sure there used to be more berries before. There are some now, but not the way they used to grow in certain areas.”

—John Jararuse

“Berries seem to be decreasing [in number], and I also noticed in the last few years that the berries even taste different; they’re not as tasty as they were before.”

—Elizabeth Ittulak

"ᑕᐃᔅᓱᒪᓂ, ᓄᓂᕙᕆᐊᕌᖕᒐᒪ [...] ᐊᖏᔪᐊᓘᔮᖅᐸᓚᐅᖅᑐᑦ, ᐊᐅᐸᖅᑐᑦ, ᑭᒍᑕᖏᕐᓇᑦ ᐸᐅᖖᒐᐃᑦ ᐊᕐᕌᒍᑕᒫᑦ ᒥᑭᓐᓂᖅᓴᐅᔫᔮᖅᐸᓕᖅᑐᑦ. ᑕᐃᔅᓱᒪᓂ ᐊᖏᔪᐊᓘᕙᓚᐅᕐᒪᑕ ᕿᓚᒥᐊᓗᒃ ᑕᑕᑦᑎᓐᓇᖅᐸᓚᐅᖅᑐᖅ."

—ᐃᓕᓴᐱ ᐃᑦᑐᓚᒃ

"ᑕᐃᔅᓱᒪᓂ ᐸᐅᖖᒐᖅᑲᓕᖅᐸᓚᐅᖅᑐᒍ ᔫᓂᒥ, ᓯᓚ ᐅᖅᑰᓕᖅᑎᑦᓗᒍ, ᒫᓐᓇ ᔪᓚᐃᒥ ᐋᒡᒋᓯᒥ ᓯᑎᐱᕆᒥᑦ ᓘᓐᓃᑦ ᓄᓂᕙᒃᐸᓕᖅᑐᒍᑦ."

—ᑳᑎ ᐊᒻᒪ ᐅᐃᓕᐊᒻ ᐅᐃᓐᑐᔅ

"In the past, when I picked those berries [...] they all seemed to be large, but these years, the redberries, the blueberries, and the blackberries, they all seem to be getting smaller every year. In the past they were so big you could fill up your bucket real quick."

—Elizabeth Ittulak

"Before, we used to get berries in June, when the weather was warm and good, but now you have to wait until July or August or September to pick."

—Katie and William Winters

ᐱᕈᖅᑐᑦ ᐊᓯᖏᑦ

"ᐅᔾᔨᕈᓱᒃᑐᖓ ᐊᒥᓱᐊᓗᖕᓂᒃ ᒥᑭᑦᑐᓂᒃ ᓄᑖᓂᒃ ᓇᐹᖅᑐᓂᒃ ᐱᕈᖅᑐᖅᑲᖅᐸᓕᕐᓂᖓᓂᒃ ᓇᓂᑐᐃᓐᓇᖅ. ᑕᑯᖅᑲᑦᑕᖅᓯᒪᔭᒃᑲ. ᑕᒪᑐᒪᓂ ᐅᔾᔨᕆᓕᑕᐃᓐᓇᖅᑕᒃᑲ. ᐊᓯᒃᑲᓗ ᐅᖅᑲᖅᑐᑦ ᑐᓴᖅᓯᒪᔭᒃᑲ, 'ᑕᑯᒋᑦ ᐊᒥᓱᓂᒃ ᓄᑖᓂᒃ ᓇᐹᖅᑐᕋᓛᓂᒃ ᐱᕈᖅᑐᖅᑲᓕᖅᑐᖅ.'"

—ᑯᕆᔅᑏᓐ ᐸᐃᑭ

"ᐊᕐᕌᒍᓂ ᐊᓂᒎᓴᖅᑐᓂ ᐅᖅᐱᒐᐃᑦ ᐱᕈᕐᓂᖅᓴᐅᕙᓕᖅᑐᑦ ᓱᒃᑲᓕᔪᒥᒃ."

—ᓴᐃᕋ ᐃᑦᑐᓚᒃ

Other Plants

"One thing I notice is that there are a lot of little new trees growing in different places. I've seen that a lot. Only this year I noticed it. And I've heard other people commenting, 'My, what a lot of little new trees are growing around.'"

—Christine Baikie

"In the last few years we have noticed that the bushes have been growing very rapidly."

—Sarah Ittulak

“ᐊᕐᕌᒍᓂ ᐊᓂᒍᓵᕐᒃᑐᓂ ᑕᑯᓯᒪᓕᕐᒃᑐᖕᒐ ᐱᕈᕐᒃᑐᓂᒃ, ᕐᑲᐅᔨᒪᖖᒋᑕᓐᓂᒃ. [...] ᕐᑲᓄᕐᒃ ᑕᐃᔭᐅᕙᖕᒪᖕᒑᑕ ᕐᑲᐅᔨᒪᖖᒋᑕᒃᑲ.”

—ᔪᓚᐊᔅ ᐃᒃᑯᓯᕐᒃ

“ᐊᐅᔭᕐᒃᑕᒫᑦ ᐃᕚᑦ ᑕᑭᓂᕐᒃᓴᐅᕙᑦᓕᐊᖕᒋᓐᓇᐅᔭᓕᕐᒃᑐᑦ, ᐅᕐᒃᐲᑦᑐ ᓇᓂᓖᕐᒃ ᓄᓇᓕᖕᒥ ᐱᕈᕐᒃᐸᓕᕐᒃᓲᑎᒃ ᓱᒃᑲᔪᐊᓗᖕᒥᒃ ᐱᕈᕐᒃᐸᓕᕐᒪᑕ.”

—ᕚᕈᓇ ᐃᑦᑐᑦᓚᒃ

“ᑕᑯᓯᒪᔪᖕᒐ ᑎᖕᒐᐅᔭᑦ ᐊᓯᔾᔨᕐᓂᖕᒋᓐᓂᒃ; ᐱᑕᕐᒃᑲᖖᒋᓐᓂᕐᒃᓴᐅᓕᕐᒃᓲᑎᒃ. ᓄᖕᒍᑉᐸᑦᓕᐊᕌᓚᓴᓕᕐᒃᑐᑦ.”

—ᐃᓕᓴᐱ ᐃᑦᑐᑦᓚᒃ

“ᓇᐹᕐᒃᑐᑦ ᐳᕐᒃᑐᓂᕐᒃᓴᐅᕙᓕᕐᒃᑐᑦ ᓯᓕᖕᓂᕐᒃᓴᐅᓕᕐᒃᓲᑎᒡᓗ ᓄᓇᓕᑦᑎᓐᓂ; ᒫᓐᓇ ᓇᐹᕐᒃᑐᒐᑦ ᐊᕐᕌᒍ 15–20-ᐅᓚᐅᕐᒃᑐᓂᑦ ᑕᑭᓂᕐᒃᓴᒻᒪᖑᐅᓕᕐᒃᑐᑦ.”

—ᑎᒧᑎ ᑕᐅᓐᓕ

“ᑕᐃᔅᓯᒪᓂ, ᒫᓐᓇᓕᒫᕐᒃ ᓄᓇᐃᓐᓇᐅᓚᐅᕐᒃᓯᒪᔪᕐᒃ, ᓄᓂᕙᒡᕚᒃᓴᒻᒪᖑᐅᓚᐅᕐᒃᓯᒪᔪᕐᒃ. ᒫᓐᓇᓕ ᐅᕐᒃᐲᑦ ᐊᕚᓚᕐᑭᐊᑦᑐ ᐱᕈᕐᓂᕐᒃᓴᐅᓕᕐᒃᑐᑦ. ᓱᒃᑲᔪᒻᒪᖑᖕᓂᒃ ᐱᕈᕐᒃᐸᓕᕐᒪᑕ. [...] ᐱᓱᔪᒡᕚᒃᓴᐅᑦᑎᐊᕈᓐᓃᒻᒪᖑᒃᑐᕐᒃ ᐅᕐᒃᐱᒐᕐᓄᑦ.”

—ᐄᑐᐊᕐᒃ ᐊᒻᒪ ᓗᐃᓴ ᕙᓚᐅᕗᔅ

“In the last few years I have seen a few plants, unknown plants. [...] I don’t know what their names are.”

—Julius Ikkusek

“Every summer the grass grows higher and higher, and the shrubs seem to be taking over the community because they’re growing so fast.”

—Verona Ittulak

“I have seen a big change in lichens; there is not as much around as before. They’re declining fast.”

—Elizabeth Ittulak

“The trees are growing taller and wider in the town; the trees that we have now were not as tall 15 or 20 years ago.”

—Timothy Townley

“In the past, all of this area was mostly open, and that’s where we used to get a lot of berries. But these years the willows and *avâlaKiat* [birches] are getting in the way of things. They’re growing so fast. [...] These days it’s even hard to go for a walk because of the shrubs.”

—Edward and Louisa Flowers

“ᐅᖅᐲᑦ ᓇᐹᖅᑐᑦᑐ
ᐱᕈᖅᓯᒪᔪᒻᒪᕆᐅᓕᖅᑐᑦ
ᑕᐃᒪᙵᑦ ᑕᒪᐅᖓ
ᓅᓚᐅᖅᓯᒪᒐᑦᑕ. [...] ᐊᖅᖀᑐᑖᑦ
ᑕᑭᓂᖅᓴᐅᓕᖅᐸᓪᓕᐊᖏᓐᓇᖅᑐᑦ.”

—ᒥᓂ ᐊᒻᒪ ᔭᐃᑯ ᒥᖅᑯᕋᑦᓱᒃ

“The willows and the trees have grown immensely since the time we moved here. [...] They’ve been growing taller every year.”

—Minnie and Jacko Merkuratsuk

ᐆᒪᔪᑦ

“ᐅᔾᔨᕈᓱᒃᓯᒪᓕᖅᑐᖓ ᐊᔾᔨᐅᙱᑦᑐᓂᒃ
ᑎᖕᒥᐊᓂᒃ. ᐊᖅᖀᓂ ᐅᖓᑦᑕᕐᒥ
ᑎᖕᒥᐊᖅᑲᓚᐅᖅᐳᑦ; ᑕᒪᐅᖓ
ᑎᑭᑦᑕᓚᐅᖅᓯᑎᒃᑐ. ᐹᓂ
ᐃᒃᑐᒪ ᖁᑖᓂ ᐃᕙᓚᐅᖅᑐᖅ.
[...] ᐊᓯᖏᓐᓂᒃᑐ ᑎᖕᒥᐊᓂᒃ
ᑕᑯᓯᒪᓕᖅᑐᖓ ᖃᐅᔨᒪᙱᑕᓐᓂᒃ;
ᐊᕙᓂ ᕿᑭᖅᑕᒥ ᑕᑯᓚᐅᖅᑐᖓ,
ᓄᑖᓂᒃᑐ ᒥᑭᑦᑐᓂᒃ ᑎᖕᒥᐊᓂᒃ
ᑕᑯᓯᒪᑦᑐᖓ—ᖃᐅᔨᒪᙱᑕᒃᑲ
ᑭᓲᖕᒪᖔᑕ.”

—ᑯᓇᔅᑏᓐ ᐸᐃᑭ

“ᐆᒐᖅᑕᖃᕈᓐᓃᖅᑐᖅ.
ᐆᒐᖃᑐᐊᕈᓐᓃᖅᑐᖅ. ᒥᑭᑦᑑᓕᖅᑐᑦ
ᐱᔭᒃᓴᐅᔪᓐᓃᖅᓱᑎᒃᑐ.”

—ᔮᓐ ᔭᖃᕈᔅ

Animals

“I’ve noticed different birds. And we had mourning doves here a couple of years ago; it was the first time ever they were here. There was one that had a nest up in the eave of my house the year before last. [...] And I have seen other birds that I didn’t recognize. One was about the size of a ptarmigan or a partridge; I saw that over on Tabor Island, and I’ve seen [new] little birds—I don’t know what they are.”

—Christine Baikie

“There’s no more codfish. Hardly any codfish. They’re small now and you don’t get them.”

—John Jararuse

“ᐱᕈᖅᓴᑎᑦᑐᖕᒐ ᓄᑕᕐᒥ
ᓂᕐᔪᑎᖅᑲᑦᑎᐊᓚᐅᖅᓯᒪᔪᖅ.
ᐆᒃᑑᑎᒋᑐᒍ, ᓄᑕᕐᒥ
ᒥᑎᖅᑲᑦᑎᐊᖅᐸᓚᐅᖅᓯᒪᔪᖅ, ᑭᓯᐊᓂ
ᑭᖕᒍᓪᓕᖅᐹᕐᒥ ᑕᐅᕗᖖᒐᕋᒪ ᐊᑕᐅᓯᕐᒥᒃ
ᒪᕐᕉᖕᓂᓪᓘᓐᓃᑦ ᑕᑯᓚᐅᖅᑐᖕᒐ.
ᐆᒐᐃᑦ ᐱᑕᖅᑲᕈᓐᓃᑦᑎᐊᖅᑐᑦ,
ᐃᖅᑲᓗᖅᑲᕈᓐᓃᖅᐸᓪᓕᐊᓪᓗᓂ
ᑲᓇᔪᖅᑲᕈᓐᓃᖅᐸᓪᓕᐊᓪᓗᓂᓗ.
ᑲᓇᔪᓕᒻᒪᕆᐊᓘᕙᓚᐅᖅᓯᒪᔪᖅ—
ᓯᒡᔮᖅᐸᓚᐅᖅᓯᒪᒻᒪᕆᒃᑐᑦ ᐸᓂᖅᖢᑎᒡᓗ
ᓯᒡᔭᒥ, ᐊᕝᕗᒐᒃᓴᒻᒪᕆᐅᕙᓚᐅᖅᑐᑦ [...]
ᐱᑕᖅᑲᕈᓐᓃᖅᑐᑦ.”

—ᕓᕈᓇ ᐃᑦᑐᓪᓚᒃ

“When I was growing up in Nutak there was an abundance of animals. For example, Nutak was full of eider ducks, but last time I was over there I saw only one or two. The cod is completely gone; the char is declining in numbers; and the capelin is gone as well. There was so much capelin before—they used to drift ashore and dry up, they were easy picking [...] but that's all gone.”

—Verona Ittulak

“ᓇᓄᖅᑲᕐᓂᖅᓴᐅᔮᓕᖅᑐᖅ
ᐊᒃᖤᖅᑲᕐᓂᖅᓴᐅᓕᖅᖢᓂᓗ, ᐊᒃᖤᐃᑦᑐ
ᓴᓗᖕᓂᖅᓴᐅᖅᑯᔮᓕᖅᖢᑎᒃ
—ᑳᒃᑐᐊᓘᔮᖅᑐᑦ. [...]
ᐃᓱᒫᓗᓇᒻᒪᕆᒃᑐᖅ.”

—ᑯᕆᔅᑏᓐ ᐸᐃᑭ

“There seem to be more polar bears and black bears around, and the black bears look so much skinnier—as if they're hungry. [...] I'm concerned about that.”

—Christine Baikie

“ᑕᐃᔅᓱᒪᓂ
ᐃᒍᑦᓴᓕᒻᒪᕆᐊᓘᓚᐅᖅᓯᒪᔪᖅ,
ᐱᓗᐊᖅᒥ ᐅᖅᐱᒐᓕᖕᓂ [...]
ᒫᓐᓇᓕ ᑕᑯᒐᔪᖕᓇᕈᓐᓃᖅᑐᑦ,
ᐅᓄᕈᓐᓃᕐᓂᖅᓴᐅᕙᓪᓕᐊᓕᖅᑐᒃᓴᐅᖕᒪᑕ.”

—ᔪᓕᐊᔅ ᐃᒃᑯᓯᖅ

“In the past [bumblebees] were very, very plentiful, especially in the bush, [...] but nowadays you only see one once in a very long while, so they're declining, I guess.”

—Julius Ikkusek

“ᑕᐃᔅᓱᒪᓂ ᑭᒃᑯᓕᒫᖕᒍᔮᖅᑐᑦ
ᖅᑭᑦᑐᕆᐊᖅᑕᐅᔅᔭᐃᒃᑯᑎᓂᒃ
ᔭᐃᑲᒃᓯᒪᕙᓚᐅᖅᑐᑦ, ᒫᓐᓇᓕ
ᖅᑭᒃᑐᕆᐊᖅᑲᓗᐊᖅᑑᔮᕈᓐᓃᖅᑐᖅ.”

—ᒫᑕ ᐅᒃᑯᐊᖅᑦᓯᐊᖅ

“In years before, you could hardly see people without bug jackets, but mosquitoes seem to be getting less [abundant].”

—Martha Okkuatsiak

“ᐊᕐᕌᒍᓂ ᐱᖕᒐᓱᓂᒃ
ᑎᓴᒪᓂᒡᔫᓐᓃᑦ ᐊᓂᒍᖅᑐᓂ,
ᕐᑯᐱᕐᕈᕐᑲᕐᓂᖅᓴᐅᓕᕐᓂᖕᒐᓂᒃ,
ᑕᒪᑐᒪᓂ ᐊᕐᕌᒍᒥ ᐱᔪᐊᖅᑐᖅ.”

—ᑕᐃᔭᐅᔦᒪᙱᑦᑐᖅ

“ᑭᑦᑐᓯᕐᑲᐃᓐᓇᖅᐸᒃᓯᒪᔩᒐᓗᐊᖅ,
ᑭᓯᐊᓂ ᒫᓐᓇᔫᙱᑦᑐᖅ,
ᐅᓄᖅᓯᕙᓪᓕᐊᓕᖅᑐᑦ.”

—ᔫᓐ ᔭᕋᕈᔅ

“ᐊᑕᐅᓯᑐᐊᕐᒥᒃ ᐊᐅᔭᒃᑯᑦ ᓇᓄᕐᒥᒃ
ᑕᑯᕙᓚᐅᖅᑐᒍᑦ, ᐃᓛᓐᓂᒃᑯᑦ
ᑕᑯᓚᐅᖅᓯᒪᙱᑦᑎᐊᒻᒪᕆᒃᐸᒃᓱᑕ;
ᒫᓐᓇ ᐊᔾᔨᒋᔦᓐᓃᖅᑕᖓ.
ᐅᓄᖅᓯᕙᓪᓕᐊᓕᖅᑐᑦ. ᐊᒥᓱᐊᓗᖕᓂᒃ
ᓇᓄᕐᑲᖅᐸᓕᖅᑐᖅ. ᑭᓪᓕᓂᓕᐊᓚᐅᕋᑦᑕ
20-ᐸᓗᖕᒋᓐᓂᒃ ᓇᓄᕐᓂᒃ
ᑕᑯᓚᐅᖅᑐᒍᑦ ᐃᑲᕐᕋᐅᑉ ᓇᑉᐸᖕᒐᓂ
[...] ᐅᓄᖅᓯᕙᓪᓕᐊᓪᓚᕆᒃᑐᑦ
ᓯᒡᔭᖅᐸᓯᖕᓂ.”

—ᔫᓐ ᔭᕋᕈᔅ

“ᐊᒥᓱᐊᓗᖕᓂᒃ ᐊᒃᓴᕐᑲᖅᐸᓕᖅᑐᖅ. [...]
ᐅᑭᐅᑦ ᑕᓪᓕᒪᑦ ᓯᕗᕐᖕᒐᒍᑦ ᐊᒃᓴᓂᒃ
ᑐᕐᑳᑦᑎᔭᕆᐊᕐᑲᖅᐸᓚᐅᙱᑦᑐᑦ
ᓄᓇᓕᖕᒧᑦ ᕐᑲᓪᓕᓗᐊᖅᑐᓂᒃ. ᒫᓐᓇ
ᑕᐃᒪᐃᑉᐸᓕᖅᑐᑦ. [...] ᐊᒃᓴᓂᒃ
ᑐᕐᑳᑦᑎᕙᓕᖅᑐᑦ ᐹᖕᒋᓐᓄᑦ
ᓯᓚᑎᑦᑎᐊᖕᒋᓐᓄᑦ ᑎᑭᑉᐸᓕᕐᒪᑕ.
ᑭᓱᒎᖕᒪᖕᒑᑦ ᕐᑲᐅᔨᒪᙱᑦᑐᖕᒐ ᑭᓯᐊᓂ
ᐊᒃᓴᐃᑦ ᐅᓄᕐᓂᖅᓴᒻᒪᕆᐅᓕᖅᑐᑦ.”

—ᑕᐃᔭᐅᔦᒪᙱᑦᑐᖅ

“In the past three or four years, I’ve noticed the bugs are much worse, and certainly this year.”

—Anonymous

“There used to be some [dragonflies], but not a lot like now; they’re increasing.”

—John Jararuse

“If we were lucky we could see one polar bear in a summer, and sometimes we never saw a polar bear; now it is different. They’re increasing. There’s a lot of polar bears now. We travelled to the Killinik area and saw almost 20 polar bears within half an hour, [...] so they’re increasing in number for sure, mostly on the coast.”

—John Jararuse

“There’s been a lot of black bears. [...] Up until the last five years I’ve never known anybody to go out shooting the bears because they were in the community, but now they do. [...] They’re taking out these bears because they’re on people’s doorsteps and in their backyards. I don’t know what it is, but the number of bears in the community has really risen.”

—Anonymous

Nunatsiavut / Nain

ᓄᓇᑦᓯᐊᕗᑦ / ᓇᐃᓂ

“ᐅᐱᕐᖔᑕᒫᑦ ᐊᒃᓴᓂᒃ ᑕᑯᕙᓕᖅᑐᖓ, ᓄᓇᓕᖕᒧᓪᓗ ᖃᓂᓐᓂᖅᓴᐅᕙᓕᖅᓱᑎᒃ, ᐃᒻᒪᖄ ᓄᓇᐃᓚᕐᒥ ᓂᕆᔭᒃᓴᖃᑦᑎᐊᖖᒋᓕᕐᓂᖕᒋᓐᓄᑦ.”

—ᒫᑕ ᐅᒃᑯᐊᑦᓯᐊᖅ

“I’ve seen more black bears every spring, and they’re getting closer to the town, maybe because of a lack of food in the interior.”

—Martha Okkuatsiak

“ᐃᓚᖕᒋᑦ ᐆᒪᔪᑦ ᐊᓯᖕᒋᓐᓂᒃ ᓂᕆᕙᓕᖅᑐᑦ, ᓲᕐᓗ ᐊᒃᓴᐃᑦ, ᓇᐅᔭᐃᑦ, ᑐᓗᒃᑲᓪᓗ—ᓂᕆᔭᒃᓴᖅᓯᐅᖅᐸᓕᖅᑐᑦ ᐊᒃᑕᖅᕕᖕᒥᑦ.”

—ᕕᕈᓇ ᐃᑦᑐᓪᓚᒃ

“Some species of animals have changed their eating habits, like, for instance, the black bear, the gull, and the crow—they all eat from the dump.”

—Verona Ittulak

“ᐅᕙᓐᓄᑦ ᑐᒃᑐᐃᑦ ᓴᓗᖕᓂᖅᓴᐅᔫᓕᖅᑐᑦ. ᐃᓛᓐᓂ ᑐᓐᓄᖃᖖᒋᑦᑎᐊᖅᑑᔫᖅᓱᑎᒃ, ᐱᓗᐊᖅᑐᒥ ᐸᖕᓃᑦ—ᐊᖑᓯᓪᓗᐃᑦ. ᓯᓕᒃᑐᐊᓗᖕᓂᒃ ᑐᓐᓄᖃᖅᐸᓚᐅᖅᑐᒍᑦ, ᑕᐃᒪᐃᑉᐸᒍᓐᓃᖅᑐᖅ.”

—ᔮᓐ ᔭᕋᕈᔅ

“To me [the caribou] seem a lot skinnier than they were before. Sometimes it seems like they have no fat, especially stags—males. We used to have thick ones before, but it’s not like it used to be anymore.”

—John Jararuse

“ᐊᒥᓱᐊᓗᖕᓂᒃ ᑎᖕᒥᐊᓂᒃ ᑕᑯᓚᐅᖅᓯᒪᖖᒋᑕᑦᑎᓐᓂᒃ ᑕᑯᕙᓕᖅᑐᒍᑦ, ᖃᓄᖅ ᑕᐃᔭᐅᕙᖕᒪᖔᑕ ᖃᐅᔨᒪᖖᒋᒻᒪᕆᒃᑕᕗᑦ.”

—ᐃᓕᓴᐱ ᐃᑦᑐᓪᓚᒃ

“There’s a lot of species of birds that we had never seen before, and don’t even know what they are called.”

—Elizabeth Ittulak

“ᐊᔾᔨᒌᖖᒋᒻᒪᕆᒃᑐᓂᒃ ᑕᕐᕋᓕᑭᓵᓂᒃ ᑕᑯᖅᑲᑦᑕᖅᓯᒪᔪᖓ. ᐊᒥᓱᐊᓘᓕᖅᑐᑦ ᒫᓐᓇ. ᑕᐃᔅᓱᒪᓂ ᖁᖅᓱᖅᑐᐃᓐᓇᐅᕙᓚᐅᖅᑐᑦ, ᑭᓯᐊᓂ ᐊᓯᖕᒋᓐᓂᒃ ᑕᖅᓴᓕᖕᓂᒃ ᑕᑯᕙᓕᖅᑐᒍᑦ; ᑕᕐᕋᓕᑭᓵᑦ ᐊᔾᔨᒌᖖᒋᑦᑐᐊᓗᖕᓂᒃ ᑕᖅᓴᖅᑲᖅᐸᓕᖅᑐᑦ.”

—ᑎᒧᑎ ᑕᐅᓐᓕ

“I’ve seen a lot of difference in [butterflies]. There are a lot now. In the past they were more of a yellowish colour, but in the last few years the colours have been changing; there’s a lot of colours in the butterflies now.”

—Timothy Townley

“ᑎᖕᒥᐊᖑᕙᓚᐅᖅᑐᑦ
ᓯᒡᔭᕐᒦᑉᐸᓚᐅᖅᑐᑦ—ᑕᑯᕙᖑᓐᓃᖅᑕᕗ.
ᓯᒡᔪᐊᔭᑦ, ᐊᒪᐅᓕᒐᑦ—
ᑕᑯᕙᖑᓐᓃᖅᑕᕗᑦ.”

—ᑳᑎ ᐊᒻᒪ ᐅᐃᓕᐊᒻ ᐅᐃᓐᑐᔅ

“ᓇᑦᑏᑦ ᐅᓄᖅᓯᕙᑦᓕᐊᓕᖅᑐᑦ.
ᐊᒥᓱᒻᒪᕆᐊᓘᔪᑦ ᑕᒪᑐᒪᓂ ᐅᐱᖅᖄᒥ
ᓯᑯᒥ.”

—ᑳᑎ ᐊᒻᒪ ᐅᐃᓕᐊᒻ ᐅᐃᓐᑐᔅ

“ᐅᔅᐱᕈᓱᒃᑐᖕᒐ ᐊᕐᖄᒍᓂ
ᐊᓂᒍᖅᑐᓂ ᑐᒃᑐᐃᑦ ᐸᑎᖕᖏᑕ
ᑎᐱᖕᖏᑦ ᐊᔅᐱᕆᔭᓐᓃᕐᓂᖕᖏᓐᓂᒃ;
ᑕᖕᖏᖅᑲᖖᖏᓐᓂᖅᓴᐅᔮᓕᖅᑐᑦ.”

—ᔪᓕᐊᔅ ᒥᖅᑯᕋᑦᓱᒃ

“ᐊᒃᕙᐃᑦ ᓄᓇᓕᖕᓄᑦ
ᑎᑭᖀᖅᐸᓚᐅᖖᖏᑦᑎᐊᖅᑐᑦ.
ᓄᓇᓕᖕᓄᑦ ᖅᑲᒡᓕᑕᐃᓐᓇᖅᐸᓚᐅᖅᑐᑦ
ᑰᖕᒍᐊᖀᖕᒐᒥᒃ, ᐃᖅᑲᓗᐃᑦ
ᒪᔪᐊᓕᖅᑎᓪᓗᑎᑦ.
ᖅᑲᒡᓕᓐᓂᑐᐊᕆᕙᓚᐅᖅᓯᒪᔭᖕᒐᑦ, ᒫᓐᓇ
ᓄᓇᓕᖕᓄᑦ ᑎᑭᖀᖅᐸᓕᖅᑐᑦ.”

—ᐄᑐᐊᖅ ᕙᓚᐅᕗᔅ

“Those birds that used to be on the beach—we don’t see them anymore. *Sijjariat* [beach birds], *amauligait* [snow buntings]—we don’t see them anymore.”

—Katie and William Winters

“Seals seem to be increasing [in number]. There were a lot this spring out on the ice.”

—Katie and William Winters

“I’m noticing in the last few years that the marrow of the caribou is tasting a lot different; it’s more like a watery taste.”

—Julius Merkuratsuk

“The black bear never used to come to town. The only time they really got close to the communities was when they went to the brook, when the chars were going upriver. That was the only time they were close to the community, but now they are all around town.”

—Edward Flowers

“ᑐᒃᑐᐃᑦ ᑎᐱᖕᒋᑦ
ᐊᔾᔨᒋᙱᒋᒻᒪᑎᓕᖅᑕᖕᒋᑦ.
ᓄᓇᐃᑦᓚᖅᒦᑎᑦᑐᒋᑦ
ᑎᖕᒐᐅᔭᓂᒃ ᓂᕆᕙᒃᑐᑦ, ᒫᓐᓇ
ᓯᒡᔭᖅᐸᓯᒃᑐᒍᐊᖅᐸᓕᖅᒪᑕ
ᓇᐹᖅᑐᓂᑦᑐ ᓂᕆᕙᓕᖅᖁᑎᒃ,
ᑐᒃᑐᑦ ᓂᖅᑭᖕᒋᑦ ᓂᕆᑦᑐᒋᑦ
ᓇᐹᖅᑐᖅᓂᖅᐸᓕᖅᑐᑦ. [...]
ᓄᓇᐃᑦᓚᒦᕝᐸᒃᑎᑦᑐᒋᑦ
ᑎᖕᒐᐅᔭᖅᑐᖅᑎᑦᑐᒋᑦ
ᒪᒪᖅᓂᖅᓴᐅᓚᐅᖅᑐᑦ ᑎᒥᑦᑎᓐᓄᑦᑐ
ᐱᐅᓂᖅᓴᐅᑦᑐᑎᒃ.”

—ᐄᑐᐊᖅ ᕙᓚᐅᕗᔅ

“The caribou taste is really different now. When they’re in the country, way inland, they eat lichen, but these years they go more on the beach and they’re eating the trees here, and when you eat [caribou meat] the taste is like the trees. [...] When they were up in the country eating lichen they were more healthy to eat.”

—Edward Flowers

“ᖅᑭᒃᑐᕆᖅᑲᙱᓐᓂᖅᓴᐅᔮᓕᖅᑐᖅ,
[...] ᐃᒻᒪᖅᑳ ᓄᓇ
ᐸᓂᖅᓯᒪᓂᖅᓴᐅᓕᖅᓂᖕᒐᓄᑦ.”

—ᐄᑐᐊ ᐊᒻᒪ ᓗᐃᓴ ᕙᓚᐅᕗᔅ

“The mosquitoes seem to be getting less [abundant], [...] maybe because of the dryness of the land.”

—Edward and Louisa Flowers

“ᑐᒃᑐᐃᑦ ᑎᐱᖕᒋᑦ
ᐊᔾᔨᒋᔪᓐᓃᖅᑕᖕᒋᑦ; ᑭᓯᓂᒃ
ᓂᕆᕙᓕᖅᒪᖕᒫᑕ ᖅᑲᐅᔨᒪᙱᑦᑐᖕᒐ.
ᐃᓚᖕᒋᑦ ᐋᓐᓂᐊᖅᑲᖅᑰᔮᖅᑐᑦ;
ᐸᑎᖅᑲᙱᓐᓂᖅᓴᐅᓕᖅᖁᑎᒡᑐ.”

—ᐋᓂ ᓕᑦ

“[The caribou] taste different; I don’t know what they’re eating. Some of them are sick or something; there’s not enough *patik* [marrow] in their bones.”

—Annie Lidd

ᐅᑭᐅᒥ ᐊᓯᔾᔨᖅᑕᕐᓂᖅ

Seasons

"ᐅᑭᐅᒃᑯᑦ ᒪᕐᑯᒃᐸᓕᖅᑐᖅ. ᐊᐱᔭᕌᖕᒐᑦ ᐊᐳᑕᐅᖏᓐᓇᓕᖅᐸᓚᐅᖅᓯᒪᔪᖅ, ᒫᓐᓇᓕ ᐊᐱᕆᐊᖅᐸᒃᑲᓗᐊᖅᖢᓂ, ᐃᓱᒪᓇᖅᖢᓂ, 'ᑕᐃᒪ ᐅᑭᐅᕐᓂᐊᓕᖅᐳᖅ,' ᐊᐅᑐᐃᓐᓇᖅᐸᒃᖢᓂ. ᑕᐃᒪᐃᑉᐸᓚᐅᖅᓯᒪᙱᑦᑑᒐᓗᐊᖅ. ᐊᐱᔭᕌᖕᒐᑦ, ᖃᐅᔨᒪᓇᖅᐸᓚᐅᖅᑐᖅ ᐅᑭᐅᖑᓂᐊᓕᕐᓂᖕᒐ. ᒫᓐᓇᓕ ᓇᓗᓇᖅᐸᓕᖅᑐᖅ."

—ᑕᐃᔭᐅᔪᒪᙱᑦᑐᖅ

"ᑕᐃᔅᓯᒪᓂ, ᑕᐃᔅᓯᒪᓂᐅᓗᐊᖏᑦᑐᖅ, ᐊᐱᕙᓚᐅᖅᓯᒪᔪᖅ ᐅᑐᐱᕆᒥ, ᑭᓯᐊᓂ ᒫᓇ ᓇᓂᓕᒫᖅ ᐊᐱᓇᓵᖅᐸᓕᖅᑐᖅ. [...] ᐊᐱᕙᓕᖅᑐᖅ ᖁᕕᐊᓯᒡᕕᒃ ᖃᒡᓕᒻᒪᓇᐅᑎᑦᑐᒍ, ᖃᓐᓂᕋᓗᐊᖅᐸᑦ ᖃᓐᓂᑯᑖᒐᓂ; ᔭᓚᐃᒥᓗ ᖃᓐᓂᙱᑦᓗᓂ ᒪᕐᑯᐃᓐᓇᓇᐊᕐᑲᖅᖢᓂ."

—ᒫᑕ ᐅᒃᑯᐊᑦᓯᐊᖅ

"ᐅᑭᐅᓂ ᐊᓂᒍᖅᑐᓂ, ᐃᒪᐃᓐᓇᐅᓕᖅᑲᑦᑕᓚᐅᖅᑐᖅ ᔪᓚᐃᒥ, ᑕᒪᑐᒪᓂ ᐊᕐᕌᒍᒥ ᓯᑯᓕᒻᒪᓇᐊᓘᓕᓚᐅᖅᑐᖅ ᒪᐃᒥ, ᑲᒪᓇᖅᑐᐊᓘᓚᐅᖅᑐᖅ ᐅᕙᑦᑎᓐᓄᑦ ᔫᓂᒥ ᓯᓚ ᖃᒧᑕᐅᔭᒃᑯᑦ ᐊᑐᕈᓐᓇᓚᐅᕋᑦᑎᒍᑦ [...] ᓯᓚᐅᑉ ᐊᓯᔾᔨᖅᐸᒡᓕᐊᓂᖓ ᐊᓯᔾᔨᖅᐸᒡᓕᐊᕌᓚᓴᕋᓯᕆᔭᕋ."

—ᐃᓕᓴᐱ ᐃᑦᑐᓪᓚᒃ

"It rains in the winter. Snow used to come and stay, and now we'll get a couple of starts to the winter, and we think, 'Okay, this is it,' and then it disappears. It never used to do that. When we had snowfall, that was it; we were into winter. None of this guessing."

—Anonymous

"In the past, not too long ago, the snow would come sometime in October, but it seems to be getting later everywhere. [...] We get our snow now just a few days before Christmas, and even if it does snow it's going to stop snowing; come January you might get rain instead of snow."

—Martha Okkuatsiak

"In past years, [we would only have] open water [...] in the month of July, and [this year] this ice started getting bad around the month of May, which is strange for us, because we would still be going on skidoo in June, [...] so I feel the climate is changing and is changing fast."

—Elizabeth Ittulak

Nunatsiavut / Nain

“ᓯᓚ
ᓂᒡᓚᖅᐸᓪᓕᐊᖕᒋᓐᓇᐅᕗᖅᑰᔨᓕᖅᑐᖅ.
ᒫᓐᓇ ᔫᓂᐅᓕᖅᑐᖅ ᓱᓕ ᖃᖅᑲᓂ
ᐊᐳᑎᖄᖅᑐᖅ, ᑲᒪᓇᒻᒪᕆᒃᑐᖅ.
[...] ᓯᓚ ᓂᒡᓚᓱᓗᐊᓕᕐᓂᖕᒐᓄᑦ
ᐊᐅᒃᐸᒍᓐᓃᖅᑐᖅ.”

—ᑎᒧᑎ ᑕᐅᓐᓕ

“ᐊᐱᓇᓵᖅᐸᓕᖅᑐᖅ; ᐅᑐᐱᕆ
ᐱᒋᐊᕐᓂᖕᒐᓂ ᐊᐱᕙᓚᐅᖅᓯᒪᔪᖅ. [...]
ᐅᑭᐅᓂ ᐊᓂᒍᓵᖅᑐᓂ ᖁᕕᐊᓱᒡᕕᐅᑉ
ᓯᕗᓂᑦᑎᐊᖕᒐᓂ ᐊᐱᕙᓕᖅᑐᖅ. [...]
ᐊᐳᑎᓕᐊᓘᕙᓚᐅᖅᓯᒪᔪᖅ ᓇᐃᓂᒥ
ᐅᑭᐅᑦ 10 ᐊᓂᒍᖅᑐᓂ, ᑭᓯᐊᓂ
ᒫᓐᓇ ᐊᐳᑎᖃᓗᐊᖅᐸᒍᓐᓃᖅᑐᒍᑦ.
ᐃᓛᓐᓂ ᐊᐱᐊᕐᔪᑐᐃᓐᓇᖅᐸᓕᖅᑐᖅ,
ᐊᐳᑎᒐᓛᒃᐳᑦᓗ ᐊᓄᕆᒧᑦ
ᑎᒃᑕᐅᑐᐃᓐᓇᖅᐸᓕᖅᑐᑦ; ᓄᓇ
ᐊᐱᓯᒪᕙᒍᓐᓃᖅᑐᖅ. ᐄ, ᒫᓐᓇ
ᐊᓄᕆᓂᖅᓴᐅᕙᓕᖅᑐᖅ.”

—ᑎᒧᑎ ᑕᐅᓐᓕ

“ᐊᐱᓇᓵᕐᓂᖅᓴᐅᕙᓕᖅᑐᖅ
ᐅᑭᐊᒃᑯᑦ, ᑎᓯᒃᑎᑉᐸᒍᓐᓃᖅᑐᖅ,
ᑰᑦᓗ ᖁᐊᖅᐸᒍᓐᓃᖅᑐᑦ. ᓯᑯᓗ
ᓵᑦᑐᒻᒪᕆᐅᓕᖅᔪᓂ.”

—ᐃᓚᐃ ᒥᖅᑯᕋᑦᓱᒃ

“The weather seems to be getting cooler all the time. It is now June and you can still see snow on top of the hill; that is strange. [...] The weather is so cold now that it doesn’t melt.”

—Timothy Townley

“The snow comes later now; it used to come in early October. [...] But these last few years we’ve seen snow only a few days before Christmas. [...] There was a lot of snow here in Nain before, maybe 10 years ago, but these years we’re hardly getting any snow. Sometimes it’s only a few centimetres at a time, and mostly the snow we get now is just blown away by the wind; it’s not sticking to the ground. Yes, there’s a lot more wind now.”

—Timothy Townley

“[Snow comes] later in the fall, [and] it doesn’t get hard enough, the river doesn’t freeze. The ice is really thin.”

—Eli Merkuratsuk

“ᐅᑭᐅᖅᑲᖅᑑᔮᖖᒋᓚᒍᑦ ᑕᒪᑐᒪᓂ ᐅᑭᐅᒥ. ᑕᐃᔅᓯᒪᓂᑑᖖᒋᑦᑐᖅ ᓱᖕᒋᐅᖕᒐᔭᑦᑎᑐᑦ. ᐅᑐᐱᓇᒥ ᐃᓄᐃᑦ ᖃᒧᑕᐅᔭᖅᐸᓚᐅᖅᑐᑦ, ᑕᐃᒪᐃᑉᐸᒍᓐᓃᖅᑐᑦ.”

—ᔪᓕᐊᔅ ᒥᖅᑯᕋᑦᓱᒃ

“We never had winter this past winter. Not like the winters we had before, [that] we’re used to. In October [in the past] people would be going by skidoo around town, but they don’t do that anymore.”

—Julius Merkuratsuk

ᓯᓚ/ᐊᓄᓇ

“ᐅᑭᐅᑦ 10-15 ᐊᓂᒍᖅᑐᓂ ᐊᐳᑎᖃᓗᐊᖅᓯᒪᖖᒋᑦᑐᒍᑦ. [...] ᓯᓚᑎᑦᑎᓐᓂ ᐹᕿᑦ ᐳᐊᕐᓇᔭᓇᐊᖅᑲᖅᐸᒍᓐᓃᖅᑕᒃᑲ ᐅᑭᐅᒃᑯᑦ.”

—ᑯᓇᔅᑏᓐ ᐸᐃᑭ

Climate/Weather

“We haven’t had what you would call a lot of snow for the last 10 or 15 years. [...] I never have to shovel snow away from the doorways anymore in the winter.”

—Christine Baikie

“ᒫᓐᓇᐅᓕᖅᑐᖅ, ᖃᓐᓂᓕᑐᐊᕐᒪᑦ, ᐅᑭᐊᒃᔮᑎᑐᑦ, ᐊᐱᔩᒐᓗᐊᑦ ᐊᓄᓇᒧᑦ ᑎᒃᑕᐅᑐᐃᓐᓇᖅᐸᓕᖅᑐᑦ. ᑕᐃᔅᓯᒪᓂᓕ ᖃᓐᓂᖅᐸᑦ ᐊᐱᒃᐸᑦ ᐊᐳᑎᖅᑲᖅᐸᓚᐅᖅᑐᒍᑦ ᐅᓪᓗᓄᑦ ᐱᖕᒐᓱᓄᑦ ᑕᓪᓕᒪᓄᓪᓘᓐᓃᑦ. ᖃᓐᓂᖅᐸᓚᐅᖅᑐᖅ ᐊᓄᕌᕋᓂ, ᐊᒡᓛᑦ ᐅᓪᓗᑦ ᐱᖕᒐᓱᑦ ᑕᓪᓕᒪᓪᓘᓐᓃᑦ ᐊᓂᒍᕋᓗᐊᖅᑎᓪᓗᒋᑦ ᐊᐳᑎᖃᐃᓐᓇᖅᐸᓚᐅᖅᓯᒪᔪᖅ.”

—ᔪᓕᐊᔅ ᐃᒃᑯᓯᖅ

“These days, as soon as it starts to snow, like in the fall, all the snow that comes down just blows off the roads and the land right away. But in the past we had snow [on the ground for] maybe three or four days. It kept snowing without any wind at all, and even after three or four days the snow didn’t blow away.”

—Julius Ikkusek

“ᑕᐃᔅᓱᒪᓂ ᐃᓄᑐᖅᑲᕗᑦ
ᖃᐅᔨᒪᑦᑎᐊᖅᐸᓚᐅᖅᓯᒪᔪᑦ ᓯᓚ
ᖃᓄᐃᑦᑑᓂᐊᕐᒪᖕᒑᑦ, ᒫᓐᓇᓕ
ᓇᓚᐅᑦᑖᒐᕐᓂᑭᓐᓃᖅᑐᖅ.
ᓯᓚᑦᑎᐊᕙᓪᒪᕆᐊᓘᓗᓂ
ᓯᓚᓗᓕᑳᓪᓚᒍᓐᓇᖅᓯᔪᖅ.”

—ᔪᓕᐊᔅ ᐃᒃᑯᓯᖅ

“ᐊᓄᕆᓂᖅᓴᐅᕙᓕᖅᑐᖅ,
ᐱᓗᐊᖅᑐᒥᒃ ᑲᓇᖕᓂᖅᑐᓂᒃ.
ᓯᓚ ᓇᓗᓇᖅᑐᐊᓘᓕᖅᑐᖅ. [...]
ᐊᓄᕌᖕᒋᓐᓇᐅᔭᖅᑰᔨᓕᖅᑐᖅ,
ᓇᓗᓇᖅᑐᐊᓘᓕᖅᑐᖅ.”

—ᑕᐃᔭᐅᔪᒪᙱᑦᑐᖅ

“ᐊᕐᕌᒍᓕᒫᖅ
ᐊᓄᕆᑐᓂᖅᓴᐅᖕᒋᓐᓇᐅᔭᖅᑰᔨᓕᖅᑐᖅ.”

—ᕗᕈᓇ ᐃᑦᑐᓪᓚᒃ

“ᐅᔾᔨᕈᓱᒃᑐᖕᒐ ᓄᓇᐃᑦ ᐃᓚᖕᒋᓐᓂ
ᐸᓂᖅᐸᓪᓕᐊᓕᕐᓂᖕᒋᓐᓂᒃ.
ᑕᓯᕋᓛᖅᑕᖅᑲᖅᐸᓚᐅᖅᑐᖅ,
ᐃᒪᖃᕈᓐᓃᖅᑐᑦ.”

—ᔮᓐ ᔭᕋᕈᔅ

“In the past our Elders knew exactly [...] what the weather was going to be like, but today it’s so unpredictable. In one day it could be really, really sunny, and then the weather will come down all of a sudden, out of nowhere.”

—Julius Ikkusek

“I find that there’s a lot more wind, especially from the north and northeast. I just find the weather is really unstable. [...] You almost always have wind, and it’s not predictable.”

—Anonymous

“The wind seems to be a lot stronger all year round.”

—Verona Ittulak

“I have noticed that [...] the ground is drying up in different areas. There used to be a lot of ponds, but they’re drying out; they don’t have water anymore.”

—John Jararuse

“ᐃᓄᑐᖃᐃᑦ ᐅᖃᖅᐸᓚᐅᖅᓯᒪᔪᑦ
ᖃᓪᓗᓈᑦ ᓄᓇᖓ
ᓛᐸᑐᐊᑎᑑᓯᓛᕐᓂᐊᕐᓂᖓᓂᒃ,
ᓛᐸᑐᐊᒥᓗ ᖃᓪᓗᓈᑦ ᓯᓚᖓᓐᓂᒃ
ᐊᔾᔨᖃᑎᓛᖅᑐᑕ.”

—ᐃᓕᓴᐱ ᐃᑦᑐᓪᓚᒃ

“ᒪᖁᖃᑦᑕᕐᓂᖓ ᐊᔾᔨᒋᔪᓐᓃᒻᒪᕆᒃᑕᖓ
ᖃᖕᒐᑲᓪᓛᖅᔪᒃᑎᑐᑦ.
[...] ᖃᖕᒐᑦᑎᐊᓵᖅᓂᑦ
ᒪᖁᖕᓂᖅᓴᓪᒪᕆᐅᕙᓕᖅᑐᖅ ᐅᑭᐅᒃᑯᑦ,
ᐃᒻᒪᑦᑎᐋᓴᖅ ᐱᓇᓱᐊᕈᓯᓕᒫᐸᓗᐊᓗᒃ
ᒪᖁᓚᐅᖅᑐᒍᑦ, ᓄᖅᑲᙱᑕᐃᓐᓇᖅᑐᓂ,
ᑕᒪᓐᓇ ᐊᓯᔾᔨᖅᓯᒪᒻᒪᕆᒃᑐᖅ.”

—ᐃᓕᓴᐱ ᐃᑦᑐᓪᓚᒃ

“ᓄᓇ ᐃᔾᔪᖅᓗ
ᐊᓯᔾᔨᖅᓯᒪᒻᒪᕆᒃᑐᖅ ᑕᑯᓯᒪᔭᕋ.
[...] ᓯᒡᔭᒥ ᒪᕋᐃᓐᓈᓘᑎᖅᑐᖅ—
ᑭᕙᐊᓚᓴᕐᓇᖅᓯᑦᑐᓂᓗ.”

—ᐃᓕᓴᐱ ᐃᑦᑐᓪᓚᒃ

“ᓯᓚ ᐊᓯᔾᔨᖅᐸᓪᓕᐊᕌᓚᓴᓕᖅᑐᖅ.
[...] ᐊᖓᔪᖅᑳᕗᑦ ᖃᐅᔨᒪᔪᓚᐅᖅᑐᑦ
ᓯᓚ ᖃᐅᒃᐸᑦ ᖃᓄᐃᑦᑑᓂᐊᕐᒪᖔᑦ—
ᒫᓐᓇ ᓇᓗᓇᒻᒪᕆᓕᖅᑐᖅ ᐅᓪᓛᒃᑯᑦ
ᒪᖁᓚᐅᕐᓗᓂ ᐅᓐᓄᒃᓴᒃᑯᑦ
ᓯᕿᓐᓂᑦᑎᐊᓕᕈᓐᓇᕐᒪᑦ.”

—ᑎᒧᑎ ᑕᐅᓐᓂ

“[The Elders] had predicted before that the white man’s land would become more like Labrador, and in Labrador we will get their weather instead.”

—Elizabeth Ittulak

“There’s a big difference in the rainfall compared to a few years ago. [...] Just recently we’ve been getting a lot of rain in the middle of winter, and just a few weeks ago we had rainfall for almost a whole week, nonstop, so that’s a big change.”

—Elizabeth Ittulak

“I’ve seen quite a change in the ground and the soil itself. [...] When you go to the beach it’s all like soft mud—you sink so fast.”

—Elizabeth Ittulak

“The weather is changing so rapidly. [...] Our parents could predict what the weather was going to be like tomorrow or for the next few days—today it is impossible to do that because it could be raining in the morning and sunny in the afternoon.”

—Timothy Townley

“ᐱᕈᖅᓴᑎᑦᓗᑕ, ᐃᒡᓗᕗᑦ
ᐊᐳᒻᒧᑦ ᐊᖁᓗᔭᐅᓯᒪᖁᓚᐅᖅᑐᑦ.
ᑐᒃᓯᐊᕐᕕᒃ ᐳᖅᑐᕙᒻᒪᓇᒃ,
ᐳᐊᕐᓇᔭᖅᑕᐅᔭᓇᐊᖅᑲᖅᐸᒃᓗᓂ
ᐊᐳᑎᖅᑲᓗᐊᕐᓂᖓᓄᑦ. ᑭᓯᐊᓂ
ᒫᓐᓇ ᐊᐱᓗᐊᖅᐸᒍᓐᓃᖅᑐᒍᑦ.
ᐊᒡᓛᑦ ᐊᐳᑎᖅᑲᖖᒋᓗᐊᖅᐸᓕᖅᑐᖅ
ᐱᓱᒡᕕᒃᓴᓂᒃ ᖃᒧᑕᐅᔭᕐᕕᒃᓴᓂᒡᓗ.”

—ᑎᒧᑎ ᑕᐅᓐᓕ

“ᑰᑦᑑᓐᓃᑦ ᑰᒍᓐᓃᖅᑐᑦ.
ᐃᒪᖅᑲᕈᓐᓃᖅᑐᑦ. ᐹᓂ
ᑕᓯᖅᑕᖅᑲᖅᐸᓚᐅᖅᑐᖅ [...]
ᐊᑖᑕᑦᑎᐊᓇᓚᐅᖅᑕᕋᓗ
ᐅᖅᑲᖅᐸᓚᐅᖅᑐᖅ ᑯᓛᑯᑕᐃᕙᒥᒃ
ᑕᓯᕐᒦᑦᑐᖅᑲᓚᐅᖅᓯᒪᓂᖓᓂᒃ.
[...] ᖃᖓᒐᑦᓚᐅᓗᐊᖏᑦᑐᖅ
ᑕᐅᕗᖖᒐᐅᓚᐅᖅᑐᒍᑦ
ᑕᑯᔭᖅᑐᑐᐃᓐᓇᖅᓗᑕ ᐊᑖᑕᑦᑎᐊᒪ
ᐅᖅᑲᐅᓯᓇᖁᓚᐅᖅᑕᖕᒋᓐᓂᒃ—
ᑭᓱᖅᑲᖖᒋᑦᑎᐊᖅᑐᖅ. [...]
ᐱᓱᒡᕕᒃᓴᐅᓕᖅᑐᖅ; ᐸᓂᖅᓯᒪᓕᖅᑐᖅ.”

—ᑳᑎ ᐊᒻᒪ ᐅᐃᓕᐊᒻ ᐅᐃᓐᑐᔅ

“ᐊᐳᑦ ᐱᐅᔪᓐᓃᓕᖅᑐᖅ;
ᓱᑲᑐᐃᓐᓇᐅᔫᓕᖅᑐᖅ. ᐊᐳᑦ
ᑕᐃᔅᓱᒪᓂᑐᑦ ᐊᔾᔨᖖᒋᓪᒪᓇᓕᖅᑕᖕᒐ.”

—ᑳᑎ ᐊᒻᒪ ᐅᐃᓕᐊᒻ ᐅᐃᓐᑐᔅ

“ᓇᐹᖅᑐᑦ ᓱᖅᑕᐅᖁᓕᖅᑐᑦ ᐊᐳᑎᓄᑦ
ᑎᓱᐊᔪᓄᑦ. ᖃᒧᑕᐅᔭᒃᑯᑦ ᐃᖕᒋᖅᕋᑦᓗᓂ,
ᑕᑯᓐᓇᓕᖅᑐᖅ ᓇᐹᖅᑐᓂᒃ
ᐅᖅᑯᖕᒐᔪᓂᒃ ᑕᒡᔪᐊᔪᓄᑦ ᓱᖅᑕᕕᓂᖅᓂᒃ.
ᐊᐅᒃᐸᓪᓕᐊᓕᑦᑐᖅ.”

—ᐃᓚᐃ ᒥᖅᑯᕋᑦᓱᒃ

“When we were growing up, our houses used to be buried with snow. You know, the church is pretty high, but they even used to have to shovel that out because of the snow. But these days we hardly get any snow. We'd be lucky to get [enough] snow to walk on or skidoo on.”

—Timothy Townley

“Even the brooks are dried up. No water. We had a big old pond over there [...] and my old grandfather used to tell us all that there was a crocodile in there. [...] Not too long ago we went over there just to see what my grandfather was talking about—there's nothing there. [...] You can walk on it now; it's dry.”

—Katie and William Winters

“The quality of the snow is different; it's just like sugar. The snow is not like it used to be.”

—Katie and William Winters

“The trees get broken down a lot by avalanches. When you travel on skidoo to the country, you can see the broken trees where the avalanches come down. It's melting.”

—Eli Merkuratsuk

“ᓯᑯᕙᒍᓐᓃᓪᒪᕆᒃᑐᖅ.
ᐊᑦᑕᕐᓇᐃᓐᓇᐅᔭᓕᖅᑐᖅ. […] ᐊᐳᑦ
ᖄᖕᒐᓃᑎᓪᓗᒍ ᖃᐅᔨᒪᓇᖖᒋᓪᒪᑦ
ᖃᓄᖅ ᐃᔾᔪᑎᒋᖕᒪᖕᒫᑦ. […]
ᒫᓐᓇ ᐊᑦᑕᕐᓇᖅᑐᓪᒪᕆᐅᓕᖅᑐᖅ
ᓄᒃᑕᕆᐊᒃᓴᖅ ᖃᒧᑕᐅᔭᒃᑯᑦ.”

—ᐃᓚᐃ ᒥᖅᑯᕋᑦᓱᒃ

“ᐱᖕᒐᓐᓇᔮᖅᐸᒍᓐᓃᖅᑐᒍᑦ.
[…] ᐊᑯᓂᐊᓗᖅ ᑕᐃᒪᓐᓇ
ᓯᓚᖃᕈᓐᓃᖅᑐᒍᑦ.
ᐱᖕᒐᓐᓇᔮᖅᐸᒍᓐᓃᖅᑐᖅ
ᐅᐊᖕᓂᖅᐸᒍᓐᓃᖅᓱᓂᓗ;
ᑲᓇᖕᓃᓐᓇᕐᓂᖅᓴᐅᓕᖅᑐᖅ
ᓂᒡᒋᕐᓂᖅᓴᐅᓕᖅᓱᓂᓗ.”

—ᐄᑐᐊᖅ ᐊᒻᒪ ᓗᐃᓴ ᕙᓚᐅᕗᔅ

“ᐃᓚᖕᒋᓐᓂ ᐊᕐᖀᒍᓂ ᓯᓚ
ᐅᖅᑰᓂᖅᓴᐅᕙᒃᑐᖅ, ᑭᓯᐊᓂᓗ
ᐃᓚᖕᒋᓐᓂ ᓂᒡᓚᓱᖕᓂᖅᓴᓪᒪᕆᐅᑦᑐᓂ.
ᓯᓚ ᐊᓯᔾᔨᖅᐸᑦᑕᐊᓪᒪᕆᒃᑐ.
[…] ᐃᓛᓐᓂ ᐅᑦᓗᓕᒫᖅ
ᐅᖅᑰᕙᓪᒪᕆᐅᓚᐅᕐᓗᓂ ᖃᐅᑐᐊᖅᐸ
ᓂᒡᓚᓯᓕᕐᓗᓂᐅᕙᒃᑐᖅ,
ᐊᓯᔾᔨᖅᐸᑦᑕᐊᖕᒋᓇᐅᔭᖅᑐᖅ.”

—ᒥᓂ ᐊᒻᒪ ᔭᐃᑯ ᒥᖅᑯᕋᑦᓱᒃ

“[The ice] doesn’t even freeze up. It’s dangerous all the time. […] The snow goes on top and you can’t tell [how thick the ice is]. […] It’s really, really dangerous now to travel by skidoo.”

—Eli Merkuratsuk

“We don’t get any more westerly winds. […] It’s been a long while now since we’ve had that weather. We can’t get any more westerly or northwesterly winds; it’s mainly easterly and northerly winds now.”

—Edward and Louisa Flowers

“In some years the weather is really warm, but in some years it’s really cold; that’s happening more often now than in the past. The weather is changing a lot. […] One day you will have a really, really hot day, and the next day it could be [cool], so it’s changing all the time.”

—Minnie and Jacko Merkuratsuk

“ᐅᑭᐅᓂ ᐊᑐᓕᖅᑕᑦᑎᓐᓂ
ᐊᐳᑦ ᓇᔪᓇᒻᒪᑎᓕᖅᑐᖅ.
ᐃᓛᓐᓂ ᑎᓯᔪᒻᒪᑎᐅᓪᓗᓂ
ᐅᖕᖏᓘᓐᓃᑦ ᐊᕐᑭᑦᑐᒻᒪᑎᐅᓪᓗᓂ
ᕐᑭᐅᒃᑲᖅᓯᒪᓪᓗᓂᓗ.
ᐊᑦᑕᕐᓇᕐᓂᖅᓴᐅᓕᖅᑐᖅ
ᓄᒃᑕᑎᐊᒃᓴᖅ, ᐊᒻᒪ ᖃᖓᑐᐃᓐᓇᖅ
ᐊᓄᑎᓯᑳᓪᓚᒍᓐᓇᓕᖅᕈᓂ.
ᓯᓚᐅᑉ ᖃᓄᐃᓐᓂᐊᕐᓂᖓ
ᓇᓚᐅᑦᑖᕆᓐᓇᖅᐸᓚᐅᖅᑕᕗᑦ,
ᑕᐃᒪᐃᑦᑐᓐᓇᐃᓪᓕᔪᒍᑦ.”

—ᒥᓂ ᐊᒻᒪ ᔭᐃᑯ ᒥᖅᑯᕋᑦᓱᒃ

“These years the snow is so unpredictable. It could be really hard sometimes, or really soft and wet at the same time. It is more dangerous to travel, as well, because the wind could come at any time. We used to predict weather before, but we can’t anymore.”

—Minnie and Jacko Merkuratsuk

ᐃᓄᐃᑦ ᐱᖅᑯᓯᑐᖃᖓᑦᑕ ᐊᒃᑐᖅᑕᐅᓂᖓ

“ᖃᓐᓂᖅᑐᒻᒪᑎᐊᓘᔭᓐᓇᖅᑐᖅ,
ᑭᓯᐊᓂ ᑎᒃᑕᐅᑐᐃᓐᓇᕐᓂᐊᖅᑐᓂᒃ.
ᐅᑭᐅᓂ ᐊᓂᒍᓵᖅᑐᓂ
ᐊᐳᑎᓯᐊᕐᓇᕈᓐᓇᐃᓪᓕᖏᓪᓕᐊᔭᖅ
ᐃᒡᓗᓕᔅᔪᑎᒃᓴᓂᒃ. [...] ᒫᓐᓇ
ᐊᐅᓪᓛᕐᓂᐊᕌᖓᑦᑕ ᐊᖑᓇᓯᒋᐊᕐᓗᑕ
ᑐᐱᒋᑎᐊᕐᑲᖅᐸᓕᖅᑐᒍᑦ ᐊᐳᑦ
ᐃᒡᓗᓕᕝᕕᒃᓴᐅᔪᓐᓃᕐᓂᖓᓄᑦ [...]
ᐅᑭᐅᒻᒪᑎᐅᒐᓗᐊᒃᑯᑦ.”

—ᔪᓕᐊᔅ ᐃᒃᑯᓯᖅ

“ᐃᖅᑲᐅᒪᕗᖓ ᐅᑭᐅᒃᑯᑦ,
ᒪᕐᑯᒃᐸᓚᐅᖅᓯᒪᙱᑦᑎᐊᖅᑐᖅ;
ᓯᑯᓪᓗᓂ ᓯᑯᓯᒪᐃᓐᓇᖅᐳᓂᓗ
ᐅᐱᕐᖔᒧᑦ, ᒪᕐᑯᕐᑲᑦᑕᓕᑕᐃᓐᓇᖅᐳᓂ.
ᒫᓐᓇ ᒪᕐᑯᒃᐸᓕᖅᑐᖅ
ᕐᑯᐊᖅᐸᓪᓕᐊᓕᖅᑎᓪᓗᒍ

Impacts on Traditional Ways of Life

“There will be a lot of snow coming down, but it all blows off. These years it’s hard to find good snow to make a snow house, really hard. [...] These days when you go on the land or anywhere at all for hunting, you have to take a tent along, because the snow is not fit for a snow house [...] even in the middle of winter.”

—Julius Ikkusek

“I remember winters, we never got rain; the ice froze and it stayed frozen until spring, and then you got rain. Now the rain

ᑕᓇᐅᖑᙱᑦᑐᒥᒃ ᓱᑯᖃᓱᐅᔾᔭᕙᓪᓕᐊᓪᓗᓂ ᐃᒪᖅ ᓱᑯᕙᓪᓕᐊᑎᓪᓗᒍ, ᑕᓇᐅᑉ ᓱᑯᔫᑉ ᐊᔾᔨᙱᑦᑎᒻᒪᒍ, [...] ᐊᐅᒃᐸᓪᓕᐊᓕᖅᑎᓪᓗᒍ, ᓱᒃᑲᓕᔪᒥᒃ ᐊᐅᕚᓚᕐᓂᖅᓴᐅᕙᒃᑐᖅ. ᑕᒪᓐᓇ ᐊᒃᑐᐃᓂᖃᖅᓯᒪᓪᓚᓇᒃᑐᑦ ᐃᓄᖕᓄᑦ ᐊᖑᓇᓱᒍᓯᖕᒋᓐᓄᓪᓗ ᓱᑯᒥ.”

—ᑕᐃᔭᐅᔪᒪᙱᑦᑐᖅ

“ᐊᐳᑦ ᐊᖅᑭᓐᓂᖅᓴᐅᕙᓕᖅᑐᖅ. ᑕᐃᔅᓱᒪᓂ ᐃᒡᓗᓕᕐᕕᒃᓴᐅᕙᓚᐅᖅᑐᖅ ᐊᐳᑦ ᓈᒻᒪᖕᓂᖕᒐᓄᑦ, ᒫᓐᓇᓕ ᐊᖅᑭᑦᑐᐊᓘᕙᓕᖅᑐᖅ. ᑕᖅᑮᖕᓄᒃ ᒪᖅᕉᖕᓄᒃ ᐅᑕᖅᑭᔭᓇᐊᖃᖅᐸᓕᖅᑐᒍᑦ ᐅᕝᕙᓘᓐᓃᑦ ᒥᓂᓚᐅᖅᑎᓪᓗᒍ ᑎᓯᒃᑎᑦᑎᐊᓇᓗᒍ ᐃᒡᓗᓕᔪᒪᒍᑦᑕ.”

—ᒫᑕ ᐅᒃᑯᐊᖅᑕᓯᐊᖅ

“ᐃᖅᓂᒃᑲ ᓇᒧᙵᐅᔪᓐᓇᓚᐅᙱᑦᑎᐊᖅᑐᑦ ᐅᐱᖅᖔᖅ ᓱᑯᖃᑦᑎᐊᙱᓐᓂᖕᒐᓄᑦ, ᐊᑦᑕᖅᓇᖅᑐᒻᒪᓇᐊᓘᓚᐅᖅᒪᑦ. ᐊᖑᓇᓱᒋᐊᕈᓐᓇᖅᓯᓚᐅᙱᑦᑐᑦ.”

—ᒫᑕ ᐅᒃᑯᐊᖅᑕᓯᐊᖅ

is mixing in with the freezing process and you’re getting the properties of freshwater ice, and it doesn’t have the same strength as the sea ice [used to], [...] so when it does start to melt, it melts really fast. There’s been a significant impact on people’s patterns for harvesting out on the ice.”

—Anonymous

“The snow is more powdery these years. In those years it was easy to make a snow house because the snow was good for it, but these years it’s all powdery. You would have to wait a couple of months or after it rained a little bit to make a snow house.”

—Martha Okkuatsiak

“My sons could not go anywhere over the spring because of the [lack of] ice, because it was so dangerous. They couldn’t go to their natural hunting places.”

—Martha Okkuatsiak

Nunatsiavut / Nain

“ᓱᑯᕙᖕᓂᖓ
ᐊᓯᔾᔨᖅᓯᒪᔪᒻᒪᕆᐅᓕᖅᑐᖅ, ᐊᒡᓛᓪ
ᓯᑯᐃᑦ ᑕᓯᕐᓂ. ᐃᓄᐃᓪᓘᓐᓃᑦ
ᐊᐅᓚᓴᕆᐊᖅᐸᒍᓐᓃᖅᑐᑦ
ᐃᖃᓪᓕᐊᖅᐸᒍᓐᓃᖅᑐᑦ ᑕᓯᕐᒧᑦ
ᓯᑯᑭᓗᐊᒧᑦ ᐊᑦᑕᕐᓇᓗᐊᓕᕐᓂᖓᓄᑦ.
[...] ᐊᒃᑐᐃᓂᖅᑲᓪᓚᕆᒃᓯᒪᔪᖅ
ᐊᖑᓇᓱᒃᑎᓄᑦ, ᐃᓕᖅᑯᓯᕐᒥᓂᒃ
ᐊᑐᕈᓐᓃᓕᕐᒥᒃ.
ᐊᖑᓇᓱᖕᓂᖅᐸᐅᕙᖕᒪᑕ ᐅᐱᕐᖔᒃᑯᑦ,
ᑭᖑᓂᖅᑲᓈᕈᔾᔨᓪᓗᑎᒡᓗ.”

—ᕗᕈᓇ ᐃᑦᑐᓪᓚᒃ

“ᓯᑯ ᐊᑦᑕᕐᓇᕐᓂᖅᓴᒻᒪᕆᐅᓕᖅᑐᖅ
ᐃᖏᕐᕋᕝᕕᒋᒋᐊᖓ, ᓇᓗᓇᖅᓯᖕᒪᑦ
ᐃᔾᔪᒐᓗᐊᕐᒪᖔᑦ ᓴᓗᐊᕐᒪᖔᓪᓘᓐᓃᑦ.”

—ᔮᓐ ᔭᕋᕈᔅ

“ᐃᓄᐃᓪᓗ ᐊᓯᔾᔨᓕᕐᒥᔪᑦ. ᐃᓄᐃᑦ
ᐃᓅᖅᑲᑎᒥᖕᓄᑦ ᐊᓯᔾᔨᖅᐸᓪᓕᐊᓕᕐᒥᔪᑦ.
[...] ᓇᒡᓕᒃᑎᒌᒃᐸᓚᐅᖅᑐᑦ
ᓇᒡᓕᒃᑎᒌᒍᓐᓃᓕᕐᒥᔪᑦ. ᑕᒪᓐᓇ
ᐊᓯᔾᔨᓕᖅᑐᓂᒃ ᐅᔾᔨᕆᓯᒪᖕᒥᔭᕋ. ᓯᓚ
ᐊᓯᔾᔨᖅᐸᓪᓕᐊᔪᑑᙱᑦᑐᖅ; ᐃᓄᐃᑦ
ᓂᕐᔪᑎᑦ ᐱᕈᖅᑐᓪᓗ ᐊᓯᔾᔨᓕᕐᒥᔪᑦ.”

—ᐃᓕᓴᐱ ᐃᑦᑐᓪᓚᒃ

“ᒪᓃᓛᓘᓕᕐᓂᖃᖅᑕᐅᔪᖅ.
ᐃᓄᐃᑦ ᐊᖑᓇᓱᒋᐊᖅᑑᒐᓗᐊᑦ
ᐅᑎᑐᐃᓐᓇᖅᐸᓕᖅᑐᑦ
ᒪᓂᓚᑎᒎᕈᓐᓇᙱᑦᑕᖄᖓᒥᒃ.”

—ᑳᑎ ᐊᒻᒪ ᐅᐃᓕᐊᒻ ᐅᐃᓐᑐᔅ

“There’s a big change in ice patterns, and also ice in the lakes. It’s even hard for people to go ice fishing up in the lakes because of the danger of the thinness of the ice. [...] It’s affecting the hunters quite a lot, because they’re missing their traditional ways. They usually do their hunting in the spring, and they’re missing that.”

—Verona Ittulak

“The ice is much more dangerous to travel on now, because you don’t know if it’s thick or not.”

—John Jararuse

“People are changing as well, not only the environment. People are changing toward each other as well. [...] People who used to love each other are not loving each other anymore; they’re not so close in ties anymore. That is a part of the change I have been seeing today. It’s not only the climate that has been changing; it’s also the people and the wildlife and the plants.”

—Elizabeth Ittulak

“ᐆᒪᔪᓕᒫᑦᑎᐊᑦ—ᑎᖕᒥᔾᔪᑦ
ᓂᕆᕙᒃᑕᕗᑦ, ᓇᑦᑏᑦ
ᓂᕆᕙᒃᑕᕗᑦ—ᑭᓯᓕᒫᑦ
ᓴᓗᒃᑎᑉᐸᓪᓕᐊᖏᓐᓇᐅᔪᓕᖅᑐᑦ.”

—ᐋᑐᐊᖅ ᐊᒻᒪ ᓗᐃᓴ ᕙᓚᐅᕗᔅ

“ᓄᓇᐅᑉ ᐃᑭᐊᖕᒐᓂ
ᓂᓚᖅᑲᖅᐸᓚᐅᖅᓯᒪᔪᖅ,
ᓂᓚᒃᑕᖅᐸᓚᐅᖅᑐᒍᑦ
ᐃᒪᕆᓂᐊᖅᑕᑦᑎᓐᓂᒃ. ᒫᓐᓇ
ᑕᐃᒪᐃᑦᑐᖅᑲᕈᓐᓃᖅᑐᖅ; ᓄᓇᐅᑉ
ᐃᑭᐊᓂ ᓂᓚᒃᑕᖅᑲᖅᐸᒍᓐᓃᖅᑐᖅ.”

—ᐋᑐᐊᖅ ᐊᒻᒪ ᓗᐃᓴ ᕙᓚᐅᕗᔅ

“ᖃᓐᓂᖀᖕᒑᑦ ᓇᓗᓇᖅᐸᓕᖅᑐᖅ
ᓇᐅᒃᑯᑦ ᓯᑯ ᐊᑦᑕᖅᓇᖖᒋᒻᒪᖕᒑᑦ.
[...] ᓯᑯᓚᐅᖅᑎᓪᓗᒍ ᖃᓐᓂᖀᖕᒑᑦ,
ᖃᐅᔨᒪᓇᖅᐸᒍᓐᓃᖅᑐᖅ ᓯᑯᑦ ᐃᔾᔪᔪᑦ
ᓵᓗᐊᖅᑐᓪᓗ. ᑕᐃᔅᓯᒪᓂᓕ ᓯᑯᒥᒃ
ᖃᐅᔨᒪᓂᖅᓴᐅᕙᓚᐅᖅᑐᒍᑦ.”

—ᔭᐃᑯ ᒥᖅᑯᕋᑦᓱᒃ

“They say there’s lots and lots of ice outside. People try to go hunting, but they have to come back because they can’t get through the ice.”

—Katie and William Winters

“All species of wildlife—the birds that we eat, the seals that we eat—everything is getting skinnier all the time.”

—Edward and Louisa Flowers

“There used to be ice under the ground, and we used to [gather] ice for melting water. These days there’s nothing like that; there’s no ice under the ground anymore.”

—Edward and Louisa Flowers

“When it snows it is now hard to tell which parts of the ice are good and which parts of the ice are not safe. [...] When it freezes over and then it snows, we can’t tell the difference between the good and the bad. In the past the ice was more reliable.”

—Jacko Merkuratsuk

Nunatsiavut / Nain

Baker Lake
Our land our knowledge
Caribou • Tuktut
Dwarf Birch • Avaalaqiat
Sorrel
Grasses • Iviit
Thomas Qaqimat
"The sand is covering everything. All the berries, willows and birches are smaller and shorter nowadays; it really dried up. Long time ago [...] when there was rain more often, they would grow taller and faster, the berries, the shrubs and the grasses, and the caribou would be fatter."
Martha Nukik
"The Kazan falls are really far from Baker Lake. Years ago we used to hear their noise but since the water levels are going down we don't hear them anymore. [...] I used to see lots of those mountain sorrel leaves (along the Kazan River) [...], the shore was reddish because of it but nowadays you don't see that anymore."
Basil and Julie Tuluqtuq
"There is a lot of grasses (iviit) nowadays, growing where people used to have tents. Birch (avaalaqiaq) was always there but I was really surprised to see how much grass there was."
Very special thanks are due to Qamani'tuamiut Silas Aittauq, Margaret Amauruq-Niviatsiaq, Norman Attungala, Hattie Attutuvaa, Paul Attutuvaa, Vera Avaloa, May Haqpi, Jacob Ikinilik, Winnie Ikinilik, Janet Ikuutaq, Toona Iqulik, Irene Kaluroq, Martin Kreelak, Mary Kreelak, John Nukik, Martha Nukik, Thomas Qaqimat, Lucy Qaunaq, Elizabeth Quinangnaq, Joan Scottie, Simon Tookoome, Basil Tuluqtuq, Julie Tuluqtuq and Lucy Tunguaq. It is with great respect that we would like to thank them for sharing with us their knowledge of the land and their experiences. Their generosity and welcoming manner made this work possible. Special thanks also to the interpreters Vera Avaala, Thomas Mannik and Hattie Mannik for their essential role in conducting the interviews. Finally, we want to express our gratitude for the resources at the Qamani'tuaq Heritage Centre which helped to facilitate our work.
Photo credits: Carmen Spiech (Grasses), Sylvie Ferland (Caribou, Dwarf Birch)
Contact: Dr. Alain Cuerrier, 514.872.3182, alain.cuerrier@umontreal.ca
Institut de recherche en biologie végétale
CICAT
2007-2008 IPY·API
ArcticNet
UQTR
Centre d'études nordiques
Université Laval

ᐊᑭᓐᓇᕐᒥᐅᑕᑦ

ᕐᑯᔭᒋᔪᒪᓪᓗᑎᒍᑦ ᐃᓄᑐᕐᑲᐃᑦ ᐊᐱᕐᒃᓱᕐᒃᑕᐅᕐᑲᑕᐅᓚᐅᕐᒃᑐᑦ ᕐᑲᐅᔨᒪᔨᑕᐅᔪᓪᓗ ᓄᓇᓕᖕᓂ, ᓂᕈᐊᕐᒃᓴᕐᒃᓱᑕ ᐅᕐᑲᕐᒃᑕᐅᓯᒪᔪᓂᒃ ᐱᒻᒪᕆᐅᔪᓂᒃ ᐊᐱᕐᒃᓱᓚᐅᕐᒃᑕᑦᑎᓐᓂᒃ ᐊᑭᓐᓇᕐᒥᐅᑕᒃᓴᓕᐊᕆᓯᒪᔭᕗᑦ ᑕᑯᒃᓴᐅᑎᑦᑎᓇᓱᐊᕐᒃᓱᑕ ᐊᔾᔨᒌᙱᑦᑐᓂᒃ ᐊᓯᔾᔨᕐᒃᐸᓪᓕᐊᓂᐅᔪᓂᒃ ᐃᓄᐃᑦ ᓄᓇᖏᓐᓂ ᑲᓇᑕᒥ. ᑕᒪᐃᓐᓄᑦ ᓄᓇᓕᖕᓄᑦ, ᐱᖓᓱᓂᒃ ᑎᓴᒪᓂᓪᓘᓐᓃᑦ ᐊᑭᓐᓇᕐᒥᐅᑕᒃᓴᓕᐅᕐᒃᓯᒪᔪᒍᑦ ᐅᑯᓄᖓ ᐊᖏᒃᑐᕐᒃᓯᒪᓪᓗᑎᒍᑦ ᐊᕙᑎᑦᑕ ᐊᓯᔾᔨᕐᒃᐸᓪᓕᐊᓂᖓᓄᑦ: ᓯᓚᐅᑉ ᐊᓯᔾᔨᕐᒃᐸᓪᓕᐊᓂᖓ, ᐱᕈᕐᒃᐸᒃᑐᑦ, ᓄᓂᕙᒃᓴᑦ, ᐊᒻᒪ ᐆᒪᔪᑦ. ᐅᕐᑲᕐᒃᑕᐅᓯᒪᔪᓕᒫᑦ ᐅᕐᑲᕐᒃᑑᑉ ᐊᔾᔨᙳᐊᖕᒐᓂᒃ ᐊᔾᔨᙳᐊᕐᑲᕐᒃᑐᑦ ᐊᔾᔨᙳᐊᒃᑲᓐᓂᕐᑲᕐᒃᓱᑎᒃ ᐅᕐᑲᐅᓯᓇᔭᖕᒋᑕ ᒥᒃᓵᓅᖕᒐᔪᓂᒃ. ᐅᕐᑲᕐᒃᑕᐅᓯᒪᔪᓕᒫᑦ ᐅᕐᑲᓚᐅᕐᒃᑐᒧᑦ ᐊᖕᒋᕈᑕᐅᓯᒪᔪᑦ. ᑕᒪᐃᓐᓄᑦ ᓄᓇᓕᖕᓄᑦ, ᐊᑭᓐᓇᕐᒥᐅᑕᑦ ᐅᕐᑲᐅᓯᖕᒋᓐᓄᑦ ᑐᑭᓕᐅᕐᒃᑕᐅᓯᒪᔪᑦ ᐃᓄᒃᑎᑑᓕᕐᒃᑎᑕᐅᓯᒪᔪᑦ.

Posters

To pay tribute to the participating Elders and knowledge holders in each community, we selected a number of important quotations from our interviews and presented them on posters to highlight and reflect the diversity of changes being experienced across Inuit communities in Canada. For each community, we created three or four posters that covered the following categories of environmental change: climate and weather, vegetation, berries, and animals. Each quotation was accompanied by the interviewee's photograph and an additional image related to the quotation's theme. All quotations and images were validated with the interviewees who were shown poster drafts. For each community, the posters were translated into the local dialect and the other regional languages.

ᓄᖅᑲᕈᑎᒥᔭᖕᒥᑦ

ᑖᒃᑯᐊ ᓄᐊᑕᐅᔪᑦ ᐃᓄᐃᑦ ᐃᓱᒪᒋᔭᖕᒋᓐᓂᒃ ᐊᕙᑎᐅᑉ ᐊᓯᔾᔨᖅᐸᓪᓕᐊᓂᖕᒐᓄᑦ 8-ᖑᔪᓂ ᑲᓇᑕᒥ ᐅᑭᐅᖅᑕᖅᑐᒥ ᓄᓇᓕᖕᓂ ᐅᖃᐅᓯᖃᖅᓯᒪᔪᑦ ᓄᓇᐃᑦ ᓇᓃᓐᓂᕆᔭᖕᒋᑦ ᓄᓇᖓᑕᓗ ᐱᕈᖅᑐᖅᑲᕐᓂᖕᒋᑦ ᐃᓱᒪᒋᔭᐅᓪᓗᑎᒃ ᐱᓗᐊᖅᑐᒥᒃ ᓄᓂᕙᒃᓴᓂᒃ. ᑕᒪᐃᓐᓂᑲᓴᒃ, ᓄᓇᓖᖅ ᐱᕈᖅᑐᖅᑲᕐᓂᖅᓴᐅᓕᕐᓂᖁᐃᔪᑦ, ᓇᓗᓇᐃᖅᑕᐅᓯᒪᔪᑎᒍᑦ ᓯᓚᕐᔪᐊᕐᒥ ᐅᑭᐅᖅᑕᖅᑐᒥ "ᑐᖑᔪᖅᓯᕙᓪᓕᐊ" ᓂᐅᓂᖁᖅᑕᐅᔪᒥᒃ. ᑕᒪᓐᓇ ᐱᓕᕆᐊᕆᔭᕗᑦ ᐃᓄᐃᑦ ᐆᒪᔪᐃᑦ ᒥᒃᓵᓄᑦ ᖃᐅᔨᒪᔭᖕᒋᓐᓂᒃ ᐃᑲᔫᑎᖃᕐᓂᐊᖁᓱᒋᓪᓗᑎᒍᑦ ᐅᑭᐅᖅᑕᖅᑐᕐᒥ ᓱᒃᑲᓕᔪᒥᒃ ᓯᓚ ᐊᓯᔾᔨᖅᐸᓪᓕᐊᓕᖅᑎᓪᓗᒍ. ᑖᒃᑯᐊ ᐊᑑᑎᖃᕈᓐᓇᕐᓂᐊᕐᑯᓗᐊᖅᐸᕗᑦ ᓄᓇᓕᖕᓄᑦ, ᐃᖅᑲᓇᐃᔭᕐᕕᓄᑦ, ᐃᓕᓐᓂᐊᕐᕕᓄᓪᓗ ᑎᑎᕋᐅᓯᐅᕙᓕᖅᑎᓪᓗᒍ ᓯᓚᐅᑉ ᐊᓯᔾᔨᖅᐸᓪᓕᐊᓂᖓ ᐃᓄᐃᑦ ᐃᓱᒪᒋᔭᖕᒋᓐᓂᑦ ᐅᖃᐅᑎᐅᑎᓪᓗᒋᑦ.

ᖁᕕᐊᒋᒻᒪᕆᓚᐅᖅᐸᕗᑦ ᐃᓄᐃᑦ ᐱᓕᕆᖃᑎᒋᒋᐊᖕᒋᑦ ᑐᓵᓪᓗᒋᓪᓗ ᐅᖃᐅᓯᒃᓴᖕᒋᑦ ᐅᔾᔨᕆᓯᒪᔭᖕᒋᓪᓗ ᐊᕙᑎᐅᑉ ᒥᒃᓵᓄᑦ. ᖃᐅᔨᒪᔭᑎᓐᓄᑦ ᐃᑲᔫᑎᖅᓯᒪᒻᒪᕆᒃᑐᑦ ᐊᒥᓱᑎᒍᑦ. ᐃᓄᐃᑦ ᖃᐅᔨᒪᔭᕐᒥᓂᒃ ᐅᖃᐅᓯᖃᕆᐊᒃᓴᖅ ᖁᕕᐊᓱᒻᒪᕆᒃᐸᓚᐅᖅᑐᑦ, ᐱᓗᐊᖅᑐᒥᒃ ᖃᐅᔨᒪᒍᒥᒃ ᐅᓂᒃᑳᑎᒃ ᐃᑲᔫᑎᖃᕐᓂᐊᕐᓂᖏᓐᓄᑦ ᑭᖑᕚᖏᓐᓄᑦ.

Closing Remarks

This compilation of Inuit perceptions of environmental changes among eight Canadian Arctic communities and along latitudinal and longitudinal gradients has a special focus on vegetation changes, with a particular berry flavour. In general, all communities are experiencing more vigorous vegetation growth, consistent with the circumpolar "greening" of the Arctic. We think this work contributes to promoting Inuit local ecological knowledge as a unique reference in this era of rapid changes in the Arctic regions. We hope this legacy will be useful for communities, institutions, and schools for documenting climate change impacts from an Inuit perspective.

It was a real pleasure for us to work with Inuit and listen to what they had to say as keen observers of their environment. They have contributed to our knowledge in many ways. People were very happy to share their knowledge, especially because it provided an opportunity to keep their memories alive for the benefit of future generations.

ᑖᓐᓇ ᐅᕐᑲᓕᒫᒐᖅ ᐱᓕᕆᐊᕐᑲᕐᑲᑎᒌᑦᑎᐊᕐᓂᕐᒧᑦ ᓴᖅᑭᑕᐅᔪᓐᓇᖅᓯᒪᔪᖅ, ᐃᖅᑲᓇᐃᔭᖅᑎᒌᑦ ᓄᓇᓕᖕᓂ ᐅᖕᒐᓯᒌᒃᑐᒻᒪᕆᐅᒐᓗᐊᖅᕗᑎᒃ—ᓲᕐᓗ ᓄᓇᓖᑦ ᐱᓕᕆᐊᕐᑲᕐᑲᑎᒋᓚᐅᖅᑕᕗᑦ. 7-ᓂᒃ ᐅᕐᑲᓪᓚᐅᓯᕐᑲᕐᑲᑎᒌᖖᒋᑦᑐᓂᒃ ᐱᓕᕆᐊᕐᑲᕐᑲᑎᕐᑲᖅᕗᑕ ᐊᓯᖕᒋᓐᓂᒡᓗ ᐊᒃᓱᕈᕐᓇᖅᑐᒃᑰᖅᐸᒃᕗᑕ. ᐅᕙᑦᑎᓐᓄᑦ, ᐅᔾᔨᕈᓱᓚᐅᖅᐳᒍᑦ ᐃᓄᐃᑦ ᐅᕐᑲᐅᓯᖕᒐ ᓴᖖᒋᕚᓂᖕᒐᓂᒃ ᐊᔾᔨᖖᒋᑦᑑᔪᓐᓇᕐᓂᖕᒋᓪᓗ. ᓯᕗᓂᒃᓴᑦᑎᓐᓂ, ᐅᕐᑲᓕᒫᒐᓕᐅᒃᑲᓐᓂᓛᕈᒪᕗᒍᑦ ᐊᔾᔨᖖᒋᑦᑐᓂᒃ, ᓄᓇᓕᖕᓂᓗ ᐊᓯᖕᒋᓐᓂᒃ ᐊᐱᖅᓱᕆᐊᖅᑐᕐᓗᑕ.

ᑭᖕᒍᓕᖅᐹᕐᒥ, ᕐᑯᔭᒋᔪᒪᕙᕗᑦ ᑕᒪᕐᒥᒃ ᐃᓄᑐᕐᑲᐃᑦ, ᕐᑲᐅᔨᒪᔨᑕᐃᑦ, ᑐᓵᔩᑦ ᐃᑲᔪᕐᓂᕐᑲᓚᐅᖅᑐᑦ ᐱᓕᕆᐊᕐᑲᕐᑲᑎᒌᒃᑎᑦᑐᑕ, ᐱᓗᐊᖅᑐᒥᒃ ᑕᐃᒃᑯᐊ ᐱᒋᐊᓚᐅᖅᑎᓪᓗᑕ ᑭᖕᒍᕐᖕᒐᒍᑦ ᐃᓅᔪᓐᓃᖅᓯᒪᓕᖅᑐᑦ.

This book is the result of great teamwork, with colleagues located along an east-west gradient—like the communities we worked with! Dealing with seven different languages and dialects was another challenge when getting this work published. For us, it was also a reconnaissance of the richness and diversity in Inuit language. In future, we plan to produce different versions of this book in the different dialects spoken in the communities where we conducted interviews.

Finally, we want to thank again all the Elders, knowledge holders, and interpreters who contributed to this collaborative work, especially as some of them have passed away between the beginning of this work and its completion.

ᐃᓱᒪᒋᔭᖕᒋᑦ ᓄᖅᑲᐱᑎᒋᓪᓗᒋᑦ

ᐃᓄᐃᑦ ᑎᓴᒪᐅᔪᖖᒋᒐᖅᑐᓂᒃ ᓄᓇᓖᑦ—50 ᓄᓇᓖᑦ ᑲᓇᑕᒥ ᐃᓄᐃᑦ ᓄᓇᖕᒋᓐᓂ—ᑐᓴᖅᑕᐅᑎᑦᑎᓪᒪᑕ ᐅᕙᑦᑎᓐᓂ ᖃᐅᔨᓯᒪᔭᖕᒋᓐᓂᒃ ᐃᓱᒫᓘᑎᖕᒋᓐᓂᓪᓗ ᓯᓚ ᐱᔾᔪᑎᒋᓪᓗᒍ. ᐊᓯᔾᔨᖅᓃᑦ ᓂᕆᐅᓇᖅᓂᑦ ᓱᒃᑲᓕᓯᒪᖕᒪᑦ, ᓄᓇᑦᑎᐊᕚᓗᖕᒐᓐᓂᑦ. ᓯᓚᐅᑉ ᐅᖅᑰᓯᕙᓪᓕᐊᓂᖕᒐ ᐱᑦᑐᒍ, ᓴᙱᑦᑎᑉᐸᓪᓕᐊᑐᐃᓐᓇᖅᓂᐊᖅᑐᑦ, ᐃᓄᐃᑦ ᓄᓇᖕᒋᓐᓂ ᓴᖅᑭᔮᓪᓚᑎᒃᑐᒥᒃ ᑕᐃᒪᐃᓐᓂᐊᕆᕗᖅ. ᑲᑎᖅᓱᖅᓯᒪᔭᕗᑦ ᑖᒃᑯᐊ ᐱᒻᒪᕆᐊᓗᐃᑦ, ᒐᕙᒪᖅᑲᖅᑯᖕᓂ ᐃᓱᒪᓕᐅᕆᔩᓪᓗ ᓯᕗᓕᖅᑏᓪᓗ ᐊᑏᑐᖅ ᓈᓚᑦᑎᐊᕐᓂᐊᖅᐳᑦ ᐃᓄᐃᑦ ᑐᓴᖅᑕᐅᑎᑕᖕᒋᓐᓂᒃ ᐅᕙᑦᑎᓐᓄᑦ, ᓱᓕᔪᕈᓱᓪᓗᑎᒡᓗ ᐱᓕᕆᖃᑎᖃᕋᓱᐊᕐᓂᐊᖅᐳᑦᑐᖅ ᑐᑭᓯᒃᑲᓐᓂᕐᑯᓪᓗᒋᑦ ᐊᓯᔾᔨᖅᑐᓂᒃ ᐅᑭᐅᖅᑕᖅᑐᒥ ᐊᓯᔾᔩᓂᐊᖅᑐᓂᒃ. ᐊᕐᕌᒍᐃᑦ ᐊᑐᖅᑕᕗᑦ 2000ᖕᒋᓐᓃᓕᖅᑎᓪᓗᒋᑦ; ᐃᓄᖕᓄᑦ ᐊᕙᑎᒧᓪᓗ ᑲᑉᐱᐊᓇᖅᑐᒦᑎᑕᐅᓕᕐᑕ ᑕᐃᔅᓱᒪᓂᐅᓚᐅᖅᓯᒪᔪᒥ. ᑲᓇᑕᒥᐅᑦ ᐊᑕᖕᒋᕐᓗᑎᒃ ᐱᓕᕆᖃᑎᒌᒋᐊᖃᓕᖅᑐᑦ ᖃᓄᖅᑑᕐᓗᑎᒃ ᖃᓄᖅ ᐋᖅᑮᒋᐊᕈᓇᕐᒪᖔᑦᑕ ᐅᑭᐅᖅᑕᖅᑐᒥᐅᑦ ᓯᕗᓂᒃᓴᖃᑦᑎᐊᕐᑯᓪᓗᒋᑦ. ᐱᒻᒪᕆᓪᓚᑎᐊᓘᕗᖅ ᑲᓇᑕᒥᐅᑦ ᑭᐅᑦᑎᐊᕐᑯᓪᓗᒋᑦ ᓯᓚᐅᑉ

Final Thoughts

Inuit from eight communities—a subset of the more than 50 Canadian Inuit communities—shared their perceptions and concerns regarding environmental changes with us. Such changes are happening at an unprecedented rate within their land, their beautiful Nunangat. As a result of global warming, these changes will continue to gather force, with Inuit community members at the forefront of this reality. If we reflect on the important information we have gathered, we hope that political figures and leaders will listen carefully to what Inuit have shared with us, and embrace a true partnership with them in order to learn and understand more about these changes and how they have and will continue to transform the North. We are now in the 21st century; we are exposed to greater social and environmental threats than in the past. Canadian society as a whole must collaborate and share ideas and solutions with Northerners to ensure a healthy future for all citizens. This is especially important as Canada

ᐊᓯᔾᔨᕐᓂᖓᑕ ᒥᒃᓵᓄᑦ. ᐃᓄᐃᑦ ᓇᐅᑦᑎᖅᓱᖅᑎᐅᒻᒪᑕ ᐊᓯᔾᔨᖅᑐᓂᒃ ᐅᑭᐅᖅᑕᖅᑐᒥ, ᖃᐅᔨᒪᔭᖏᑦ ᐃᓱᒪᒋᔭᖏᓪᓗ ᓄᓇᖓᑕ ᐊᓯᔾᔨᕐᓂᖓᑕ ᒥᒃᓵᓄᑦ ᐃᓚᓕᐅᔾᔭᐅᔭᕆᐊᖃᖅᐳᖅ ᑲᓇᑕᒥᐅᑦ ᑭᐅᓚᖅᐸᑕ ᓯᓚᐅᑉ ᐊᓯᔾᔨᕐᓂᖓᑕ ᒥᒃᓵᓄᑦ. ᐃᒃᓯᕚᖃᑎᒋᔭᕆᐊᖃᓕᖅᐸᒍᑦ, ᑏᑐᖃᑎᒋᓗᒋᑦ, ᖃᓄᖅ ᖃᐅᔨᓴᖅᑕᐅᑦᑎᐊᕈᓐᓇᕐᒪᖔᑦ ᐊᒻᒪ ᓯᕗᓕᑦᑎᐊᕈᓐᓇᕐᒪᖔᑦ ᑲᓇᑕ.

ᓇᑯᕐᒦᒃ
ᖁᔭᓐᓇᒦᒃ
ᒪ’ᓇ
ᖁᐊᓇ
ᓇᑯᕐᒦᒃ
Merci
Thank you

formulates our nation's response to climate change. Inuit are the sentinels of these changes in the North, and their knowledge and understanding of how their land is changing must be incorporated into how Canada addresses climate change. It is time to sit with them, have tea, and discuss avenues for better research and better leadership in Canada.

Nakurmiik
Qujannamiik
Maatna
Koana
Nakummek
Thank you
Merci

ᐃᓚᓕᐅᓯᒪᔪᖅ: ᖃᐅᔨᓴᐅᔪᑦ ᓇᐃᓈᖅᑕᐅᓯᒪᓪᓗᑎᒃ ᐅᓂᒃᑳᓕᐊᑦ

Appendix: Summary of Findings

ᓇᐃᓈᖅᓯᒪᔪᑦ ᐊᐱᖅᓱᖅᑕᐅᔪᓂᑦ 8-ᖑᔪᓂᑦ ᓄᓇᓕᖕᓂᑦ ᐊᒻᒪ 145-ᓂᒃ ᐊᐱᖅᓱᖅᑕᐅᔪᓂᒃ ᓄᓇᕗᒻᒥ, ᓄᓇᕕᖕᒥ, ᓄᓇᑦᓯᐊᕗᒻᒥᓗ
Summary of interview results from 8 communities and 145 interview participants in Nunavut, Nunavik, and Nunatsiavut

ᐊᕕᒃᑐᖅᓯᒪᔪᑦ Region	ᓄᓇᕗᑦ NUNAVUT				ᓄᓇᕕᒃ NUNAVIK			ᓄᓇᑦᓯᐊᕗᑦ NUNATSIAVUT
ᓄᓇᓕᒃ Community	ᖁᒡᓗᖅᑐᖅ Kugluktuk	ᖃᒪᓂᑦᑐᐊᖅ Baker Lake	ᒥᑦᑎᒪᑕᓕᒃ Pond Inlet	ᐸᖕᓂᖅᑑᖅ Pangnirtung	ᐅᒥᐅᔮᖅ Umiujaq	ᑲᖏᖅᓱᐊᓗᔾᔪᐊᖅ Kangiqsualujjuaq	ᑲᖏᖅᓱᔪᐊᖅ Kangiqsujuaq	ᓇᐃᓂ Nain
ᓄᓂᕐᒪᒃᓴᑦ **BERRIES**								
ᓄᓂᕐᒪᒃᓴᑦ ᐅᓄᕐᓂᖏᑦ Berry abundance	—	—	—	—	—	—	—	↓
ᓄᓂᕐᒪᒃᓴᖃᕐᓂᖓ Berry distribution	—	—	—	—	—	—	—	NA
ᓄᓂᕐᒪᒃᓴᑦ ᐱᕈᐊᓂᖏᓐᓂᖏᑦ Berry ripening	—	—	—	—	—	—	—	•
ᓄᓂᕐᒪᒃᓴᑦ ᑎᐱᖏᑦ Berry taste	—	—	—	—	—	—	—	√
ᓄᓂᕐᒪᒃᓴᑦ ᐊᖏᓂᖏᑦ Berry size	—	—	—	—	—	—	—	↓
ᐊᖅᐱᑦ ᐅᓄᕐᓂᖏᑦ Bakeapple abundance	—	—	—	—	—	↑	—	↓
ᐊᖅᐱᑦ ᐱᕈᐊᓂᖏᓐᓂᖏᑦ Bakeapple ripening	—	—	—	—	—	←	—	—

ᐊᕕᒃᑐᖅᓯᒪᔪᑦ Region	ᓄᓇᕗᑦ NUNAVUT				ᓄᓇᕕᒃ NUNAVIK			ᓄᓇᑦᓯᐊᕗᑦ NUNATSIAVUT
ᓄᓇᓕᒃ Community	ᖁᕐᓗᖅᑐᖅ Kugluktuk	ᖃᒪᓂ'ᑐᐊᖅ Baker Lake	ᒥᑦᑎᒪᑕᓕᒃ Pond Inlet	ᐸᖕᓂᖅᑑᖅ Pangnirtung	ᐅᒥᐅᔭᖅ Umiujaq	ᑲᖏᖅᓱᐊᓗᒃᔪᐊᖅ Kangiqsualujjuaq	ᑲᖏᖅᓱᔪᐊᖅ Kangiqsujuaq	ᓇᐃᓂ Nain
ᓄᓂᕙᒃᓴᑦ **BERRIES**								
ᑭᒍᑕᖕᒋᕐᓇᑦ ᐅᓄᕐᓂᖕᒋᑦ Blueberry abundance	—	—	↑	—	—	—	—	↓
ᑭᒍᑕᖕᒋᕐᓇᑦ ᐱᕈᐊᓂᖕᓂᖕᒋᑦ Blueberry ripening	—	—	—	—	—	—	—	—
ᐸᐅᕐᖓᑦ ᐅᓄᕐᓂᖕᒋᑦ Blackberry abundance	—	—	—	—	—	—	—	—
ᐸᐅᕐᖓᑦ ᐱᕈᐊᓂᖕᓂᖕᒋᑦ Blackberry ripening	—	—	—	—	—	—	—	—
ᑭᖕᒥᖕᓇᑦ ᐅᓄᕐᓂᖕᒋᑦ Cranberry abundance	—	—	—	—	—	—	—	—
ᑭᖕᒥᖕᓇᑦ ᐱᕈᐊᓂᖕᓂᖕᒋᑦ Cranberry ripening	—	—	—	—	—	—	—	—

ᓇᓗᓇᐃᔭᐅᑦ
Legend

—	ᐊᓯᔾᔨᕐᓂᖃᖕᒋᑦᑐᖅ no change
√	ᐊᓯᔾᔨᖅᓯᒪᔪᖅ change
↑	ᐊᖕᒋᓂᖅᓴᒥᒃ more, higher
↓	ᒥᑭᓂᖅᓴᒥᒃ less, lower
←	ᕿᓚᒻᒥᐅᓂᖅᓴᖅ earlier
→	ᑭᖑᕚᕆᓂᖅᓴᖅ later
•	<50% ᐊᖏᖃᑎᒌᒃᑐᑦ <50% consensus
NA	ᖃᐅᔨᓴᖅᑕᐅᖏᑦᑐᖅ not addressed

ᐊᕕᒃᑐᖅᓯᒪᔪᑦ Region	ᓄᓇᕗᑦ NUNAVUT				ᓄᓇᕕᒃ NUNAVIK			ᓄᓇᑦᓯᐊᕗᑦ NUNATSIAVUT
ᓄᓇᓕᒃ Community	ᖁᕐᓗᖅᑐᖅ Kugluktuk	ᖃᒪᓂᑦᑐᐊᖅ Baker Lake	ᒥᑦᑎᒪᑕᓕᒃ Pond Inlet	ᐸᖕᓂᖅᑑᖅ Pangnirtung	ᐅᒥᐅᔭᖅ Umiujaq	ᑲᖏᖅᓱᐊᓗᔾᔪᐊᖅ Kangiqsualujjuaq	ᑲᖏᖅᓱᔪᐊᖅ Kangiqsujuaq	ᓇᐃᓂ Nain
ᐱᕈᖅᑐᑦ ᐊᓯᖏᑦ OTHER PLANTS								
ᐱᕈᖅᑐᑦ ᐅᓄᕐᓂᖏᑦ Vegetation	—	—	—	—	—	↑	—	↑
ᐱᕈᖅᑐᖃᕐᓂᖓ Plant distribution	—	—	—	—	—	√*	—	NA
ᐱᕈᖅᑐᑦ ᐱᕈᑦᑎᐊᕐᓂᖏᑦ Plant blooming	—	—	—	—	—	←	—	—
ᓇᐹᖅᑐᑦ ᐅᓄᕐᓂᖏᑦ Tree abundance	—	—	—	—	—	↑	—	↑
ᓇᐹᖅᑐᖃᕐᓂᖓ Tree distribution	—	—	—	—	—	—	—	NA
ᐅᖅᐱᒐᑦ ᐅᓄᕐᓂᖏᑦ Shrub abundance	↑	—	—	↑	↑	↑	↑	↑
ᐅᖅᐱᒐᖃᕐᓂᖓ Shrub distribution	—	—	—	—	—/√	—	—	NA
ᐃᕕᑦ ᐅᓂᕐᓂᖏᑦ Grass abundance	•	—	•	—	—	—	—	—
ᐃᕕᖃᕐᓂᖓ Grass distribution	—	—	—	—	—	—	—	NA
ᑎᖓᐅᔭᑦ ᐅᓄᕐᓂᖏᑦ Lichen abundance	—	—	—	—	—	—	—	—

ᐊᕕᒃᑐᖅᓯᒪᔪᑦ Region	ᓄᓇᕗᑦ NUNAVUT				ᓄᓇᕕᒃ NUNAVIK			ᓄᓇᕠᐊᕗᑦ NUNATSIAVUT
ᓄᓇᓕᒃ Community	ᖁᕐᓗᒃᑐᖅ Kugluktuk	ᖃᒪᓂ'ᑐᐊᖅ Baker Lake	ᒥᑦᑎᒪᑕᓕᒃ Pond Inlet	ᐸᖕᓂᖅᑑᖅ Pangnirtung	ᐅᒥᐅᔭᖅ Umiujaq	ᑲᖏᖅᓱᐊᓗᒃᔪᐊᖅ Kangiqsualujjuaq	ᑲᖏᖅᓱᔪᐊᖅ Kangiqsujuaq	ᓇᐃᓂ Nain
ᐆᒪᔪᑦ ANIMALS								
ᐆᒪᔪᑦ ᐅᓄᕐᓂᖏᑦ Mammal abundance	—	—	√	—	—	↑	↑	√
ᑎᖕᒥᐊᑦ ᐅᓄᕐᓂᖏᑦ Bird abundance	—	—	—	—	—	—	—	√
ᐃᖃᓗᐃᑦ ᐅᓄᕐᓂᖏᑦ Fish abundance	—	—	—	—	—	—	—	—
ᐊᓇᓐᖏᑦ ᐅᓄᕐᓂᖏᑦ Black fly abundance	↓	—	—	—	—	—	—	•
ᐊᓇᓐᖏᑦ ᓴᖅᑭᓐᓂᖏᑦ Black fly emergence	—	—	—	—	—	—	—	—
ᕿᒃᑐᕆᐊᑦ ᐅᓄᕐᓂᖏᑦ Mosquito abundance	•	—	—	—	—	—	•	•
ᕿᒃᑐᕆᐊᑦ ᓴᖅᑭᓐᓂᖏᑦ Mosquito emergence	—	—	—	—	—	—	—	—

ᓇᓗᓇᐃᔭᐅᑦ
Legend

—	ᐊᓯᔾᐲᕐᓂᖅᑲᖏᑦᑐᖅ no change	↑	ᐊᖏᓂᖅᓴᒥᒃ more, higher	→	ᑭᖑᕙᓯᖕᓂᖅᓴᖅ later
√	ᐊᓯᔾᐲᖅᓯᒪᔪᖅ change	↓	ᒥᑭᓐᓂᖅᓴᒥᒃ less, lower	•	<50% ᐊᖏᖅᑲᑎᒌᒃᑐᑦ <50% consensus
		←	ᕿᓚᒻᒥᐅᓂᖅᓴᖅ earlier	NA	ᖃᐅᔨᓴᖅᑕᐅᖏᑦᑐᖅ not addressed

ᐊᕕᒃᑐᖅᓯᒪᔪᑦ **Region**	ᓄᓇᕗᑦ **NUNAVUT**				ᓄᓇᕕᒃ **NUNAVIK**			ᓄᓇᑦᓯᐊᕗᑦ **NUNATSIAVUT**
ᓄᓇᓕᒃ **Community**	ᖁᒡᓗᖅᑐᖅ **Kugluktuk**	ᖃᒪᓂ'ᑐᐊᖅ **Baker Lake**	ᒥᑦᑎᒪᑕᓕᒃ **Pond Inlet**	ᐸᖕᓂᖅᑑᖅ **Pangnirtung**	ᐅᒥᐅᔭᖅ **Umiujaq**	ᑲᖏᖅᓱᐊᓗᔾᔪᐊᖅ **Kangiqsualujjuaq**	ᑲᖏᖅᓱᔪᐊᖅ **Kangiqsujuaq**	ᓇᐃᓂ **Nain**
ᓯᓚᐃᑦ **SEASONS**								
ᐅᑭᐅᖅ Winter	√	—	—	—	—	—	—	√
ᐅᐱᕐᖔᖅ Spring	√	—	—	—	—	—	—	√
ᐊᐅᔭᖅ Summer	√	—	—	—	—	—	√	√
ᐅᑭᐊᖅ Fall	√	—	—	—	—	—	—	√
ᓯᓚ/ᐊᓄᓂ **CLIMATE/ WEATHER**								
ᐊᓯᔾᔨᖅᑕᕐᓂᖓ Variability	—	↑	—	—	—	—	↑	↑
ᐆᓇᕐᓂᖓ ᓂᒡᓚᓱᙱᓂᖓ Temperature	—	—	—	—	—	↑	•	↓
ᐊᓄᓂ Wind	√	√	√	—	√	—	√	√
ᒪᕐᑯᒃᐸᙱᓂᖓ Rain	↓	↓	—	—	—	—	↑	↑
ᖃᓐᓂᖅᐸᙱᓂᖓ Snow	↓	↓	•	•	—	↓	↓	↓
ᐊᐳᑦ ᐊᐅᓚᕐᓂᖓ Snowmelt timing	—	—	—	—	—	←	←	—
ᐃᒪᕐᖃᕐᓂᖓ Hydrology	√	√	—	—	√	√	√	√
ᓯᑯᓂᖓ Sea ice freeze-up	→	—	—	—	—	—	—	→
ᓯᑯᐃᕐᓂᖓ Sea ice breakup	•	—	←	←	—	←	←	—/←

ᐊᕕᒃᑐᖅᓯᒪᔪᑦ Region	ᓄᓇᕗᑦ NUNAVUT				ᓄᓇᕕᒃ NUNAVIK			ᓄᓇᑦᓯᐊᕗᑦ NUNATSIAVUT
ᓄᓇᓕᒃ Community	ᖁᕐᓗᖅᑐᖅ Kugluktuk	ᖃᒪᓂʼᑐᐊᖅ Baker Lake	ᒥᑦᑎᒪᑕᓕᒃ Pond Inlet	ᐸᖕᓂᖅᑑᖅ Pangnirtung	ᐅᒥᐅᔭᖅ Umiujaq	ᑲᖏᖅᓱᐊᓗᔾᔪᐊᖅ Kangiqsualujjuaq	ᑲᖏᖅᓱᔪᐊᖅ Kangiqsujuaq	ᓇᐃᓂ Nain
ᓯᓚ/ᐊᓄᓂ CLIMATE/ WEATHER								
ᓯᑯᑉ ᐃᔾᔪᓂᖓ Sea ice thickness	—	—	—	—	—	↓	↓	↓
ᐃᒪᐅᑉ ᓂᒡᓚᔑᖕᓂᖓ Seawater temp.	—	—	—	—	—	—	—	NA
ᑕᓯᑦ ᓯᑯᓂᖓ Lake ice freeze-up	—	—	—	—	—	—	—	—
ᑕᓯᑦ ᓯᑯᐃᕐᓂᖓ Lake ice breakup	—	—	—	—	—	—	←	—
ᑕᓯᕐᓂ ᓯᑯ ᐃᔾᔪᓂᖓ Lake ice thickness	—	—	—	—	—	—	—	—
ᑕᓯᕐᓂ ᑰᖕᓂ ᓂᒡᓚᔑᖕᓂᖓ Lake/river temp.	—	—	—	—	—	—	—	NA
ᓯᕿᓂᐅᑉ ᐆᓇᕐᓂᖓ Sun radiation	—	—	—	—	—	—	—	—

ᓇᓗᓇᐃᔭᐅᑦ
Legend

—	ᐊᓯᔾᔨᕐᓂᖅᑲᖏᑦᑐᖅ no change	↑	ᐊᖏᓂᖅᓴᒥᒃ more, higher	→	ᑭᖑᕙᓯᖕᓂᖅᓴᖅ later
√	ᐊᓯᔾᔨᖅᓯᒪᔪᖅ change	↓	ᒥᑭᓂᖅᓴᒥᒃ less, lower	•	<50% ᐊᖏᖅᑲᑎᒌᒃᑐᑦ <50% consensus
		←	ᕿᓚᒻᒥᐅᓂᖅᓴᖅ earlier	NA	ᖃᐅᔨᓴᖅᑕᐅᖏᑦᑐᖅ not addressed

ᐊᕕᒃᑐᖅᓯᒪᔪᑦ Region	ᓄᓇᕗᑦ NUNAVUT				ᓄᓇᕕᒃ NUNAVIK			ᓄᓇᑦᓯᐊᕗᑦ NUNATSIAVUT
ᓄᓇᓕᒃ Community	ᖁᕐᓗᖅᑐᖅ Kugluktuk	ᖃᒪᓂ'ᑐᐊᖅ Baker Lake	ᒥᑦᑎᒪᑕᓕᒃ Pond Inlet	ᐸᖕᓂᖅᑑᖅ Pangnirtung	ᐅᒥᐅᔭᖅ Umiujaq	ᑲᖏᖅᓱᐊᓗᔾᔪᐊᖅ Kangiqsualujjuaq	ᑲᖏᖅᓱᔪᐊᖅ Kangiqsujuaq	ᓇᐃᓐ Nain
ᓯᓚ/ᐊᓄᕆ CLIMATE/ WEATHER								
ᓄᓇᐅᑉ ᐃᑭᐊ ᖁᐊᖑᐃᓐᓇᖅᐸᒃᑐᖅ ᐊᐅᖕᓂᖕᒐ Permafrost melting	√	√	—	√	—	√	—	√
ᓄᓇ ᑎᓱᐊᕐᓂᖅ Erosion	↑	—	—	—	—	↑	—	↑
ᐊᒃᑐᐃᓂᑦ IMPACTS								
ᓄᒃᑕᕕᑦ ᐊᖅᑯᑕᐅᕙᒃᑐᑦ Travel routes	—	—	—	—	—	—	√	√
ᐃᓅᓯᑎᔭᑦ Ways of life	—	—	—	—	—	—	—	√

ᓇᓗᓇᐃᔭᐅᑦ
Legend

—	ᐊᓯᔾᔨᕐᓂᖃᖕᒋᑦᑐᖅ no change	↑	ᐊᖕᒋᓂᖅᓴᒥᒃ more, higher	→	ᑭᖑᕚᕐᓯᖕᓂᖅᓴᖅ later
√	ᐊᓯᔾᔨᖅᓯᒪᔪᖅ change	↓	ᒥᑭᓂᖅᓴᒥᒃ less, lower	•	<50% ᐊᖕᒋᖅᑎᕆᒃᑐᑦ <50% consensus
		←	ᕿᓚᒻᒥᐅᓂᖅᓴᖅ earlier	NA	ᖃᐅᔨᓴᖅᑕᐅᖕᒋᑦᑐᖅ not addressed

* ᓇᓗᓇᐃᖅᑕᐅᔪᓕᒫᑦ ᐊᑯᓂᒍᑦ ᐊᓯᔾᔨᕐᓂᐅᔪᒍᑦ 50-ᐳᓴᓐᖑᔪᑦ ᐅᖓᑖᓂᓘᓐᓃᑦ ᐊᐱᖅᓱᖅᑕᐅᔪᑦ ᐊᖕᒋᖅᑎᕆᒃᑲᖓᑕ.
* All indicators of environmental change are based on 50 percent or greater consensus among interview participants.

Photo Credits

All photographs © the editors, except:

cover/p. 248 © Freezingpictures/Dreamstime.com
p. 13 City Escapes Nature Photo/Shutterstock.com
p. 24 © Joëlle Taillon
p. 26 © Yuriy Brykaylo/Dreamstime.com
p. 36© Courtenay ClarkDan Bach Kristensen/Shutterstock.com
p. 41 Dan Bach Kristensen/Shutterstock.com
p. 46 Fufachew Ivan Andreevich/Shutterstock.com
p. 51 © David Herraez/Dreamstime.com
p. 54 Dmytro Pylypenko/Shutterstock.com
p. 60 Dimos/Shutterstock.com
p. 65 © Tyler Olson/Dreamstime.com
p. 81 Mike Garcelon/Shutterstock.com
p. 83 © Benoit Tremblay
p. 90 Wyatt Rivard/Shutterstock.com
p. 95 Coprid/Shutterstock.com
p. 96 © Kenhutchsion/Dreamstime.com
p. 101 © Carmen Spiech
p. 102 Glamorous Images/Shutterstock.com
p. 105 © Benoit Tremblay
p. 117 Flore/Shutterstock.com
p. 131 © Emi/Shutterstock.com
p. 132 MVPhoto/Shutterstock.com
p. 143 jps/Shutterstock.com
p. 149 Dennis W. Donohue/Shutterstock.com
p. 155 © Benoit Tremblay
p. 163 © Christine Demers
p. 164 © Clickrchick/Dreamstime.com
p. 167 © Naïm Perreault
p. 178 GreenLandStudio/Shutterstock.com
p. 186 John Wollwerth/Shutterstock.com
p. 188 © Guillaume Rheault
p. 191 Sophia Granchinho/Shutterstock.com
p. 199 © Brandon Alms/Shutterstock.com
p. 200 © Dreamstime.com

p. 203 © Alice and Charlie Tooktoo
p. 209 Bildagentur Zoonar GmbH/Shutterstock.com
p. 221 © Thorsten Mauritsen c/o ArcticNet
p. 222 Trevor Jones/Shutterstock.com
p. 225 Footage.Pro/Shutterstock.com
p. 237 Collette3/Shutterstock.com
p. 238 Jim H. Walling/Shutterstock.com
p. 247 © Kamchatka/Dreamstime.com
p. 251 Tony Campbell/Shutterstock.com
p. 255 Tyler Olson/Shutterstock.com
p. 274 Vladimir Melnik/Shutterstock.com
p. 286 © Elias Obed
p. 289 © Romanchuck/Dreamstime.com
p. 296 © Achim Baque/Shutterstock.com
p. 299 AndreAnita/Shutterstock.com